POLYMERIC FOAMS

Innovations in Processes, Technologies, and Products

Polymeric Foams

series editor Shau-Tarng Lee

INCLUDED TITLES

POLYMERIC FOAMS SERIES

POLYMERIC FOAMS

Innovations in Processes, Technologies, and Products

Edited by
Shau-Tarng Lee

CRC Press
Taylor & Francis Group
Boca Raton London New York

CRC Press is an imprint of the
Taylor & Francis Group, an **informa** business

CRC Press
Taylor & Francis Group
6000 Broken Sound Parkway NW, Suite 300
Boca Raton, FL 33487-2742

First issued in paperback 2021

© 2017 by Taylor & Francis Group, LLC
CRC Press is an imprint of Taylor & Francis Group, an Informa business

No claim to original U.S. Government works

ISBN-13: 978-0-367-78275-7 (pbk)
ISBN-13: 978-1-4987-3887-3 (hbk)

Library of Congress Cataloging-in-Publication Data

Names: Lee, S.-T. (Shau-Tarng), 1956- editor.
Title: Polymeric foams : innovations in processes, technologies, and products
/ [compiled by] Shau-Tarng Lee.
Other titles: Polymeric foams (CRC Press : 2017)
Description: Boca Raton : Taylor & Francis, CRC Press, 2017. | Series:
Polymeric foams series ; volume 6 | Includes bibliographical references
and index.
Identifiers: LCCN 2016020297 | ISBN 9781498738873 (alk. paper)
Subjects: LCSH: Plastic foams.
Classification: LCC TP1183.F6 P6475 2017 | DDC 668.4/93--dc23
LC record available at https://lccn.loc.gov/2016020297

Visit the Taylor & Francis Web site at
http://www.taylorandfrancis.com

and the CRC Press Web site at
http://www.crcpress.com

Dedicated to the Lord, who is above all and in all, that we can understand

His wonderful creation. His kindness is beyond our comprehension.

Contents

Preface

There is no question that foamed plastics is quite a fascinating subject. It involves various components in three states processed through a highly unstable nucleation into expansion to generate a product with a cellular structure possessing unique properties for a wide variety of applications. It has enriched our lives significantly since its commercialization as early as the 1930s. For thermoplastic foam, its lightweighting after expansion lent itself directly into flotation and packaging segments in the beginning, then the food tray and insulation board, and nowadays, structural and aviation industries. The global low-density foam market exceeded US$100 billion in 2015. Although blowing agent emission and landfill degradation issues remain, the main driver for growth is innovation. As the population distribution continues to change, the concomitant demand changes as well. For instance, in the United States, baby boomers are close to retiring or are retired already. Living quality (i.e., air and water quality, sound absorption), medical devices, protective designs, and energy management become clear customer voices, which surely drive development efforts. Polymeric foams are a solution, or, at least, a vital component in the solution to meet or exceed the customer's expectation.

In 2014, *Foam Update* organized an international foam conference in Shanghai with a clear focus on innovation. It drew wide and deep support to speed up the cradle to adult process or research to commercialization. Quite a few interesting presentations were presented and generated genuine onsite discussion, as well as follow-up actions. A valuable network was created. Invitations were made to authors who are relevant to innovation topics to revise their presentation for a chapter format to benefit a wider audience. CRC Press/Taylor & Francis Group was willing to commit to the publishing. We do not believe that this is the end of foam development, but it is certainly a vital part of the autodynamics. It not only directs and encourages development but also makes the valuable ones stay and strengthen the whole infrastructure for foam.

This volume begins with an introduction in Chapter 1, followed by product developments. Polypropylene, polyethylene terephthalate, and blends are addressed in Chapters 2 through 4, respectively. Chapters 5 and 6 are focused on polyactide foam in extrusion and molded beads. PLA foam is a good stepping stone into sustainable production on account of social responsibility. Nanocellular foam is a product that would open up many more application doors. Chapter 7 has a very comprehensive coverage of its development detail. After that, the rest is devoted to applications: Chapter 8 on the ever-growing structural applications in sandwich composites with foam cores, electromagnetic interference foam in Chapter 9, innovative foam productions in Chapter 10, and injection molding foam insulation and practice in Chapter 11. It has been quite a pleasure working with so many global foam experts who were dedicated to completing each chapter with updated and confirmed materials to make this book a precious collection for continued progress in polymeric foams.

Acknowledgments

It can be a challenge to edit a book written by well-known experts whose schedules today are much busier than yesterday. Over 20 authors committed themselves to the challenge of finishing the assigned chapters and I truly appreciate their dedication for the genuine efforts that they put into this comprehensive work. Although some of them could not deliver on time due to unexpected circumstances, my appreciation still goes to them as well, and I wish them continued success in foam.

Special thanks go to Richard Gendron, who retired from the Canadian National Research Council in September 2015, where he had served as an officer since 1990. He is a prominent researcher for foam. His publications are a good testimony for him. His commitment after retirement and ever-refreshing enthusiasm in the polylactide (PLA) foam chapter are very impressive. I wish him nothing but my best. Dr. Stéphane Costeux of Dow Chemical also stepped forward to secure company permissions to prepare a very collective and comprehensive chapter on nanocellular foams. Professor Park of the University of Toronto, a great contributor to foam for so many years and a good friend of mine, had his group prepare the chapter on PLA beads on short notice. Their faithful efforts are simply beyond what my words can express. May God grant me the opportunity to pay back more.

I would also like to deliver my deep appreciation to Ken Chrisman, president of Product Care, Sealed Air Corp. He delivered a resounding keynote speech at the Poly-Foam Conference Shanghai, December 3–4, 2014, highlighting the importance of innovation for foam. He also encouraged his business colleagues to continue on his keynote speech to make the foam chain more business valuable. Moreover, without the support of my direct reports, Tim Denninson (process platform executive director) and George Wofford (vice-president of core research and development), it would not have been possible to finish the editing of this book. Both of you are much appreciated! Lastly, although my wife, Mjau-Lin, is not that aware of polymeric foams, I learned a lot in the kitchen from cake mix, egg-white bubbling, yeast-induced expansion, rice growth in the cooker, etc. More importantly, her soft and kind character makes me short of thanks before the Lord.

Shau-Tarng Lee
Sealed Air Corporation
Saddle Brook, New Jersey

Editor

Dr. Shau-Tarng (S.-T.) Lee received his PhD from Stevens Institute of Technology Hoboken, New Jersey, and joined Sealed Air Corporation in foam research, foam product/process, and technology development as a development engineer, a research-and-development director, and currently as a senior research Fellow. He has over 100 publications, including 28 patents. He was elected to Fellow by the SPE in 2001. He was inducted into Sealed Air's Inventors Hall of Fame in 2003. He has authored and edited five foam books that are published by Technomic and CRC Press. At present, he serves as a foam series editor for Taylor & Francis Group, a co-editor-in-chief for the *Journal of Cellular Plastics*, and the principle editor for *Foam Update*.

Contributors

Rong-Yeu Chang
CoreTech System (Moldex3D)
Co. Ltd.
Hsinchu, Taiwan

Stéphane Costeux
The Dow Chemical Company
Midland, Michigan

Kurt Feichtinger
Tech Services Manager
Baltek Inc.
High Point, North Carolina

Richard Gendron
National Research Council
of Canada
Quebec, Canada

Peng Guo
SINOPEC Beijing Research Institute
of Chemical Industry
Beijing, China

Chao-Tsai Huang
Department of Chemical and
Materials Engineering
Tamkang University
New Taipei City, Taiwan

Han-Xiong Huang
Lab for Micro Molding and Polymer
Rheology
The Key Laboratory of Polymer
Processing Engineering of the
Ministry of Education
South China University
of Technology
Guangzhou, China

Tatsuyuki Ishikawa
JSP Corporation
Kanuma, Japan

Xiulei Jiang
Xinhengtai Advanced Materials
Co. Ltd.
Zhejiang, China

Shau-Tarng Lee
Sealed Air Corporation
Saddle Brook, New Jersey

Yang Li
Ningbo Key Lab of Polymer
Materials
Ningbo Institute of Material
Technology and Engineering
Chinese Academy of Science
Zhejiang, China

Tao Liu
East China University of Science
and Technology
Shanghai, China

Wenguang Ma
Director of Technology
Vixen Composites LLC
Elkhart, Indiana

Mihaela Mihai
National Research Council
of Canada
Quebec, Canada

Mohammadreza Nofar
Faculty of Chemical and
    Metallurgical Engineering
Metallurgical and Materials
    Engineering Department
Istanbul Technical University
Maslak, Istanbul

Chul B. Park
Microcellular Plastics
    Manufacturing Laboratory
Department of Mechanical and
    Industrial Engineering
University of Toronto
Toronto, Ontario, Canada

Alireza Tabatabaei
Microcellular Plastics
    Manufacturing Laboratory
Department of Mechanical and
    Industrial Engineering
University of Toronto
Toronto, Ontario, Canada

Tomoo Tokiwa
JSP Corporation
Kanuma, Japan

Zhenhao Xi
East China University of Science
    and Technology
Shanghai, China

Tian Xia
East China University of Science
    and Technology
Shanghai, China

Lin-Qiong Xu
Lab for Micro Molding and Polymer
    Rheology
The Key Laboratory of Polymer
    Processing Engineering of the
    Ministry of Education
South China University of
    Technology
Guangzhou, China

Wentao Zhai
Ningbo Key Lab of Polymer
    Materials
Ningbo Institute of Material
    Technology and Engineering
Chinese Academy of Science
Zhejiang, China

Ling Zhao
East China University of Science
    and Technology
Shanghai, China

Wenge Zheng
Ningbo Key Lab of Polymer
    Materials
Ningbo Institute of Material
    Technology and Engineering
Chinese Academy of Science
Zhejiang, China

1

Introduction

Shau-Tarng Lee

CONTENTS

1.1 Introduction

Foam has long been known as a useful product in making our way of life better. Low-density foam was made in the laboratory before World War II, after which it took off in civilian applications. In a few years, polyurethane (PU) and polystyrene foams became strong performance platforms to deliver a variety of products in floatation and packaging industries. They expanded into furniture and food segments, then into construction, medical, sports, automotive, aerospace, electronics, etc. A recent market research report indicated a steady growth in consumption to exceed 50 billion lb. by 2019 [1]. There is no doubt that technology plays a critical role in the development history. When communication basically revolutionized our way of living and doing business, the push for new technology has become ever critical in the last decade. As a result, many resources have been spent on new foam technology. It was estimated that the average number of good publications in foam was over 2000 per month in 2014. It is quite diversified. However, the market is changing rapidly. In fact, finding applications for new technology and how to turn an application into a successful business became equally or more challenging [2]. At present, innovation evidently becomes a lifeline for business success. Fundamental research is still an essential element for new technology development. It is polarized, yet the good news is that constant resources are being committed to development, and there is no lack of innovation in foam. Although the balance between research and application is changing, there is no question: keeping the balance point in control is necessary for a sustainable growth.

1.2 Blowing Agent

Since the late 1980s, ozone holes identified in the Arctic areas have alarmed the whole world. The Montreal Protocol was proposed to curb the consumption of halogenated physical blowing agents [3]. After two and a half decades, this issue seemed to be under control, yet during this time, global warming, smog (smoke + fog), acid rain, and a level of suspended 2.5-μ particles have raised more concerns. It simply makes the blowing agent qualification list much longer and more complicated. Even so, blowing agent development showed solid strides forward. Table 1.1 summarizes the five different generations in environmentally friendly blowing agent development and hydrocarbon (HC) blowing agents.

It is worth mentioning that a hydrofluoroolefin (HFO) blowing agent possesses intriguing characteristics, such as low ozone and global warming impact, non-flammable, and low thermoconductivity. Since chlorofluorocarbons (CFCs) are generally very stable and inert, they stay in the foam for a very long time so that its low thermoconductivity brings forth solid insulation benefits. Many alternatives are lighter with higher thermoconductivity, which cause direct and indirect financial burdens in making the same-density foam products while trying to match performance. HFO, as illustrated in Table 1.2, shows a much desired thermoconductivity in addition to environmental qualities. Honeywell and Du Pont have committed their manufacturing to capture its insulation value [4,5]. Figure 1.1 presents the operation of PU paneling insulation by a nozzle spray. The energy saving is a tremendous incentive for use in construction.

In contrast to inorganic physical blowing agents, such as carbon dioxide and nitrogen, HFO exhibits a much desired solubility and volatility to general thermoplastic polymers. This means that it can be processed in the conventional extrusion process when making an insulation board. When HFO is further accepted, more demand will increase volume production and reduce its unit cost to make more impact in the construction market.

TABLE 1.1

Summary of Generations of Physical Blowing Agent Development

Time Frame	1940s	1960s	1980s	1990s	2000s	2010s
Kind	Chlorofluorocarbon (CFC)	Hydrocarbon (HC)	Inorganic	Hydrochlorofluorocarbon (HCFC)	Hydrofluorocarbon (HFC)	Hydrohalogenolefin (HHO)
Blowing agent	CFC-11, 12, 114, 115	HC: pentane and butane	SFC CO_2	HCFC-22, 141b, 142b	HFC-134a, 152a	HHO-1234, 1233, 1336
Characteristics	Safe, stable, easy to process	Low cost and easy to process	Safe and microcellular	Low ozone, high GWP	Zero ODP, high GWP	Zero or low ODP and GWP
Issue	Ozone and global warming	Flammable, causes smog	Tough to process	Flammable	Higher gas thermal conductivity	Expensive

Note: GWP, global warming potential; ODP, ozone depletion potential.

TABLE 1.2

Characteristics for Selected Blowing Agent

Chemical	Formula	Boiling Point °C	Ozone Depletion Potential	Global Warming Potential[a]	Gas Thermal Conductivity W/m-°K	Flammability
CO_2	CO_2	−78.5	0	1	0.0146	No
CFC-11	CCl_3F	23.8	1	4750	0.008	No
CFC-12	CCl_2F_2	−29.8	1	10,900	0.01	No
HCFC-22	$CClHF_2$	−40.8	0.055	2310	0.011	No
HCFC-141b	$CFCl_2CH_3$	32	0.11	725	0.009	Yes
HCFC-142b	CF_2ClCH_3	−9.8	0.065	2310	0.011	Yes
HFC-134a	CH_2FCF_3	−26.1	0	1430	0.013	No
HFC-152a	CH_3CHF_2	−24.7	0	124	0.014	Yes
HFO-1234ze	$CHFCHCF_3$	−19	6	0	0.013	No
HFO-1233zd	$CF_3CHCClH$	19	4.7–7	0	0.01	No
HFO-1336mzz	$CF_3CHCHCF_3$	33	0	9	0.01	No

[a] On 100 years.

FIGURE 1.1
Spray PU foam for wall paneling.

1.3 Microcellular Foaming

Microcellular is, by large, characterized by its cell size, which is less than 100 μ. The initial success was reported from the Massachusetts Institute of Technology (MIT) by employing a supercritical carbon dioxide blowing agent [6–8]. Since then, numerous papers have been published that explore this foaming technique. The Axiomatics Corp. was formed to focus on commercial application; it was later renamed Trexel, Inc., under exclusive intellectual property rights from MIT in 1995. Over the past two decades, it seems clear that the major success of microcellular is in injection molding industry. Although it is of a low foaming percentage (i.e., under 20% cell content), it generally helps reduce the cycle time, surface quality, and mold filling. As a result, most major injection molding machinery companies are licensees of this technology. On the extrusion side, the progress seems very limited. It was concluded that, at the same expansion ratio, small cell size is a huge benefit in ascetics and skin contact yet inconclusive in major mechanical properties, which cast a limit on the application spectrum. Considering the high volatility that is associated with carbon dioxide, its die design and operation window are more challenging than the conventional process using an HC blowing agent in making a low-density thin sheet. Nonetheless, it is worth mentioning that its commercialization in polystyrene sheets for food trays and thin polypropylene (PP) sheets is quite encouraging in low-density foam extrusion. Although it is not likely to make an extruded low-density board with supercritical carbon dioxide, this blowing agent can alter the morphology of certain polymers to help understand cell nucleation and/or making foam in nanocell sizes [9,10].

For polyester, the ester group shows an interesting carbon dioxide affinity. Yet, it exhibits a much higher melting point that makes high-dissolution foaming prohibitive. However, medium-density extrusion is a good way to reap the benefits of density reduction for a high-cost polyester. For instance, polyethyleneterephthalate (PET) can soak the carbon dioxide in foam into thin sheets for a light-emitting diode (LED) reflector, as shown in Figure 1.2 [11]. The microcellular cell structure helps diffuse reflectivity to over 96% to improve LED efficiency and is good for human eyes. PET can also be extruded into a medium-density sheet that can be formed into a tray for microwavable meals, which a polystyrene (PS) tray, due to its low T_g, cannot be qualified for. It was reported that supercritical nitrogen can be processed in a specially designed corotating twin screw to make a medium-density PET sheet [12]. It should be pointed out that amorphous PET and amorphous polylactic acid (PLA) can be made into a low-density sheet in the extrusion process, since crystallization can be avoided during cooling to build up enough melt strength to manage dynamic foaming at the die exit. It is also noted that amorphous PET and PLA have limited performance virtues to make them good only in light and cold container applications.

A medium-expansion thin PET sheet was also commercialized through a soaking and oven-foaming technology by Microgreen in 2002. It takes advantage of high solubility at low temperature by foaming a PET solid roll stock with carbon dioxide at room temperature. The trade-off is the long soaking time, due to low diffusion. The foamed sheet can be thermoformed into a cup for hot beverages, as shown in Figure 1.3, and a tray [13]. It is quite an intriguing technology. Yet, it requires high capital to make up the

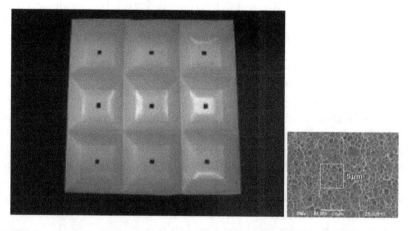

FIGURE 1.2
Furakawa's thermoformed microcellular PET as reflector for lighting fixtures. Cell size is less than 1 µ.

FIGURE 1.3
Microgreen's microcellular PET cup with cell structure in the inset. (From V. Kumar, Polyethylene terephthalate foams with integral crystalline skins, US Patent: 5,182,307, Jan., 1993.)

productivity owing to the long processing soaking time; for instance, for a less-than-1-mm sheet, over 24 h is needed for saturation in a CO_2 reservoir under room temperature. It is always a challenge to secure enough volume in the beginning to justify investment.

It was logical for supercritical carbon dioxide to foam into medium to high density to reduce weight in the extrusion and profile process. Even with a 50% microcellular void at the core of coextrusion process, it could be a significant savings, especially since recycled material is used in the core [14]. Considering that rigidity is a function of thickness to its third power, less material for a more rigid product can be made to make a flimsy bag into an elegant self-stand bag. Another example in Japan, as shown in Figure 1.4, was to build microcellular into PET bottle manufacturing to demonstrate its lightweighting feasibility in the extrusion molding process. It is understood that chemical blowing agents can be applied for the exact process and applications. The bottom line is health, environment, and cost benefits in physical blowing agents.

The microcellular cell structure in rotational and pressure mold technology has become increasingly important in making performance products, such as fuel tanks and boards [15,16]. Although its productivity cannot compete against the extrusion process, it can make a unique product by utilizing material strength through crystallization to accomplish microcellular cell morphology at high expansion for board products. When application can

FIGURE 1.4
General PET bottle and Fi-Cell PET bottle. (Courtesy of Prof. M. Oshima, Kyoto University.)

justify volume, a pressure mold can certainly be expanded into a formidable production platform.

1.4 Performance

Low-density foam has evolved from packaging and flotation to food trays, furniture, and sports and now into smart packaging and performance. Not only shock absorption and floating but also many internal surface areas with controlled orientation could be valuable in exotic applications. As demand continues to increase, it drives foam into high-end performance segments. Aerospace is a good example. According to a recent market report, aerospace foam showed $5.5 billion revenue in 2014 and is projected to reach $7.97 billion by 2019 with about 7.7% annual growth between 2014 and then. Asia-Pacific, the largest market with 39% share in 2013, is among the fastest-growing areas. PU foam is the largest type of foam in this industry. It accounted for over 34% of the total market in 2013. Flexible flame retardant (FR) PUs and rigid foam core materials have been used in flight decks and interior cabins, as shown in Figure 1.5. Integral-skin PUs are used as flight deck pads, overhead stow bins, high-end arm rests, and seating. Zotefoams, a performance foam supplier from the United Kingdom, offers polyvinyldiene fluoride (PVDF) foam out of their proprietary X-linked high-pressure two-stage foaming process while using an N_2 blowing agent. PVDF foam exhibits

FIGURE 1.5
Aircraft interior.

very good insulation properties, FR characteristics, and thermal stability up to 160°C, which is above the melting point of common thermoplastic polymers. Along with its ultraviolet resistance, it has been used in airframes, such as inside the cockpit, cargo floors, and pipe insulations.

Sabic offers polyetherimide (PEI) polymers under the trade name Ultem® and, now, an extruded PEI foam plank. PEI is an amorphous, transparent, amber polymer with high temperature deflection, excellent dimensional stability, chemical resistance, inherent flame resistance, and creep strength. Its chemical structure is presented in Figure 1.6. Sabic offered a PEI honeycomb first, as shown in Figure 1.7. Recently, a PEI foam plank was developed on a tandem extrusion for foaming into 10–30 times expansion while using a chemical blowing agent [17]. It shows the flame–smoke–toxicity characteristics of PEI resin with the light weight and low dielectric properties that are required for military and aerospace composite structures. Its unique radar transparency makes it a choice of radome applications. However, its high modulus/weight ratio lends itself to the core material of a sandwich composite that is suitable for the fast-growing aerospace market.

Rigid, tiny fibers can be introduced into the foam board in such a way that compressive modulus can be greatly enhanced [18,19]. Needles can be arranged to pick up the fibers so that they can stay in the body while the needles go all the way through. After the ends are cut off, the needle can be ready for the next round. Figure 1.8 illustrates the technique. In this way, a

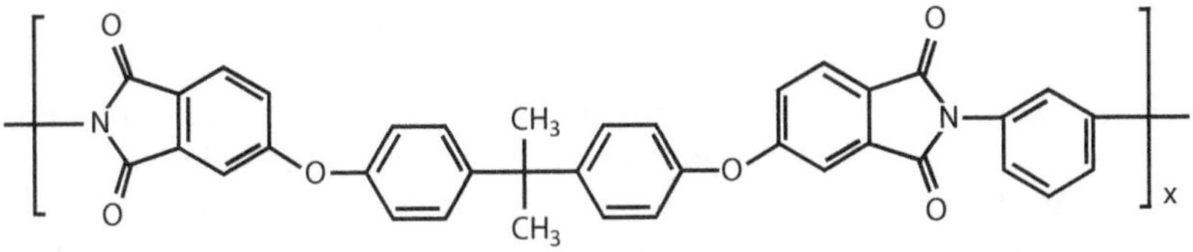

FIGURE 1.6
Polyetherimide's chemical structure.

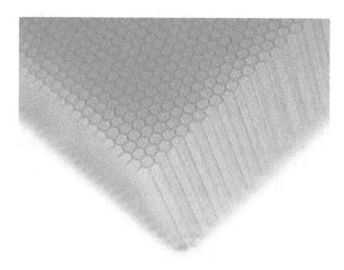

FIGURE 1.7
PEI honeycomb structure.

FIGURE 1.8
Extruded PEI plank at 20× expansion and 2.5 cm in thickness.

foam board with fiber reinforcement can serve as the core for a sandwich composite for structural flight components.

For construction, high insulation or low thermoconductivity is highly desirable from an energy management perspective. Aerogel demonstrates very low density (foam update), such as 0.007 g/cm³, and very low thermo-conductivity, around 14 mW/°K-min. This generated a tremendous amount of interest in energy management. Aerogel is made out of gelation, followed by solvent exchange and escape. The gel possesses properties that are close to liquid, yet with a shape like a solid. Since the pore size is too small, there is ample capillary force to hold the liquid inside to complete the gelation networking or cross-linking reaction. However, the solvent exchange could

FIGURE 1.9
Aerogel construction blanket from Aspen.

be a lengthy process, days or longer, depending on the pore size and dimension of the final product. Solvent exchange into liquid CO_2 seemed attractive, since it can be quickly removed when a supercritical point is reached. However, more effort is underway to make it more commercially viable.

Silica the aerogel blanket has been on the market for some time; it has a low thermoconductivity of 14 mW/°K-min relative to a PS insulation board of above 30 mW/°K-min. Figure 1.9 shows the installation of aerogel in panel insulation. Since the aerogel involves a tiny inorganic fiber, personal protection is required when handling this material. Lately, more efforts are underway to make it convenient to handle. Aerogel is proposed to be encapsulated by PS foam, which improves insulation and is easy to handle and can make it more feasible [20,21]. Also, aerogel particles can be dispersed in the PU foam matrix to obtain the similar benefits [22].

1.5 Degradable Foam

Every solid material will degrade or decompose, as a matter of environment and time, and its life-span difference may be over a few magnitudes. Degradability is the last component to complete the polymer material cycle. It has been a burden to plastics for over two decades. It is more so for high-expansion foam, due to its high consumption rate and visibility at landfills. However, it is the plastic and foam industries' social responsibility. On degradability, moisture or water solubility may be the first solution that comes to mind to tackle this challenge. Hydroxyl group–rich polymers can surely absorb water to build up enough hydrogen bonding to break apart its backbone to decompose. Polyvinyl alcohol is a good example. Due to rich hydroxyl groups, it is prone to form high crystallinity, which requires a higher melting load. Yet, it is too close to the thermal decomposition point of the base polymer to make it useful. Air Products found internal and external revisions to enlarge its processing window to make PVOH hygiene bags useful in hospitals and nursing homes to prevent unnecessary contact before cleaning in the washing system. Foaming the right-grade PVOH with water or an ethanol blowing agent became possible with the extrusion system [23]. Yet, the high cost of PVOH poses a concern in the marketplace.

Sugar and starch appear to contain quite a few hydroxyl groups. Sugar can be readily dissolved in water, but it lacks adequate structural strength. Starch contains strength in amylose and amylopectin. It drew devoted research and development efforts to make starch foam since the 1970s [24]. Amylose has hundreds of repeating unit $C_6O_5H_{10}$, and it is not water soluble. Over 45% amylose is preferred to provide strength for foaming in extrusion. However, general starch has a lower percentage of amylose, which became a big issue in extrusion processing. The water–starch slurry is not like a viscoelastic polymeric melt exhibiting gradual solidification. Instead, it cakes up rapidly after threshold cooling is reached. Adding polyvinyl alcohol becomes a common practice in increasing its strength [25]. Even so, starch foam is more brittle than most polymeric foams. Even with a certain amount of polymer (e.g., polypropylene), the foam rod product still exhibits certain brittleness and is not recommended for repeated usage [26]. Nonetheless, extruded starch foam can reach over 100 times expansion with water as the blowing agent. Figure 1.10 illustrates its cell size effects on density. It should be pointed out that hydrogen bonding formation is very rapid. In contrast to the conventional foam extrusion, it would not require a long length-over-diameter (L/D) extruder for processing. It is another advantage for starch foam in equipment investment.

Grains are known for their rich oil and hydroxyl content. They are inherently friendly to water. A good example for foaming is steamed rice out of the rice cooker. It is soaked in water and can absorb water at an elevated temperature to complete expansion in less than 30 min. It has been a human food for thousands of years. Some grains like sorghum can be mixed with moisture as the blowing agent into a short L/D extruder [27]. Under high speed, it generates high friction to break the skins for the immediate absorption of moisture. It also creates high pressure to push the solution out of the die for dynamic expansion. Immediate cutting at the die exit generates expanded foam, which is about the size of a big thumb. It is known as *loose fill* and is used to fill the space in a box for protection during transportation. The notable difference is that these expanded beads can dissolve in hot water.

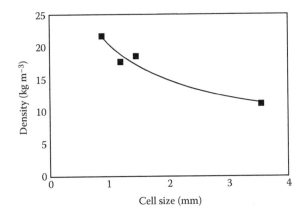

FIGURE 1.10
Cell size effects on density for starch foam.

Another degradable candidate is cellulose acetate, which is prepared by acetylating cellulose, the most abundant natural polymer. It is rigid, and its melting point is above 180°C. Cellulose acetate foam can be prepared by dissolving it in an acetone-based solvent; gelling occurs in an environment at 0°C–10°C for some time to extract impurities. Then, after freezing the gel to solidify the skeleton structure to sublime residual solvents, the open-cell structure is formed in low-density microcellular morphology. The density is between 0.1 and 0.5 g/cm³ [28,29]. Expansion over 30 times was achieved on corotating twins with a special kneading design. Blending polyethylene glycol was found to be necessary to improve its stability while using water as the blowing agent [30]. More effort is being devoted to this technique to increase its expansion ratio to stand out in this already competitive market [31].

One more development worth mentioning is cellulose foam for propellant containers in the ammunition industry [32]. Foamed celluloid makes low residue possible. Clean operation is very critical in precision gun operation. It was made in a batch process with a physical or chemical blowing agent or both. Low-density cellulose foam can be made in a mold through a two-stage heating process [33].

1.6 Three-Dimensional Printing (Additive Manufacturing)

Nature no doubt possesses wonderful design wisdom and mysteries. It is a great teacher to guide human civilization. There are many plants in nature that have a hollow space inside. The bamboo stalk in Figure 1.11 is a typical example in which hollow spaces are separated by horizontal plates as

FIGURE 1.11
Bamboo stalk.

joints. Bamboo fibers are of interest in gas permeation, and when made into sheets has been a favorite for furniture in the Orient, especially in summer for thousands of years. However, bamboo stalk can help us envision three-dimensional (3D) printing for foam in a closed structure. With a proper design, foam can be printed out instead of foaming out. Usage of a blowing agent and the associated emission issues can be totally avoided. It is not just processing but also economics. A catalytic converter is another example. Ceramic powder and necessary solvent components can be fed to form a slurry that then forms the final structure in a slow extrusion process [34]. Even in a structured space, as shown in Figure 1.12, it offers a tremendous amount of internal surface where the unreacted gasoline can be decomposed into simple inorganics before being emitted as exhaust from the muffler. Corrugated polypropylene and paper are another continuous example of a channeled sheet generated by extrusion. It can be extended to make a useful honeycomb structure for construction usage. Three-dimensional cushioning, as shown in Figure 1.13, can be made out of a hollow polyester yarn to reduce weight. Its compression,

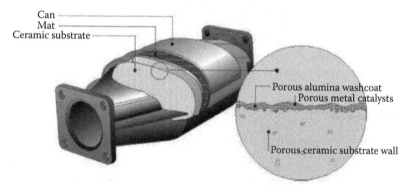

FIGURE 1.12
Ceramic catalytic converter with many tiny channels.

FIGURE 1.13
Three-dimensional cushioning product.

elasticity, and durability are as good as urethane and suitable for pillows and mattresses. According to Toyobo, its low-density range is from 30 to 150 kg/m³, and its thickness is up to 100 mm. It simply demonstrates the performance of low-density cushioning products that are made without a blowing agent.

In medical rehabilitation, a scaffold, which is an open-cell structure used to build up tissue, could be made by the 3D printing technology. Figure 1.14

FIGURE 1.14
Plastic skeleton/lattice from a 3D printer.

shows an open-cell lattice. When a multinozzle design is well advanced, it can be imagined that close cells with controlled cell size and distribution are possible. It is simply revolutionary to the whole polymeric foam world. The role of design will be increasingly important. The value of science and engineering knowledge and training will have to be reevaluated. Anyway, the combination of technology, material, and design may become the core for future innovation. Hurdles remain. But the bottom-line consideration is that 3D printing may dramatically change our lifestyle and thought process and ultimately culture. Although it could be a long shot, with many materials, processing, and design hurdles to overcome, the good news is that polymers are the choice material for this technology. In other words, material challenge could push polymers into their next spring in the future.

1.7 Hardware Development for Lower Density

Hardware is the body of the foam industry, without which science is nothing but good talking and theories. A common thermoplastic foam development path clearly shows that a batch process is in the beginning to prove the concept, then to a semicontinuous process for quality and performance, and eventually into a continuous extrusion process for productivity. A PS insulation board is a good example. However, in the high-volume commodity market, more expansion from extrusion processes has long been a development target to save material and to enhance economics. Over three decades ago, a vacuum system was implemented at the die exit to allow more expansion. Since a seal to keep a high-level vacuum is not a simple design, a huge pool system was proposed, as illustrated in Figure 1.15, by Union Carbide (UC) Industries (Parsippany, NJ) [35,36]. At present, it is still practiced in making low-density PS boards by Owens Corning.

More and more effort was spent on the cooling design, which is a flow rate and expansion bottleneck for foam plank extrusion. Enough cooling and even temperature distribution are critical parameters in controlling foam expansion and its cell integrity and cell size distribution. A cooling screw design has shown good benefits. In a screw-driven flow, the outer flow is always beneficial in contacting the cooling barrel, yet the inner flow circulates inside to form hot spots. How to break the hot core is a challenge. Anyway, some interesting designs were acknowledged [37,38]. A summary table is presented in Table 1.3 [39–42]. Into the twenty-first century, the design concept was switched to a heat exchanger to improve melt temperature distribution

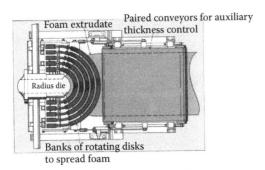

FIGURE 1.15

Vacuum design to induce more expansion for PS board; top: radius die, and bottom: vacuum chamber sealed by pool. (From Vacuum cuts PS foam extrusion costs, *Modern Plastics*, 62, 5, May, 1985.)

TABLE 1.3

Key Cooling Design Practices in Foam Extrusion

Name	Concept	Application	References
Turbo screw	Small windows on the flight of cooling screw	PS single screw extrusion	Fogarty [39]
45° offset	Break hot spot	Single screw	Rauwendaal [40]
Elongation	Prevent hot spot formation	Single and twin screw	Rauwendaal [41]
Barr screw	Force hot spot to contact cooling barrel	Single screw with special barrier design	Chung, C. I. [38]
Melt cooler	Prevent cold spot formation during cooling	Replace extruder	DE 202005001985 [42]

by reducing a big flow profile into small ones, as presented in Figure 1.16. A melt cooler with an adequate heat transfer area and dividing-and-combining design has gained attention [41]. Figure 1.17 shows a design to focus on heat exchange and streamline flow. The new extrusion system has this design to increase its capacity without increasing its size.

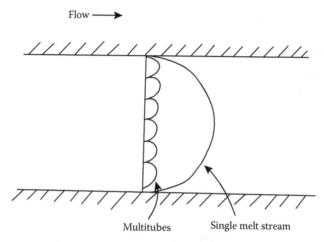

FIGURE 1.16
Temperature distribution between single melt stream and multitubes.

1.8 Nanocelullar Foaming

Nanocellular foaming has been a popular research subject in the past decade [43–45]. When the cell size is in the neighborhood of 100 nm, or 10^{-5} cm, gas molecules could be trapped in the cell to greatly reduce gas activity to alter the foam properties, such as very low thermal conductivity, dielectric constant, and improved light transmission. It can be imagined that molecular transformation from a dispersed gas to a nucleating gas is much more dynamic than microcellular foaming. It requires a much higher saturation pressure and a fast pressure release to create a cell density above 10^{12} cells/cm^3. How to transform so many gas-filled microvoids to grow against the surrounding matrix slightly relative to microcellular expansion is no wonder a great research topic. Since nucleation is a highly unstable thermodynamic phenomenon, polymer strength and its management are critical factors in making the unstable phenomenon into a very useful product.

Interesting progress has been made in cell size and porosity percentage over the past few years. Figure 1.18 shows that foaming temperature management is a viable way to generate cell size from microcellular to nanocellular [46,47]. Polymethylmethacrylate is inherently friendly to inorganic carbon dioxide. As a result, it has become the most used polymer in nanocellular investigation. High pressure, or high loading of CO_2, along with fast pressure release and polymer solidification, have been studied for a while. Template nucleation and copolymers were found to be good ways to generate nanocellular products in different laboratories. Semicrystalline PP was also found to be possible in making nanocellular or microcellular morphology

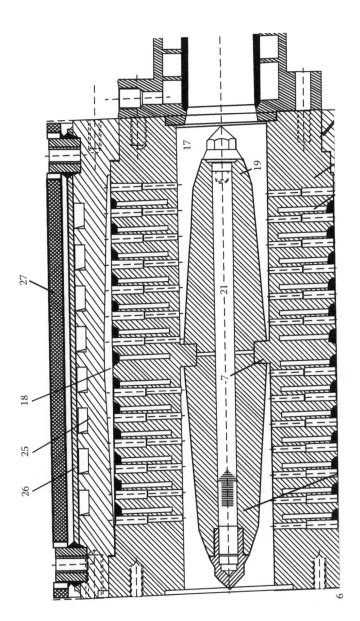

FIGURE 1.17
Melt cooler design (Reference 42: DE 202005001985).

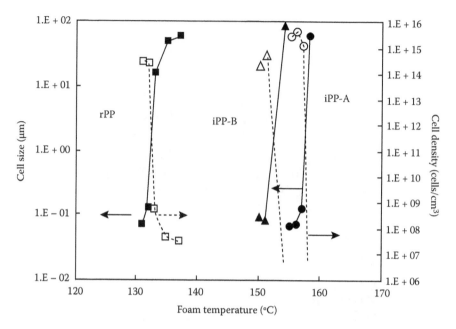

FIGURE 1.18
Cell size and its density variation for different PP resins at different foaming temperatures. (Reprinted from H. Shikuma and M. Ohshima, Preparation of polypropylene nanocellular foams by controlling their crystallinity with CO_2 and crystal nucleating agent, *Foams 2011,* spon. *Soc. Plas. Eng.*, Iselin, NJ, 2011. With permission.)

by precise temperature control. It simply suggests that crystalline formation is a good source of polymer strength. In recent research in semicrystalline nanocellular PLA at various temperatures from above to under its T_g, needle nanopores, as presented in Figure 1.19, were observed probably due to crystalline lamellae formation [48]. The needle pores could merge together to form a nucleating site for nanocellular foaming.

Nanocellular has been an interesting research subject in quite a few leading foam laboratories around the world. More pilot works are underway to generate a sample that is large enough to test out its properties. When key virtues are proved out, industrial support will be multiplied for product and technology development to further benefit our quality of life.

Since the amount of effort in research, development, and application is quite significant around the world, the progress of polymeric foam is envisioned. It is how to make each element useful in the cohesive and growing industry to continually improve our quality of life and to make global business more prosperous.

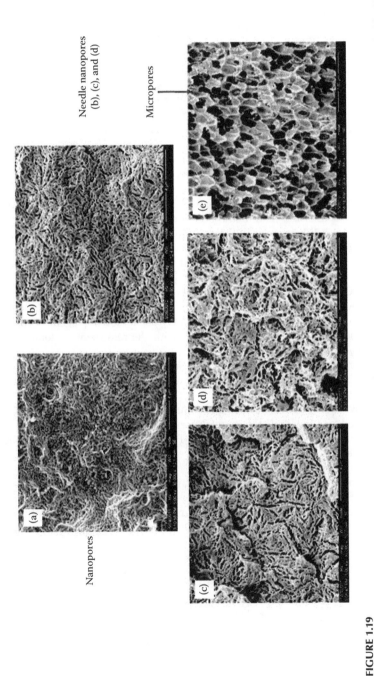

FIGURE 1.19
Effect of crystal lamellae formation on semicrystalline PLA nanoexperiments at various temperatures. (a) 80°C, (b) 90°C, (c) 100°C, (d) 105°C, and (e) 115°C (20 MPa). (From X. Liao, Foam Seminar series, Sichuan University, Sep., 2015. With permission.)

References

1. 2013. Plastics Foams: US Industry Study with Forecasts for 2017 & 2022. Freedonia Report #3114, pub. Dec.
2. K. Chrisman. 2014. Foam business development trends. *Poly-Foam Conference Shanghai*, spon. by *Foam Update*.
3. Alternative Fluorocarbons Environmental Acceptability Study (AFEAS). 1998. *Production, Sales and Atmospheric Release of Fluorocarbons through 1996*. AFEAS Program Office, Washington, D.C.
4. J. A. Creazzo, M. J. Nappa, A. C. Sievert, and E. N. Swearingen. 2014. Blowing agents for forming foam comprising unsaturated fluorocarbons. US Patent: 8,633,339, Jan.
5. B. B. Chen, M. Y. Elsheikh, P. Bonnet, B. L. Van Horn, and J. S. Costa. 2012. Blowing agent composition of hydrochlorofluoroolefin. US Patent: 8,314,159, Nov.
6. J. E. Martini-Vvedensky, N. P. Suh, and F. A. Waldman. 1984. Microcellular closed cell foams and their method of manufacture. US Patent: 4,473,665.
7. N. P. Suh. 1996. *Innovation in Polymer Processing Molding*. ed. by J. Stevenson, pub. by Hanser Publishers, Munich.
8. P. C. Lee and C. B. Park. 2014. Extrusion of high-density and low-density microcellular plastics. Ch. 14 in *Foam Extrusion, 2nd ed.*, ed. by S. T. Lee and C. B. Park, pub. by CRC Press, New York, pp. 435–488.
9. H. Yokoyama and K. Sugiyama. 2005. Nanocellular structures in block copolymers with CO_2-philic blocks using CO_2 as a blowing agent: Crossover from micro- to nanocellular structures with depressurization temperature. *Macromolecules*, 38, 10516–10522.
10. R. Miyamoto, S. Yasuhara, H. Shikuma, and M. Oshima. 2014. Preparation of micro/nanocellular polypropylene foam with crystal nucleating agent. *Pol. Eng. Sci.*, 24, 9, 2075–2085.
11. Furakawa Electric. 2012. Develop and commoditize reflective sheet MCPOLYCA optimized for LED light equipment. News release, Apr. 12.
12. S. T. Lee, M. Xanthos, and S. Dey. 2014. Foam extrusion of polyethylene terephthalate. Ch. 15 in *Foam Extrusion 2nd ed.*, ed. S. T. Lee and C. B. Park, CRC Press, New York.
13. V. Kumar. 1993. Polyethylene terephthalate foams with integral crystalline skins. US Patent: 5,182,307, Jan.
14. M. Lindenfelzer. 2014. Equipment and material considerations for physically foamed film. *Poly-Foam Conference Shanghai*, Dec. 3–4, 2014, spon. by *Foam Update*.
15. R. Pop-Live, K-H. Lee, and C. B. Park. 2006. Manufacture of integral skin PP foam composites in rotational molding. *J. Cell Plas.*, 42, 2, 139–152.
16. X. Jiang. 2014. Supercritical CO_2-assisted compression foaming of polypropylene. *Poly-Foam Conference Shanghai*, Dec. 3–4, 2014, spon. by *Foam Update*.
17. S. T. Lee. 2015. Polyetherimide Foam. Editorial of *Foam Update*, Aug.
18. K. Wang, F. Wu, W. Zhai, and W. Zheng. 2013. Effect of polytetrafluoroethylene on the foaming behaviors of linear polypropylene in continuous extrusion. *J. App. Poly. Sci.*, 129, 4, 2253–2260.

19. J. J. Childress. 1999. Interlaced Z-pin sandwiched structure. US Patent: 5,935,680, Aug.
20. S. Perkowitz. 2001. Practical foam. Ch. 4 in *Universal Foam*, pub. by Anchor Books, New York.
21. L. Lotti, G. Viro, and M. J. Skowronski. 2013. Method for making organic foam composites containing aerogel particles. WO patent: 2013000861, March.
22. G. J. Bleys. 2015. *Innovations in Polyurethanes 2014*. PDF e-Book, pub. Feb.
23. N. Malwitz and S. T. Lee. 1992. Thermoplastic compositions for water soluble foams. US Patent: 5,089,535, Feb.
24. D. Loudrin, G. D. Valle, and P. Colonna. 1995. Influence of amylose content on starch films and foams. *Carbohydrate Polymers*, 27, 261–270.
25. S. T. Lee. 2014. Starch foam. Editorial in *Foam Update*, March.
26. Z. Yang, D. Gravier, and R. Narayan. 2013. Extrusion of humidity-resistant starch foam sheet. *Polym. Eng. & Sci.*, 33, 4, 857–867.
27. L. Sorrentino, S. Iannace, S. T. Lee, and R. Pantani. 2015. Foaming technologies for thermoplastics. Ch. 11 of *Biofoams*, ed. by S. Iannace and C. B. Park, pub. by CRC Press, New York, pp. 287–314.
28. J. A. Rinde. 1978. Method of making a cellulose acetate low density microcellular foam. US Patent: 4,118,449, Oct.
29. H. Mori, M. Yoshida, M. Matsui, and M. Nakanishi. 2001. Biodegradable cellulose acetate foam and process for its production. US Patent: 6,221,924, Apr.
30. H. G. Franke and D. R. Bittner. 2001. Sorghum meal-based biodegradable formulations, shaped products made therefrom, and methods of making said shaped products. US Patent: 6,176,915, Jan.
31. S. Zepnik, A. Kesselring, C. Michels, C. Bonten, and F. van Luck. 2010. Cellulose acetate foams. *Bioplastics*, 10, 5, 26–27.
32. M.-W. Young, C. G. Gogos, N. Faridi, L. Zhu, P. Bonnett, H. Shimm, E. Caravaca, and J. Park Jr. 2013. Foamed celluloid mortar propellant increment containers. US Patent: 8,617,328, Dec.
33. N. Faridi, L. Zhu, M.-W. Young, C. G. Gogos, F. Shen, E. Caravaca, M. Elalem, V. Panchal, and D. Conti. 2014. Foamed celluloid process using expandable beads. US Patent: 8,696,838, Apr.
34. R. F. Beckmeyer and S. F. Grubber. 1998. Method for making a ceramic catalytic converter open cell substrate with rounded corners. US Patent: 5,731,562, March.
35. 1985. Vacuum cuts PS foam extrusion costs. *Modern Plastics*, 62, 5, May.
36. R. Pagan. 1988. Method and apparatus for foam extrusion into a vacuum chamber having a liquid baffle with control of the liquid level. US Patent: 4,783,291, Nov.
37. C. Yang and D. Bigio. 2014. Mixing design for foam extrusion. Ch. 12 of *Foam Extrusion, 2nd ed.*, ed. by S. T. Lee and C. B. Park, pub. by CRC Press, New York.
38. C. I. Chung. 2000. Ch. 6 of *Extrusion of Polymers*, pub. by Hanser, Munich.
39. J. Fogarty. 2000. Thermoplastic foam extrusion screw with circulation channels. US Patent: 6,015,227, Jan.
40. C. Rauwendaal. 2004. Screw design for cooling extruders. *Ann. Tech. Conf. (ANTEC)*, spon. *Sci. Plas. Eng.* preprint pp. 278–282.
41. C. Rauwendaal, T. Osswald, P. Gramann, and B. Davis. 1999. Design of dispersive mixing devices. *Intl. Poly. Proc.*, 14, 1, 28–34.
42. 2005. Melt Cooler. German Patent: DE 202005001985, June.

43. H. Guo, N. Nicolae, and V. Kumar. 2015. Solid-state poly(methyl methacrylate) (PMMA) nanofoams. Part II: Low-temperature solid state process space using CO_2 and the resulting morphologies. *Polymer*, 70, 231–241.
44. S. Costeux. 2014. CO_2-blown nanocellular foams. *J. Appl. Polym. Sci.*, 131, 23.
45. H. Guo, A. Nicolae, and V. Kumar. 2015. Solid-state poly(methyl methacrylate) (PMMA) nanofoams Part I: Low-temp. solid-state process space using CO_2 and the resulting morphologies. *Polymer*, 70, 231–241.
46. H. Shikuma and M. Ohshima. 2011. Preparation of polypropylene nanocellular foams by controlling their crystallinity with CO_2 and crystal nucleating agent. *Foams 2011*, spon. *Soc. Plas. Eng.*, Iselin, NJ.
47. P. Cassagnau. 2015. Nanofoaming of PMMA using a batch foaming process. *Poly-Foam Conf. 2015*, spon. *Foam Update*, Shanghai.
48. X. Liao. 2015. Foam Seminar series. Sichuan University, Sichuan, China. Sep.

2

Microcellular Polypropylene Foam

Peng Guo and Xiulei Jiang

CONTENTS

2.1 Introduction

Thermoplastic foams are applied widely in many industrial fields, such as packaging, cushioning, insulation, and construction, for their excellent processability, recyclability, and weak influence on environment. Among thermoplastic foams, polypropylene (PP) foam has several advantages over other foams. PP foam is resistant to chemicals and abrasion. PP foam offers higher strength as compared to polyethylene (PE) foam and better compact strength as compared to polystyrene foam. PP foam also provides a higher service temperature and good temperature stability as compared to other polyolefin foams.

    Currently, the PP foam products available on the market are high foaming ratio expanded polypropylene (EPP) foam beads, low foaming ratio EPP foam sheets, and high foaming ratio expanded cross-linked PP foam thin sheets. Normally, the thickness of the latter two is smaller than 5 mm.

EPP beads are fabricated by either autoclaving foaming or extrusion foaming. Both thermoplastic PP foam sheets and cross-linked PP foam sheets are mainly fabricated by extrusion foaming. The applications and production volume of PP foams both have been expanding in recent years. In some application cases, such as automotive, electrical devices, and buildings, PP foams play a critical role as functional parts.

Two of the obstacles to using PP foaming is its low melt strength and melt elasticity since PP is a linear semicrystalline polymer [1], which causes the processing temperature range to be narrow and the cell to collapse easily, typically in the extrusion foaming and injection molding processes, where PP is in a melt state at the stage of cell nucleation and growth. Therefore, it is essential to modify PP to enhance its melt strength and melt elasticity for the purpose of foaming.

Carbon dioxide is cheap, clean, and nonflammable. It is a notable trend to use carbon dioxide to replace conventional chemical and physical foaming agents in the foam industry. The foaming behaviors of PP, with carbon dioxide as a foaming agent in different foaming processes, were extensively studied.

PP becomes foamable at the deformation temperature, which is below its melting temperature. In a batch foaming process, PP could be impregnated with carbon dioxide at either low temperature (usually room temperature) or high temperature (usually below the PP melting temperature). The cell nucleation is induced by a rapid temperature increase if the impregnation temperature is low or by a rapid depressurization if the impregnation temperature is high. The carbon dioxide–induced plasticization leads to the melting temperature depression of PP. The foaming temperature should be chosen according to the PP melting temperature under CO_2 pressure, which could be detected with a high-pressure differential scanning calorimeter.

The impact of PP crystallization on the PP foaming behavior was studied [2–6]. It is confirmed experimentally that CO_2 can only diffuse into the amorphous phase but not the crystal phase. Thus, cells can only appear in the amorphous region. The dominant cell nucleation in a batch PP foaming process would be heterogeneous, due to the structurally heterogeneous nature of PP and the concomitantly nonuniform dispersion of carbon dioxide in the PP matrix.

A number of fillers, such as montmorillonite, $CaCO_3$, and some other polymers, such as PE and polyoxyethylene, were added to the PP to form PP base composites or blends to improve its foamability. Also, the foaming behavior of general PP without modification was also studied. In the solid-state batch foaming process, general PP can also exhibit good foamability and form a uniform cell structure. By forming a PP/rubber blend and PP/styrene-ethylene-butylene-styrene (SEBS) blend, nanocellular PP foam could even be obtained.

Due to the obstacles of PP resin, foaming facility scaling, and process control, the commercial development of microcellular PP foam is relatively slow, especially with respect to microcellular PP foam, which has a high thickness and high expansion ratio. Here, we introduce a novel foaming process to produce microcellular PP foam, which applied a newly developed PP resin, and the foaming facility is developed on the batch foaming mechanism.

2.2 Development of PP Resin for Foam

2.2.1 Background

Polypropylenes are homopolymers of propylene or copolymers of propylene with up to 10% copolymerized ethylene. Since Giulio Natta invented isotactic polypropylene (iPP) in 1954, it has been a very attractive research and industrial point for its excellent solid-state properties such as high modulus and tensile strength, rigidity and excellent heat resistance, and low cost [7]. It is estimated that the demand and capacity will reach 82 M and 90 M tons until the end of 2016, respectively [8]. However, commercial iPP, produced with Ziegler–Natta or metallocene catalysts, is of a linear structure, which is unlike amorphous polymers, such as polystyrene with a region having a property that is similar to the rubber elasticity in a wide temperature range. Thus, PP cannot be thermoformed in a wide temperature range. Meanwhile, the softening point of PP is close to its melting point. When the temperature is higher than the melting point, the melt strength and melt viscosity of PP will decrease rapidly, causing such problems as edges curling and shrinkage that would easily appear during extrusion, coating, and rolling; uneven wall thickness of the products during thermoforming; and foam collapse during extrusion foaming. Therefore, the use of PP in the fields of blow molding, foaming, and themoforming is limited. As a result, the development of a PP with a high melt strength and good ductility is always an interesting issue. The main factor that affects the PP melt strength is the molecular structure of the polymer, which comprises the size of the molecular weight, the molecular weight distribution, whether the molecular chain comprises long branched chains or not, and the length and distribution of the branched chains.

Four kinds of techniques have been developed to improve the melt strength of PP: (1) direct polymerization, (2) reactive extrusion, (3) electron beam irradiation, and (4) compounding. Considering that the larger the molecular weight of PP is, the higher the melt strength of PP, the first way is to increase the ratio of the molecular weight component through direct polymerization. However, the larger the molecular weight of PP is, the more unfavorable it is for the postprocessing forming performance of propylene polymers. In addition, the LCB introduced on the PP skeleton during polymerization is realized via a unique catalyst and an additional monomer. However, relative polymerization processing is sophisticated, and the operation is inconvenient. The last three methods introduce branches into linear polypropylene (LPP) to produce a *high-melt-strength polypropylene* (HMSPP) with enhanced processability in the presence of a cross-linker (such as peroxide and coagent), electron beam irradiation, and macromonomer grafting [9,10]. Nevertheless, the HMSPP obtained by postreactor treatment displays unstable properties and residual non-environment-friendly chemical agents.

2.2.2 Definition and Properties of HMSPP

2.2.2.1 Definition

HMSPPs are characterized by a content of long-chain branches (LCBs) of 0.05–0.15 branched/1000C and a polydispersity that is higher than 4. The presence of LCBs and the high polydispersity have an enormous influences on the melt properties of PP [10]. To improve the melt strength of PP, three methods can be used: (1) increase molecular weight, (2) broaden molecular weight distribution, and (3) introduce a branched structure [11,12]. Melt strength is the strength of the plastic while in the molten state. The melt strength, also defined as the resistance of a melt to drawdown, is an important parameter describing the extensibility of a polymer melt, which is of great importance in all processing technologies where elongational flows and the stretching of polymer melt occur [10].

2.2.2.2 Measurement

For PP, the melt strength is usually measured using a four-wheeled Göttfert Rheotens 71.97 tester in combination with a Göttfert Rheograph 25 high-pressure capillary rheometer [13]. The Rheotens was located close to the exit of the capillary. The additional pair of pull-off wheels was integrated to prevent the sticking of the elongated polymer strand once it had passed the first pair of pulleys during the test. PP pellets were melted in the heated test cylinder and pressed with a test piston at constant speed out through a capillary. The melt strand was continuously drawn down at a linear exponentially accelerating velocity between the two counterrotating wheels of the device, which were mounted on a balanced beam. The drawdown was continuously measured as a function of the angular speed of the wheels as the melt strand was pulling. The force and velocity resulting from the melt strand breakage were deemed as *melt strength* (M_s) and *velocity at break* (Br_{Vel}) for the tested condition.

Melt strength is also obtained by the melt flow rate and the corresponding equation qualitatively. In brief, when melt is extruded from the Capilograph, the diameter of melt will change due to dead load. Thus, we can measure this variation to gain the M_s indirectly. Furthermore, the sag resistance can determine the melt strength.

2.2.2.3 Elongational Viscosity

Elongational viscosity is a viscosity coefficient when applied stress is extensional stress. This parameter is often used to characterize polymer solutions.

It provides a method to measure melt flow, which is used to evaluate the consistency of materials or determine the extent of degradation of the plastic. Elongational viscosity can be measured using rheometers that exert extensional stress.

HMSPP exhibits a distinct stress–strain behavior from commercial PP. The elongational viscosity of commercial PP will decrease with the enhancement of stress. When stress attains a critical value, the fracture of melt will occur suddenly. It exhibits a ductile fracture behavior. However, HMSPP displays a different stress–strain behavior. When the strain rate is constant, elongational viscosity is kept stable. At the same time, the flow stress of melt increases gradually. Subsequently, the flow stress of melt increases exponentially, which indicates a strain-hardening behavior.

2.2.3 Preparation

2.2.3.1 Direct Polymerization

HMSPP is obtained by direct polymerization mainly through the following two methods: (1) adding an initiator and olefin monomer lending to the growth of LCB on the PP linear skeleton during the polymerization and (2) broaden molecular weight distribution through blending the low and high molecular weight fraction from two reactors. Langston et al. [14] prepared long-chain-branched polypropylene (LCBPP) with high molecular weights and well-controlled structures via a combination of rac-Me2Si(2-Me-4-Ph-Ind)ZrCl2/MAO catalyst and a T-reagent, such as p-(3-butenyl) styrene. LCBPPs indicated a regular increase in Arrhenius flow activation energy and zero-shear viscosity with the increase of branch density. The strain-hardening LCBPPs were observed in extensional flow. Wang et al. [15] synthesized a series of polypropylene-g-poly(ethylene-co-1-butene) (PP-g-EBR) graft copolymers with well-defined LCB molecular structures through the coordination and anionic polymerization. Differential scanning calorimetry (DSC) results showed that the crystallization temperature of PP-g-EBR increased with the enhancement of branch length and branch density. However, further increasing the LCB level would reduce the crystallization temperature when the LCB level exceeded a certain degree.

Direct polymerization is a main routine for the industrial production of HMSPP. There are currently two different postreactor technologies on the market: one that was developed by Montell and the other by Borealis. In 1990, Himont (Montell's predecessor) applied for a patent for HMSPP [16]. Since 1994, Montell (now named after LyondellBasell) has started the development of LCBPP by using a beam radiation process that is named after PF814 [10]. The target was the conversion of the linear iPP structure into LCB material. The experiences and know-how with the grafting of monomers during the radical reactions of PP with peroxides led the Borealis research and development group to come to a new way of creating LCBPP. By choosing small amounts of monomers with a special resonance of their radicals' stability, the Borealis Company was the first to recognize the possibility of increased stabilization of the C-radicals so that they could realize a recombination of these monomer radicals at a higher temperature. They developed a series of HMSPP resin utilizing the Daploy process. These PP resins could be characterized by an aftertreatment of the native PP granules out of the polymer synthesis by a combination of peroxides and monomers at temperatures well below the melt temperatures for PP processing [10]. Among them, Daploy WB140HMS is a structurally isomeric-modified propylene homopolymer, and Daploy WB260HMS is a propylene-based, structurally isomeric polymer. Both are designed for foamed films and sheets.

Furthermore, it is worth noting that enabling PP to have a wider molecular weight distribution is significant for processing applications. In order to obtain PP with this best performance, the ideal polymer product should comprise a small amount of polymer fraction with very high molecular weight. Generally, a certain amount of polymer fraction with relatively high molecular weight provides a high melt strength, and a large amount of polymer fraction with low molecular weight provides better processability. Sinopec Zhenhai Refining and Chemical Company produces HMSPP by direct polymerization through controlling the species and ratios of the external electron donors in the Ziegler–Natta catalyst system at different reaction stages. The Beijing Research Institute of Chemical Industry provides this innovative technique. This PP contains a wide molecular weight distribution and exhibits excellent mechanical properties, especially with very high melt strength. HMSPPs named as HMS20Z, E02ES, and E07ES are produced in a PP plant, which comprises a prepolymerization reactor, a first tubular loop reactor, and a second tubular loop reactor. The polymerization process and its steps are as follows: the first step is the prepolymerization process. The main catalyst contains titanium with a content of 2.5 wt% Ti, 8.2 wt% Mg, and 14 wt% dibutyl phthalate. The main catalyst, the cocatalyst (triethylaluminium), and the first external electron donor (dicycylpentyl dimethoxy silane [DCPMS]) are precontacted with each other at 9°C for 30 min and then continuously added into the prepolymerization reactor. The prepolymerization is carried out in the propylene liquid-phase bulk, wherein the temperature is 16°C. The residence time is about 4 min. The flow rate of triethylaluminum, DCPMS, and the main catalyst into the prepolymerization reactor are 6.15, 0.37, and 0.02 g/h, respectively. The second step is propylene homopolymerization. The catalyst flows into the two tubular loop reactors that are connected in series. The reaction temperature of the two tubular loop reactors is 71°C, and the reaction pressure is 3.9 MPa. The productivity ratio of the first tubular loop reactor to that of the second is about 44:56. No hydrogen is added in the feed to the first tubular loop reactor. A certain amount of hydrogen is added into the second tubular loop reactor, and the hydrogen content measured by the online chromatography is 4500 ppmV. Both propylene and catalyst components are added directly into the first tubular loop reactor after prepolymerization. Tetraethoxysilane with a flow rate of 0.67 g/h is supplemented into the second tubular loop reactor. After the propylene is separated by flashing from the polymer from the second tubular loop reactor, the activity of the catalyst in the reactor is eliminated by wet nitrogen. Then, polymer powders are obtained after heating and drying the polymer.

JSP realized that the low melt temperature of copolymer is useful for the preparation and thermal forming of EPP beads. First developed and commercialized in the 1980s, EPP is produced in a closed-cell structure by JSP. The autoclave process for JSP's ARPRO EPP uses carbon dioxide, water, and steam, meaning no hazardous blowing agents are applied. The material's density ranges from 18 to 250 g/L, which displays high energy absorbability and is still four times lighter than a compact component. Conventionally, PP resins are polymerized by the use of the Ziegler–Natta catalyst. In recent years, it has been found that PP resins prepared by the use of a metallocene polymerization catalyst have a lower melting point than that of the PP resin that is polymerized using the Ziegler–Natta catalyst so that the resultant expanded beads can be molded at a lower steam pressure, and cell diameter can be more uniform. Thus, PP resin polymerized by metallocene catalyst is recommended as a base resin. Through investigating the characteristic properties of the propylene homopolymer obtained by a metallocene catalyst, it has problems with moldability. Among them, a specific type of homopolypropylene exhibiting specified physical properties is used as a base resin; a molded article can be produced with good moldability by subjecting the expanded beads of the homopolypropylene to molding. Moreover, the obtained molded article exhibited better properties in rigidity compression strength and energy absorption efficiency than its prior counterparts. Moreover, a propylene homopolymer is generally poor in impact resistance at low temperatures. The resin copolymerized with a small amount of ethylene or α-olefin, is preferably used as the PP resin. Commercially available homopolypropylene resins prepared by metallocene catalyst display a 10°C lower melting point (150°C) than that of an ordinary homopolypropylene resin that is polymerized by the use of the Ziegler–Natta catalyst. In addition, a copolymer, which is prepared by random copolymerization of propylene with a small amount of ethylene or by use of a metallocene polymerization catalyst, decreases the melting temperature when an amount of ethylene is added in the polymerization system. The molded article obtained from the resultant expanded beads significantly lowers in heat resistance. JSP has made a series of intensive studies in the PP resins that are used for the preparation of EPP beads. As a result, it is found that a molded article obtained from expanded beads using a random copolymer as a base resin exhibits a higher heat resistance compared with other common commercial copolymers. This kind of random copolymer contains one or two or more in admixture of the comonomers that are selected from the group consisting of ethylene and α-olefins and is polymerized by a metallocene catalyst.

2.2.3.2 Reactive Extrusion

Reactive extrusion is a type of extrusion process through chemical reactions between different functional groups on the polymer chains [17]. Twin-screw extruders are particularly suitable for the reactive extrusion process due to their excellent mixing properties. Generally, the first step is blending the initiator, polyfunctional monomers, a cross-linking agent (silane or perox-ide), and PP resin in a high-speed mixer. The second step is the removal of solvent, reactive extrusion, pelletizing, and desiccation. The reactive extrusion possessed many merits, including simple operation, low cost, and high productivity.

Grafting is one common method of reactive extrusion. The grafting mono-mer mainly contains maleic anhydride (MAH) and trimethylolpropane triac-rylate (TMPTA). Tang et al. [18] prepared an HMSPP via a twin-screw reactive extruder from a commercial iPP through two steps. First, MAH is grafted to PP to obtain a maleic anhydride-grafted polypropylene (PP-g-MA), and then the grafted polymer is reacted with epoxy to extend the branched chain. The melt flow rate and sag resistance test exhibited that the melt strength of the HMSPP improved considerably. A DSC test showed that the LCB acts as a nucleating agent in the crystallization of the HMSPP resulting in a high crystallization temperature and crystallinity. Mabrouk et al. [19] pre-pared HMSPP by two approaches. Single-step processes generate bimodal products with highly differentiated chain populations through radical-mediated addition of PP to triallyl phosphate. Two-step methods generate more uniform architectures involving PP addition to vinyltriethoxysilane followed by moisture-curing. Sequential functionalization/cross-linking strategies can create more uniform branching distributions than single-step, coagent-based techniques. Zhang et al. [20] investigated the effect of copper N,N-dimethyldithiocarbamate (CDD) on melt reactions during preparing LCBPP via free radical grafting. The results indicated that CDD could effi-ciently control two side reactions including the degradation of PP backbone and the homopolymerization of multifunctional monomer (TMPTA) in the presence of peroxide.

A cross-linking reaction is another common path for reactive extrusion. The cross-linking favors LPP, a form of three-dimensional network leading to the enhancement of melt strength. Cross-linking agents mainly contain benzoyl peroxide (BPO) and dicumyl peroxide. Su et al. [21,22] added LCB to LPP via reactive extrusion using selected polyfunctional monomers and BPO [23]. They also prepared HMSPP in an extruder by a melt grafting reaction in the presence of various peroxides and a polyfunctional mono-mer of 1, 6-hexanediol diarylate. Through these experiments, some rules can be obtained as follows. Peroxides with more stable radicals and lower decomposition temperatures promoted the branching reaction and contrib-uted to the enhancement of the branching level and melt strength for mod-ified PPs. The type and concentration of peroxide used for modification could influence and control the branching number by peroxide properties and structure.

PP made by metallocene catalyst could also be basic resin for the preparation of HMSPP. Pivokonsky et al. [11] utilized dimethyl-2, 5-(i-butylperoxy) hexane at a concentration of 0.01 wt.% as a peroxide reaction with PP to obtain HMSPP. On the basis of the activation energy data, it is suggested that processed mPPs may have a star-branched structure.

The use of esters and a coupling agent is significant for the properties of HMSPP obtain by reactive extrusion. Cao et al. [24] produced LCBPP using ethylenediamine (EDA) as a coupling agent with the reactive extrusion of PP-g-MA in a corotating twin-screw extruder. LCBPP obtained with EDA exhibited better dynamic shear, transient extensional rheological characteristics, significantly enhanced melt strength, and strain-hardening behaviors. Augier et al. [25] demonstrated that 2-furyl acrylates are very active in maintaining the high molecular weight of PP during the various functionalization processes that are carried out in the melt in the presence of peroxides.

Furthermore, due to the degradation of macromolecules using grafting at a high temperature, a cross-linking agent should be used. The cross-linking agent is usually a polyfunctional monomer. Although reactive extrusion processing operates easily, the quality and branching number of the resultant could not be controlled precisely due to the addition of peroxide or monomer in error. The industrial preparation of HMSPP via reactive extrusion could not be realized easily.

2.2.3.3 Electron Beam Irradiation

In order to improve melt strength, long-chain branching is generated by irradiation of PP under different atmospheres, with or without polyfunctional branching agents. Generally, long-chain branching favors the improvement of melt strength, leading to the enhancement of processability. It is beneficial for specific fields needing higher melt strengths. The irradiation ways contain irradiation in an oxygen-free or a reduced-oxygen atmosphere, in molten state, and with the addition of a branching agent [26]. In order to minimize the oxidative degradation when irradiated in air, the irradiation of PP should be finished in an *oxygen-free* or a reduced-oxygen atmosphere. Moreover, oxidative degradation has been competing with cross-linking or branching, when PP is treated by ionizing radiation. Although an oxygen-free atmosphere is significant and necessary to keep the balance between the growth of long-chain branching and degradation, this atmosphere needs professional radiation equipment and processing techniques, leading to high cost.

Cerreda et al. [27] investigated the effect of electron beam irradiation on metallocene iPP in a wide electron beam dose range. The researches on films and pellets demonstrated the existence of branch structure and the absence of gel fraction in this metallocene PP. Otaguro et al. [28] analyzed the degradation of high-molecular-weight iPP that is subjected to gamma-ray irradiation up to 100 kGy in an inert atmosphere. Results indicated that the chain scission is the prioritized process creating free macroradicals that are neutralized, which seriously decreased the molecular weight of the polymer at low irradiation doses, i.e., below 12.5 kGy. The concentration of neoformed free macroradicals is too great to be neutralized when the dose level is enhanced above 12.5 kGy. A few macroradicals attack neighboring chains, resulting in chain branching and finally even cross-linking. Oliani et al. [29] synthesized HMSPP in grains by the gamma irradiation of PP under a cross-linking atom sphere of acetylene. The obtained PP is subjected to thermal treatment for the radical recombination and annihilation of the remaining radicals. With the increase of the irradiation dose in the synthesis of HMSPP visbreaking products, the frequency of cracks increases. Compared to PP, HMSPP displays more sensitivity to accelerated ageing (ultraviolet). It indicates that the previous degradation has an influence in the postuse of HMSPP.

During the irradiation process, the kind of monomer and structure of PP resin exert important and remarkable influences on the properties of the obtained HMSPP. Zarch et al. [30] investigated the rheological properties of electron beam–irradiated propylene homopolymer containing a polybutene (PB) resin TMPTA multifunctional monomer. The addition of PB resin in the PP can promote the movement of PP macromolecules, reduce the degradation/chain scission in irradiation process, and increase the branching efficiency. By evaluating the effect of electron beam irradiation on rheological properties of a poly (propylene-co-ethylene) heterophasic copolymer, Koosha et al. [31] found an increase in the melt flow rate behavior and shifting the crossover point to higher frequencies and increase in melt strength when PP was irradiated at a higher frequencies. At the same time, the ethylene–propylene phase of the PP copolymer cross-links due to irradiation, and a significant effect on the rheological behavior of samples was observed. Mousavi et al. [32] compared chemical and electron beam irradiation methods for adding a branched structure into PP and a propylene–ethylene copolymer. The chemical method utilized an initiator and a TMPTMA monomer in an internal mixer. The irradiation method is realized through irradiating a polymer by electron beam under air and nitrogen atmosphere. The branching appeared efficiently by irradiation under air, but degradation occurred remarkably in chemical method. A little introduction of ethylene in the propylene copolymer facilitated branching over degradation.

As mentioned in Section 2.2.3.1, Montell (LyondellBasell) produced LCBPP using an electron beam radiation process [10]. Montell developed a very interesting method of introducing LCBs on PP by radical reactions below 80°C. By choosing small amounts of monomers with a special resonance of their radicals' stability, the Borealis Company utilized a recombination of these monomer radicals at a higher temperature to produce HMSPP with a long branched structure.

2.2.3.4 Compounding

Compounding is a useful path to enhance melt strength. Compounding is defined as a process of incorporating additives, modifiers, or other resins into PP for achieving uniformity on a scale that is appropriate to the quality of the articles that are subsequently made from the compound. It is also known as hot or melt blending. Usually, the elastomer and a copolymer with a low melting point are used to decrease the melting point and enhance the melt strength. A blending component mainly contains the long-chain-branched PP/PE, HMSPP, and ultrafine rubber powder.

Ni et al. [33] investigated the crystallization behaviors and crystallization kinetics of iPP, LCB-iPP, and their blends. The results indicated that the incorporation of an LCB structure reduced the fold-free energy of nucleation and increased ΔE for the transport of crystalline units across the phase. Fang et al. [34] investigated the effect of blending an LCBPP with an LPP on the processability and properties of blown films. The rheological data revealed that blending an LCBPP with an LPP improved the elongational properties and the bubble stability, but a severe drop in the mechanical strength was observed for the blends. The most deteriorating effect was the reduction in the elongation at break in tensile tests that are carried out in the transverse direction, where no yielding behavior was observed for the blends. Tabatabaei et al. [35] prepared blends of LCBPP and four kinds of LPP having different molecular weights via a twin-screw extruder. The linear viscoelastic properties exhibited the immiscibility of LPP-based blends with the high molecular weight and the miscibility of LPP-based blends with the low molecular weight. Moreover, the Palierne emulsion model well suggested the predictions of the linear viscoelastic properties for both miscible and immiscible PP blends. An et al. [36] prepared HMSPP by an in situ heat induction reaction. Therein, pure PP powders were used as a basic resin, whereas low-density PE and TMPTA were added as a blending resin and as a cross-linking agent, respectively. The results indicate that the impact strength and thermal stability of HMSPP were noticeably enhanced. Furthermore, HMSPP displays good processing performance and foaming properties. The foams produced by HMSPP showed uniform, closed, and independent cells. Liu et al. [37] utilized ultrafine full-vulcanized polybutadiene rubber (UFBR) in particle sizes of ca. 50–100 nm to improve the mechanical and processing performances of PP. They also used PP-g-maleic anhydride (PP-MA) as a compatibilizer to promote the interfacial characteristics between the two components. This PP/UFBR obtains highly branched PP when the addition of UFBR attains 10 wt%. In contrast, a lower introduction of UFBR with a small amount of PP-MA facilitates the material with the rheological properties of high-melt-strength materials like LCBPP.

2.2.3.5 Summary

In order to illustrate the four technologies for enhancing melt strength, as mentioned in Section 2.2.1, a comparison of their advantages and disadvantages is summarized, as shown in Table 2.1.

From Table 2.1, the four techniques are totally divided into two types: (1) postreactor treatment (reactive extrusion, electron beam irradiation, compounding) and (2) polymerization in situ (direct polymerization). The former approach aims to add a branching structure onto an LPP chain to fabricate HMSPP with enhanced processability in the presence of a cross-linker (such as peroxide and a coagent) [18,22], electron beam irradiation [26,27], blending [34], and macromonomer grafting [15]. However, HMSPP obtained by postreactor treatment exhibits unstable properties, residual

TABLE 2.1

The Comparison of Four Techniques to Improve PP's Melt Strength

Techniques	Advantages	Disadvantages	Applications
Direct polymerization	Stable properties; better control of molecular structure; environment-friendly and recyclable	Complex control of monomer, polymerization, and catalyst systems; only obtained by mass production and not in small batches	Blowing film, physical/chemical extrusion foam (beads, sheets, and board), thermoformable, extrusion sheet/board, extrusion draw blow molding, stretch blow molding
Reactive extrusion	Controllable addition of functional groups on the polymer chains according to special needs; the optional length of chain	Unstable properties, residual non-environment-friendly chemical agents and unrecyclable cross-linking structure	Physical/chemical extrusion foam (beads, sheets and board)
Electron beam irradiation	The easy enhancement of polymer melt strength for further application	Needs special irradiation equipment; unstable properties, residual non-environment-friendly chemical agents and unrecyclable cross-linking structure	Physical/chemical extrusion foam (beads, sheets and board)
Compounding	Easy to implement; simple operation	Unstable properties, residual non-environment-friendly chemical agents	Blowing film, extrusion draw blow molding, stretch blow molding

non-environment-friendly chemical agents, and unrecyclable cross-linking structure [38]. Therefore, direct polymerization is considered as the most significant and promising method to achieve high melt strength in industry applications. LyondellBasell, Borealis, and Sinopec have exerted mass efforts and work on the in situ polymerization of HMSPP, as mentioned.

2.3 Development of Microcellular PP Foaming

As discussed, PP foams with microcellular structures can be obtained easily by a batch foaming process. The autoclave, always used in the experimental foaming process, can be scaled up to produce EPP beads. However, it is rather difficult to enlarge an autoclave to produce a PP foam board. Based on the concept of static-state batch foaming, we propose a novel foaming method using a press mold machine with CO_2 as a foaming agent to produce a microcellular PP foam board [39]. A foaming mold is used to replace the autoclave. Our previous work confirms that it is an effective route to fabricate a microcellular PP foam board.

A press machine can provide the basic prerequisites for batch foaming. The hydraulic system can provide the supercritical fluid-sealing pressure and enable the rapid mold opening for foam expansion. The temperature control system enables the foaming mold to be controlled at a desired temperature. The schematic of press mold foaming is shown in Figure 2.1. In a typical foaming process, the foaming mold is heated to a temperature below PP's melting temperature. Thereafter, the PP motherboard is put into the mold. The mold inner space is a little bigger than the PP board volume. The rest of mold space is reserved for the CO_2-induced swelling of the PP board and enabling the full cell nucleation in mold. Normally, the mold cavity is at

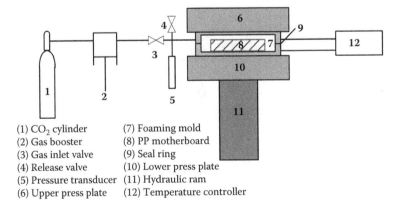

(1) CO_2 cylinder   (7) Foaming mold
(2) Gas booster   (8) PP motherboard
(3) Gas inlet valve   (9) Seal ring
(4) Release valve   (10) Lower press plate
(5) Pressure transducer   (11) Hydraulic ram
(6) Upper press plate   (12) Temperature controller

FIGURE 2.1
The schematic of press mold foaming.

least 1.5 times larger than the volume of the PP motherboard. The hydraulic system claps the foaming mold and builds the sealing pressure. CO_2 gas is injected into the foaming mold using a gas booster until it reaches the target CO_2 pressure. After the sorption of CO_2 in the PP board reaches a sort of equilibrium, the CO_2 gas in the foaming mold is released rapidly to induce cell nucleation and growth. The expansion suspends after the foam fills the mold and then continues after the mold is open. Thus, the expansion process, which always happens instantaneously in other foaming processes, is divided into two stages: (1) the expansion inside the mold and (2) the expansion outside. There is an expansion suspension before the mold is open. Figure 2.2 shows the foam expansion during the mold opening process.

The two-stage foam expansion brings about several advantages. First, both mold volume and mold area are saved. Consequently, the consumption of CO_2 and the machine hydraulic pressure are significantly saved. Obviously, it would be very costly to allow the PP motherboard to fully expand in the mold cavity. The saving of CO_2 also contributes to the increase of pressure drop rate, which significantly determines the cell nucleation rate. Second, the foam expansion rate is reduced, which results from the cell growth suspension before the mold is open. The reduction of expansion rate is very beneficial to foam dimensional stability, especially foam thickness consistency. Generally, the slower the foam expands, the higher the foam dimensional stability could be. Third, the reduction of the expansion rate relieves the requirement of the mold opening rate. The mold opening rate should be as fast as possible to avoid hindering the foam expansion. However, it is hard to achieve a fast mold opening from the point of machinery, especially when the mold is large and

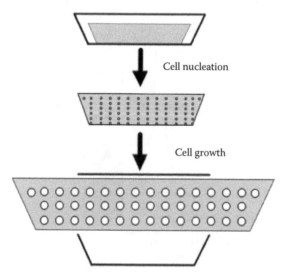

FIGURE 2.2
The schematic of the two-state cell growth.

heavy. A schematic of the two-state cell growth is shown in Figure 2.2. Figure 2.3 shows the foam expansion during the mold opening process.

One of the main challenges to press mold foaming is the relatively long saturation time, usually tens of minutes to even several hours, depending on the thickness of the PP board, which significantly restricts the production efficiency. Aside from increasing the diffusion rate by optimizing the processing parameters, shortening the diffusion route is a direct way to decrease saturation time. Generating a line of holes at the core of the PP motherboard is proposed during the continuous extrusion process [40]. With holes at the core of the PP motherboard, the CO_2 diffusion route is shortened to half of one without holes at the core, as shown in Figure 2.4. The shortening of the diffusion route would lead to a dramatic decrease of the saturation time.

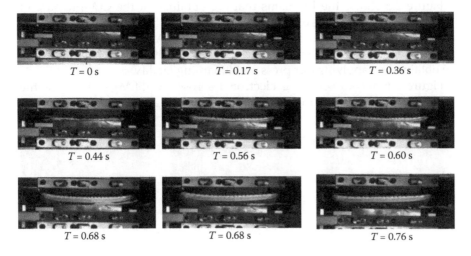

FIGURE 2.3
The foam expansion during the mold opening process.

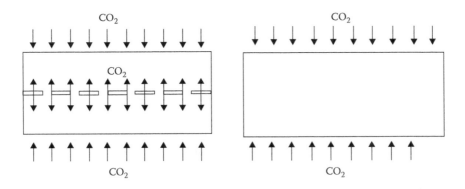

FIGURE 2.4
The comparison of CO_2 diffusion in a PP board with and without a line of holes at the core.

It is testified experimentally that about 7 h are needed to reach the diffusion equilibrium of a 20-mm-thickness board at 150°C, 10 MPa, whereas the time could be dramatically reduced to 1.5 h for a same-thickness board with a line of holes at the core.

Due to the small molecule, as well as the special affinity between CO_2 and PP, CO_2 exhibits a higher diffusion rate than other conventional physical agents in PP. The high diffusion rate of CO_2 makes it an ideal foaming agent in the press mold foaming process since the saturation time could be significantly decreased. On the contrary, the high diffusion rate of CO_2 is always regarded as a disadvantage in the molten-state foaming process because the high diffusion rate would lead to a rapid escape of CO_2 out of the polymer matrix.

Figure 2.5 shows the PP foams that are obtained at the CO_2 pressure of 10 MPa and temperatures of 148°C, 149°C, and 151°C. The corresponding foam expansion ratios are 9.6, 15.5, and 25.3, respectively. At present, PP foams with a largest expansion ratio of 30 times and a largest thickness of 80 mm could be obtained by applying our press mold foaming facility.

Figure 2.6 shows scanning electron microscopy (SEM) pictures of the microcellular PP foams that are shown in Figure 2.5. It is shown that the

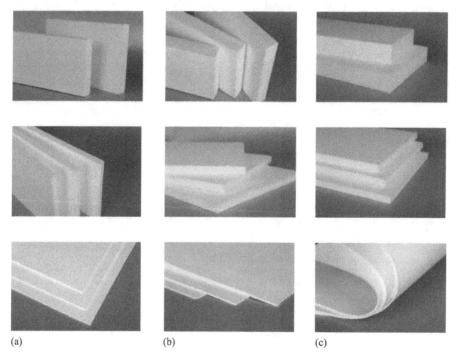

(a)   (b)   (c)

FIGURE 2.5
Microcellular PP foams obtained at a CO_2 pressure of 10 MPa and temperatures of (a) 148°C, (b) 149°C, and (c) 151°C.

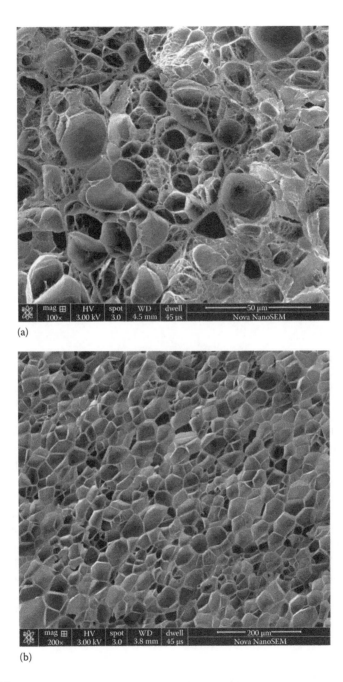

(a)

(b)

FIGURE 2.6
SEM pictures of the microcellular PP foams obtained at the CO_2 pressure of 10 MPa and the temperature of (a) 148°C and (b) 149°C as shown in Figure 2.5. *(Continued)*

(c)

FIGURE 2.6 (CONTINUED)
SEM pictures of the microcellular PP foams obtained at the CO_2 pressure of 10 MPa and the temperature of (c) 151°C as shown in Figure 2.5.

average cell size, as well as the cell uniformity, increased as the foaming temperature increased. The dependence of the foam expansion ratio on foaming temperature indicates that the processing window of E02ES is still not wide enough and should be further broadened in the future.

Interestingly, long fiber bundles, always with lengths of tens of centimeters, could be observed in the thin PP foam sheets that are sliced from the thick ones. Figure 2.7 shows pictures of the fiber bundles taken by a common camera. It is distinct that the fiber bundles consist of lots of spindly fibrils. We guess that the fibers are the extended chain crystal since the extended chain crystal is likely to appear when PP is treated at high static pressure and the temperature is close to its melting temperature. The fiber contributes greatly to the mechanical properties of the PP foams. Thanks to microcellular foaming, we can observe the long fiber bundles with naked eyes, which could only be observed in solid PP using a microscope.

2.3.1 Properties of Microcellular PP Foams

Table 2.2 shows the part mechanical properties of the microcellular PP foams with different apparent densities. All the strengths and the modulus decreased with the decrease of foam-apparent density. Compared to other

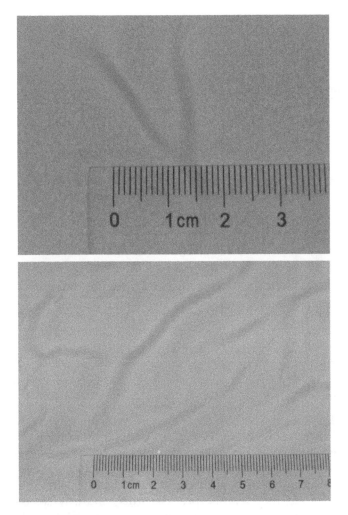

FIGURE 2.7
Long fiber bundles of extended chain crystal appears in the PP foam. Pictures were taken from a PP foam sheet with common camera.

polymer foams, microcellular PP foam exhibits both high rigidity and high toughness.

Table 2.3 shows the thermal conductivities of microcellular PP foams with different apparent densities. The thermal conductivity decreased as the foam-apparent density decreased. The microcellular structure of PP foams contributes to the thermal insulation. However, the thermal insulation property of microcellular PP foam is not as good as desired, mainly as a result of the relatively large thermal conductivity of the air, which will replace CO_2 gas in the cells after foaming.

TABLE 2.2

Part Mechanical Properties of Microcellular PP Foams with Different Apparent Densities[a]

Apparent Density, kg/m³	Compression Strength, KPa	Compression Modulus, MPa	Tensile Strength, MPa	Tensile Modulus, MPa
90	1.51	42.5	3.91	30.7
60	0.73	31.3	3.46	25.1
45	0.51	17.8	2.38	21.5
36	0.36	14.3	1.89	16.8
30	0.29	11.7	1.56	12.9

[a] The test is based on ASTM D 3575.

TABLE 2.3

Thermal Conductivities of Microcellular PP Foams with Different Apparent Densities

Foam Density, kg/m³	Thermal Conductivity, W/m.K
60	0.044
45	0.038
36	0.035
30	0.032

2.3.2 Summary

Although the consumption of PP increased with the rapid development of the global economy, PP derived from the blooming growth of alkane and shale gas technology makes excess capacity appear. The fierce market competition promotes PP production to develop diversely and functionally. Thanks to unique melt strength and fine heat resistance, the factors of weight savings, recyclability, processability, and a wider envelope of properties are driving the fast-growing market for HMSPP. As a result, HMSPP is mounting a vigorous challenge to polyurethanes, PE, and polyvinyl chloride in packaging, automotive, industrial, and food applications. Developing simple preparation technology, stable quality, and high-cost performance have been hot topics for research and industry. Some problems should be concentrated as follows: (a) high melt strength matching with high flowability, (b) good rigidity and heat resistance coupling with high transparency, (c) compatibility of HMSPP and other resin, and (d) recyclability.

Microcellular PP foam work shows that press mold foaming technology can be used to produce a microcellular PP foam thick board, which can hardly be obtained by other conventional foaming technology. The large diffusion rate of CO_2 in PP becomes a notable advantage in this foaming technology since it contributes to the shortening of the saturation time. The efficiency can also be improved by increasing the layers of mold in one press machine. Anyway, more measures should be adopted to further shorten the saturation time.

Press mold foaming is applicable not only to PP but also to other polymers. In addition, since press mold foaming technology is derived from batch pressure quench foaming, all the knowledge and results of batch pressure quench foaming experiments, which many academic researchers have published, is applicable to the press mold foaming. We can picture that many innovative experimental samples reported in earlier works could possibly become industrial-scale products in the future.

References

1. Bradley M B, Philips E M. 1990. ANTEC: Novel foamable polypropylene polymers. US: *SAE Tech*, 717–720.
2. Guo P, Liu Y, Xu Y et al. 2014. Effects of saturation temperature/pressure on melting behavior and cell structure of expanded polypropylene bead. *J Cell Plast*, 50(4): 321–335.
3. Nofar M, Guo Y, Park C B. 2013. Double crystal melting peak generation for expanded polypropylene bead foam manufacturing. *Ind Eng Chem Res*, 52(6): 2297–2303.
4. Jiang X L, Liu T, Xu Z. 2009. Metal Effects of crystal structure on the foaming of isotactic polypropylene using supercritical carbon dioxide as a foaming agent. *J Supercrit Fluid*, 48(2): 167–175.
5. Xu Z M, Jiang X L, Liu T et al. 2007. Foaming of polypropylene with supercritical carbon dioxide. *J Supercrit Fluid*, 41(2): 299–310.
6. Jiang X L. 2009. *The Control of Cell Morphology in the Process of Polypropylene Microcellular Foaming*. East China University of Science and Technology, Shanghai, China, Doctoral Dissertation.
7. Busico V, Cipullo R. 2001. Microstructure of polypropylene. *Prog Polym Sci*, 26(3): 443–533.
8. Chum P S, Swogger K W. 2008. Olefin polymer technologies—History and recent progress at the DOW chemical company. *Prog Polym Sci*, 33(8): 797–819.
9. He C, Costeux S, Wood-Adams P et al. 2003. Molecular structure of high melt strength polypropylene and its application to polymer design. *Polymer*, 44(23): 7181–7188.
10. Rätzsch M, Arnold M, Borsig E et al. 2002. Radical reactions on polypropylene in the solid state. *Prog Polym Sci*, 27(7): 1195–1282.
11. Pivokonsky R, Zatloukal M, Filip P et al. 2009. Rheological characterization and modeling of linear and branched metallocene polypropylenes prepared by reactive processing. *J Non-Newtonian Fluid Mech*, 156(1–2): 1–6.
12. Wang L, Wan D, Zhang Z et al. 2011. Synthesis and structure property relationships of polypropylene-g-poly(ethylene-co-1-butene) graft copolymers with well-defined long chain branched molecular structures. *Macromolecules*, 44(11): 4167–4179.

13. Su F, Huang H. 2010. Rheology and melt strength of long chain branching polypropylene prepared by reactive extrusion with various peroxides. *Polym Eng Sci*, 50(2): 342–351.

14. Langston J A, Colby R H, Chung T C M et al. 2008. One-pot synthesis of long chain branch PP (LCBPP) using Ziegler-Natta catalyst and branching reagents. *Macromol Symp*, 260(1): 34–41.

15. Wang L, Wan D, Zhang Z et al. 2011. Synthesis and structure property relationships of polypropylene-g-poly(ethylene-co-1-butene) graft copolymers with well-defined long chain branched molecular structures. *Macromolecules*, 44(11): 4167–4179.

16. Scheve J B, Mayfield J W, Denicola Jr., A J. High melt strength, polypropylene polymer, process for making it, and use thereof. US patent, 4,916,198 A.

17. Passaglia E, Coiai S, Augier S. 2009. Control of macromolecular architecture during the reactive functionalization in the melt of olefin polymers. *Prog Polym Sci*, 34 (9): 911–947.

18. Tang H, Dai W, Chen B. 2008. A new method for producing high melt strength polypropylene with reactive extrusion. *Polym Eng Sci*, 48(7): 1339–1344.

19. Mabrouk K E, Parent J S, Chaudhary B I et al. 2009. Chemical modification of PP architecture: Strategies for introducing long-chain branching. *Polymer*, 50(23): 5390–5397.

20. Zhang Z, Xing H, Qiu J et al. 2010. Controlling melt reactions during preparing long chain branched polypropylene using copper N, N-dimethyldithiocarbamate. *Polymer*, 51(7): 1593–1598.

21. Su F, Huang H. 2010. Rheology and melt strength of long chain branching polypropylene prepared by reactive extrusion with various peroxides. *Polym Eng Sci*, 50(2): 342–351.

22. Su F, Huang H. 2009. Rheology and thermal behavior of long branching polypropylene prepared by reactive extrusion. *J Appl Polym Sci*, 113(4): 2126–2135.

23. Su F, Huang H. 2010. Influence of polyfunctional monomer on melt strength and rheology of long-chain branched polypropylene by reactive extrusion. *J Appl Polym Sci*, 116(5): 2557–2565.

24. Cao K, Li Y, Lu Z-Q et al. 2011. Preparation and characterization of high melt strength polypropylene with long chain branched structure by the reactive extrusion process. *J Appl Polym Sci*, 121(6): 3384–3392.

25. Augier S, Coiai S, Passaglia E et al. 2010. Structure and rheology of polypropylene with various architectures prepared by coagent-assisted radical processing. *Polym Int*, 59(11): 1499–1505.

26. Cheng S, Phillips E, Parks L. 2010. Process ability improvement of polyolefins through radiation-induced branching. *Radiat Phys Chem*, 79(3): 329–334.

27. Cerrada M L, Pérez E, Benavente R et al. 2010. Gamma polymorph and branching formation as inductors of resistance to electron beam irradiation in metallocene isotactic polypropylene. *Polym Degrad Stab*, 95(4): 462–469.

28. Otaguro H, de Lima L F C P, Parra D F et al. 2010. High-energy radiation forming chain scission and branching in polypropylene. *Radiat Phys Chem*, 79(3): 318–324.

29. Oliani W L, Parra D F, Lugão A B. 2010. UV stability of HMS-PP (high melt strength polypropylene) obtained by radiation process. *Radiat Phys Chem*, 79(3): 383–387.

30. Zarch F S, Jahani Y, Haghighi M N et al. 2012. Rheological evaluation of electron beam irradiated polypropylene in the presence of a multifunctional monomer and polybutene resin. *J Appl Polym Sci*, 123(4): 2036–2041.
31. Koosha M, Jahani Y, Mirzadeh H. 2011. The effect of electron beam irradiation on dynamic shear rheological behavior of a poly (propylene-co-ethylene) heterophasic copolymer. *Polym Adv Technol*, 22(12): 2039–2043.
32. Mousavi S A, Dadbin S, Frounchi M et al. 2010. Comparison of rheological behavior of branched polypropylene prepared by chemical modification and electron beam irradiation under air and N2. *Radiat Phys Chem*, 79(10): 1088–1094.
33. Ni Q, Fan J, Dong J. 2009. Crystallization behavior and crystallization kinetic studies of isotactic polypropylene modified by long-chain branching polypropylene. *J Appl Polym Sci*, 114(4): 2180–2194.
34. Tabatabaei S H, Carreau P J, Ajji A. 2009. Rheological and thermal properties of blends of a long-chain branched polypropylene and different linear polypropylenes. *Chem Eng Sci*, 64(22): 4719–4731.
35. Fang Y, Sadeghi F, Fleuret G et al. 2008. Properties of blends of linear and branched polypropylenes in film blowing. *Can J Chem Eng*, 86(1): 6–14.
36. An Y, Zhang Z, Wang Y et al. 2010. Structure and properties of high melt strength polypropylene prepared by combined method of blending and crosslinking. *J Appl Polym Sci*, 116(3): 1739–1746.
37. Liu Y, Huang Y, Zhang C et al. 2009. Improved melting strength of polypropylene by reactive compounding with ultrafine full-vulcanized polybutadiene rubber (UFBR) particles. *Polym Eng Sci*, 49(9): 1767–1771.
38. Guo P, Xu Y, Lu M F et al. 2015. High melt strength polypropylene with wide molecular weight distribution used as basic resin for expanded polypropylene beads. *Ind Eng Chem Res*, 54(1): 217–225.
39. Jiang X L. A method to prepare microcellular polymer foams using supercritical fluid assisted press molding foaming method. Chinese patent, 201110054581.3.
40. Ji Z R. A method to produce microcellular polypropylene thick board. Chinese patent, 201210443118.2.

3

Preparation of Poly(ethylene terephthalate) Foams Using Supercritical CO_2 as a Blowing Agent

Ling Zhao, Tian Xia, Zhenhao Xi, and Tao Liu

CONTENTS

3.1 Introduction

Poly(ethylene terephthalate) (PET) foams show good mechanical properties, high temperature dimensional stability, and recyclability and have recently attracted extensive attention in academic and industrial circles. The wide applications of PET foams depend on cell morphology and matrix properties. Microcellular PETs with cell sizes less than 5 μm were developed by Furukawa Electric (Japan), and possessed the excellent property of over 99% reflectivity and 96% light diffusion rate, which increased lamps' illuminance by 40%–60% without an extra light source as the reflector. Low-density PET foams with a cell size of hundreds of microns have been widely used as packaging materials and structure materials in wind energy, marine, and transportation applications. Supercritical CO_2 has been used as a blowing agent for PET because it is inexpensive, nontoxic, and environmentally benign. CO_2 dissolved into a PET matrix causes many changes in the physical properties of PET in both the melt and solid states due to the strong plasticization effect, which enhances the free volume and chain mobility. It depresses the glass-transition temperature and the crystallization temperature, and changes the crystallization kinetics of semi-crystalline PET, which may provide many opportunities for the manipulation of foaming processes.

In a solid-state foaming process, the diffusivity of CO_2 in a PET matrix was very low owing to the low temperature, and CO_2 saturation in PET required several days or even longer. The homogeneous PET/CO_2 solution could not be obtained due to CO_2 insolubility in the crystalline region. The crystals supplied a large number of stable cell nuclei when acting as heterogeneous nucleation sites and at the same time inhibited the cell growth by increasing the stiffness of matrix [1,2]. The solid-state foaming process was usually conducted in batches. Cell nucleation was initiated by heating the saturated PET near or above its glass-transition temperature [3,4]. Kumar et al. [5] reported a semicontinuous process for the solid-state foaming of PET in which a CO_2-saturated PET roll was passed through a hot bath continuously and foamed at temperatures ranging from 50°C to 90°C. The melt-state foaming of PET could be conducted in a continuous process, such as extrusion foaming. The CO_2 diffusivity was facilitated by a high foaming temperature, and PET could be saturated by CO_2 in a short period. However, the semicrystalline PET usually had low melt strength, and elasticity resulted from the low molecular weight and well-structured backbone, which could not resist the intensive extensional deformations during cell expansion. The cells would merge into larger ones or rupture [6–8]. It is known that high molecular weight, broad molecular weight distribution, and a long-chain branch attached to the PET backbone were required to improve the polymer melt strength through enhancing the possibility of entanglements in the polymeric melt. Multifunctional chain extenders with electrophilic functional groups such as cyclic anhydride [7,9] and epoxide [10,11] were developed to react with the nucleophilic end groups of PET, which could be performed by means of *in-situ* polymerization and reactive extrusion.

Some typical PET foaming processes using supercritical CO_2 as a blowing agent, from a solid state to a melt state, will be discussed in Sections 3.2 through 3.5, including the following:

- Controllable sandwich structure of PET microcellular foams prepared by solid-state foaming process
- Melt foamability of reactive extrusion-modified PET
- Melt-foaming behavior of *in-situ*–modified PET in batch and continuous processes
- Integrated process of supercritical CO_2-assisted modification and foaming of PET

The PET foams involved in this chapter cover the range from microcellular foam that is characterized by closed cells with cell sizes that are lower than 10 μm and cell densities that are higher than 10^9 cells/cm^3 to macrocellular foam (the typical cell size of hundreds microns). The polymer matrix included PET with linear molecular chain, PET/clay nanocomposites, PET/SiO$_2$ nanocomposites, and branched PET that is prepared by both reactive extrusion and *in-situ* modification.

3.2 Controllable Sandwich Structure of PET Microcellular Foams Based on the Coupling of CO$_2$ Diffusion and Its Induced Crystallization

The sandwich structure of PET microcellular foams was fabricated by the solid-state foaming process, as illustrated in Figure 3.1. The gradient of CO$_2$

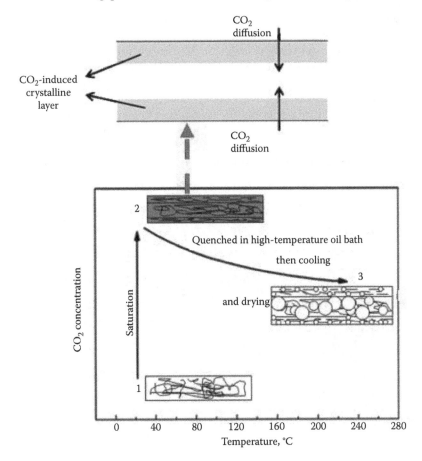

FIGURE 3.1
Schematic of the preparation process of sandwich-structure microcellular PET foams.

concentration developed gradually in amorphous PET sheets during the saturation stage. The CO_2-induced crystallization took place when the CO_2 concentration was high enough. Therefore, the layers near the sheet surfaces crystallized earlier, but PET near the central areas remained amorphous. The foaming temperature was chosen to be 235°C, just 20°C lower than the PET melting point, and was very different from that of reported patents [5] because higher foaming temperatures would create larger thermodynamic instability, which was favorable for cell nucleation, and the difference between the foaming morphologies of the crystalline layer PET and the amorphous layer PET could be more obvious at higher foaming temperatures. The sandwich structure of PET foams was produced with larger cells in the amorphous layer inside and small cells in the crystalline layers outside, as displayed in Figure 3.2 [12].

The CO_2-induced isothermal crystallization of PET was determined by *in-situ* high-pressure Fourier Transform Infrared Spectroscopy (FTIR). No crystallinity increase could be detected under 4.5 MPa CO_2 even after 2000 min, as shown in Figure 3.3, whereas a very long crystallization period of more than 1500 min was observed under 5.0 MPa CO_2, indicating the lowest pressure, corresponding to the lowest critical CO_2 concentration in PET. $C_{critical}$ for the CO_2-induced crystallization of PET was 5.0 MPa at 25°C. The well-known Avrami equation could be used to evaluate the CO_2-induced isothermal crystallization of PET films.

The solubility and diffusivity of CO_2 in an amorphous PET sheet at 25°C and 6 MPa were measured by the Magnetic Suspension Balance (MSB) method. The distribution of CO_2 concentration was found to vary with saturation time. The CO_2-induced crystallization took place at the surface layers, which would

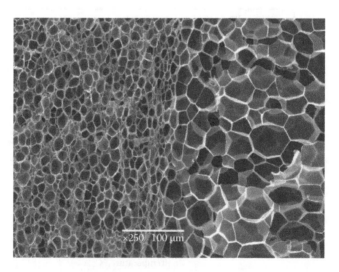

FIGURE 3.2
Typical boundary between the crystalline layer and the amorphous layer of PET foam. (From D. Li et al., *Aiche Journal*, 58, 2512–2523, 2012.)

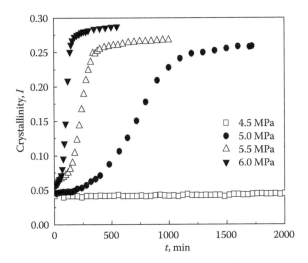

FIGURE 3.3
Changes in crystallinity of PET films during isothermal crystallization at 25°C and different CO$_2$ pressures. (From D. Li et al., *Aiche Journal*, 58, 2512–2523, 2012.)

subsequently reject CO$_2$ and could be adopted to explain the *knee-like* sorption profile, as demonstrated in Figure 3.4. The diffusion coefficient was 6.7×10^{-13} m^2/s assuming that the diffusion could be expressed by Fick's second law. The solubility of CO$_2$ in amorphous regions of PET C_0 at 25°C and 6 MPa was 11.0 wt%, and $C_{critical}$, the equilibrium concentration of CO$_2$ in amorphous PET under 5 MPa CO$_2$, was estimated to be 9.17 wt% assuming that the sorption of CO$_2$ would follow Henry's law at relatively low CO$_2$ pressure.

After a certain saturation time, the CO$_2$ concentration and crystallinity distributions in the PET sheet were controlled by CO$_2$ diffusion and

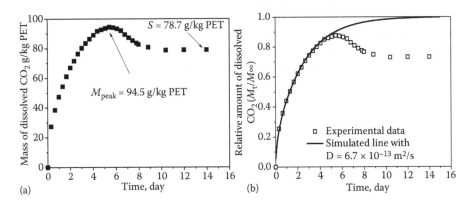

FIGURE 3.4
Sorption profiles for CO$_2$ in PET at 25°C and 6 MPa. (a) Mass of dissolved CO$_2$; (b) Relative amount of dissolved CO$_2$. (From D. Li et al., *Aiche Journal*, 58, 2512–2523, 2012.)

induced crystallization simultaneously. A model was proposed to investigate the coupled processes with assumptions that the PET sheet consisted of N well-contacted layers in the thickness direction, and the thickness of each layer was equal and so small that the CO_2 concentration distribution was uniform in each individual layer, as illustrated in Figure 3.5. In such a case, the concentration of CO_2 in each layer could be determined by the following equation:

$$\frac{C(t,i)}{C_0} = 1 - \frac{4}{\pi} \sum_{n=0}^{\infty} \frac{(-1)^n}{2n+1} exp\left[\frac{-D(2n+1)^2 \pi^2 t}{4l^2}\right] cos \frac{(2n+1)\pi x(i)}{2l} \qquad (3.1)$$

where $C(t,i)$ is the concentration of CO_2 at saturation time t in layer i, $x(i)$ is the location of layer i in the thickness dimension of PET sheet, and D is the diffusion coefficient. The crystallization rate, R_c, in each individual layer could be determined as follows:

$$R_c = \frac{dX_t}{dt} = Kt^{n-1} exp(-Kt^n) \qquad (3.2)$$

where n and K are the Avrami exponent and crystallization rate constant, respectively.

As shown in Figure 3.6, most of PET sheet was still unsaturated with one-day saturation, leading to small cell sizes in the areas near the surface and large cells in the central areas. After a five-day saturation, the distribution of CO_2 concentration had been relatively uniform in the PET sheet, which led to uniform cell morphology in most areas of the sample. Crystallinity played a major role in the microcellular foaming process on cell nucleation mechanisms, resulting in larger cell densities due to heterogeneous nucleation at the amorphous/crystalline boundaries, and cell growth mechanisms, leading to smaller cell size due to the increased stiffness of semicrystalline matrix. The thickness of crystalline layers calculated using the proposed model together with those measured from PET foams at different saturation times were

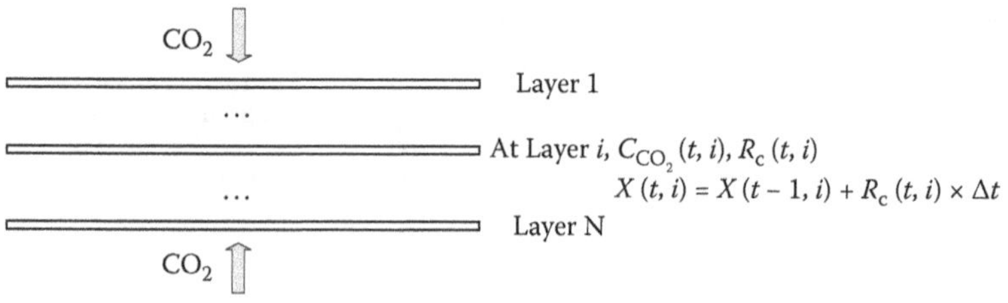

FIGURE 3.5
Schematic diagram of the proposed model. (From D. Li et al., *Aiche Journal*, 58, 2512–2523, 2012.)

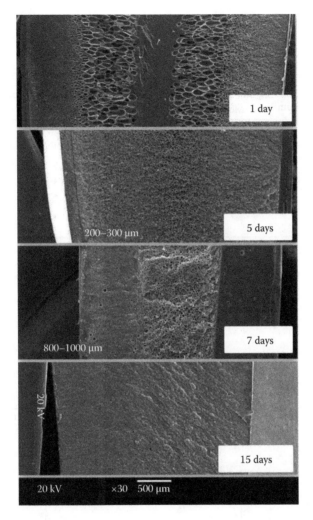

FIGURE 3.6
The overall cell morphology of PET foams prepared with different saturation times. (From D. Li et al., *Aiche Journal*, 58, 2512–2523, 2012.)

shown in Table 3.1. Basically, the model could well predict the evolution of crystalline layer against a different saturation time.

The evolution of cell size and cell density of both crystalline and amorphous layers and the expansion ratio of PET foams against saturation time are shown in Figure 3.7. The characterization of cell morphology is summarized in Table 3.3. With the increase of saturation time, the CO_2 concentration in the amorphous layer of PET specimen increased, and more CO_2 was available to support the cell growth, which led to an increase in both cell size and expansion ratio within the saturation time of five days. When the

TABLE 3.1

Comparison of the Evolution of Crystalline Layer Thickness Calculated Using the Proposed Model and in PET Foams

Saturation Time, Day		1	2	3	5	7
Thickness of crystalline layer, μm	In PET foams	0[a]	80	130	250	900
	Modeling	20	50	100	300	complete

Source: D. Li et al., *Aiche Journal*, 58, 2512–2523, 2012.

[a] The crystalline layer was not detected in the PET foams due to the overlap of crystalline and the unfoamed skin layer of the PET specimen.

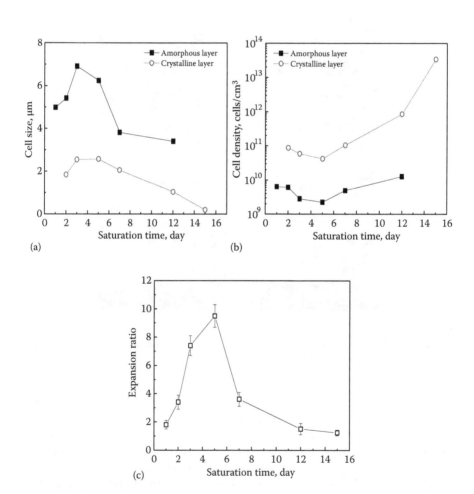

FIGURE 3.7
Characterization of PET foams obtained at different saturation times: (a) cell size, (b) cell density, and (c) expansion ratio. (From D. Li et al., *Aiche Journal*, 58, 2512–2523, 2012.)

saturation time was longer than seven days, the crystal structure of crystal-line became more perfect, and a small amount of crystals could have formed in the amorphous layer, which restricted the cell growth and increased the cell density via changing the nucleation mechanism from homogeneous to heterogeneous nucleation. Especially at the saturation time of 15 days, ultra-microcellular PET foams with an average cell diameter of 193 nm and a cell density of 3.37×10^{13} were obtained.

3.3 Melt Foamability of Reactive Extrusion-Modified PET

High melt elasticity and strength were essential in the melt foaming process of PET to avoid cell collapse and coalescence, which were acquired via the chain extension/branching reactions between the end groups of PET and the multifunctional chain extenders. And the long-chain branch was introduced to the PET backbone. In the melt foaming process of PET, cell nucleation was initiated by rapid depressurization. Cell growth was hindered, and the cell morphology was solidified by the increasing stiffness of the polymer matrix due to the nonisothermal crystallization from melts during the cooling stage, which would be influenced by the topological structure of the PET chain and CO$_2$ atmosphere. Pyromellitic dianhydride (PMDA) was often selected to react with the hydroxyl end groups of PET in extruders because the melt-ing point of PMDA, close to the PET processing temperature, together with tetrafunctionality, ensured that the modifying reactions were fast [13].

Dynamic shear rheological properties were sensitive to the topologi-cal structure of polymers. As shown in Figure 3.8a, the Newtonian zone of linear virgin polyethyleneterephthalate (VPET) was broad, and no obvious shear thinning was observed. The transition from a Newtonian plateau to a shear-thinning regime shifted to lower frequency with the increase of PMDA content, which was ascribed to the long-time-relaxation mechanism, such as entanglement couplings between high-molecular-weight fractions and those that are associated with the long-chain branch. In addition, a broad relaxation time distribution was also used to explain the shear thinning. Cole–Cole plots, in which the axes were represented as the logarithms of storage modu-lus G′ and loss modulus G″, were independent of temperature and molecu-lar weight but strongly dependent on the long-chain branch and molecular weight distribution. With the introduction of a long-chain branch or broader molecular weight distribution, G′ increased at a given G″. As demonstrated in Figure 3.8b, modified samples got a broader molecular weight distribution and a long-chain branch due to the reactive extrusion process. Figure 3.8b also included the equal-modulus line, and the position of plots with respect to this line indicated the degree of melt elasticity. It could be easily concluded that the melt elasticity of the modified sample increased with PMDA content.

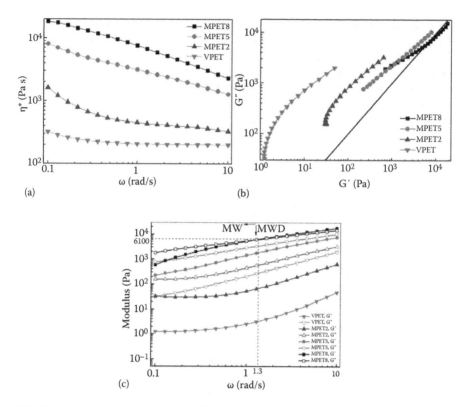

FIGURE 3.8
Dynamic rheological properties of VPET and modified PETs: (a) η^* versus ω, (b) Cole–Cole plots, and (c) G′ and G″ versus ω. (MPET2, 5, 8 represented PET modified with 0.2, 0.5, 0.8 wt% PMDA.) (From T. Xia et al., *Polymer Engineering & Science*, 55, 1528–1535, 2015.)

The cross point of G′ and G″ should shift to a lower-frequency region when the molecular weight increased and to a lower-modulus region when the molecular weight distribution became broader. It was inferred from Figure 3.8c that MPET8 had the broadest molecular weight distribution and the highest molecular weight.

The Avrami analysis was extended for the nonisothermal crystallization of a modified PET, as described by Xia et al. [13]. With the introduction of the long-chain branch, the nonisothermal crystallization rate constant Z_C became higher in general, whereas the half-crystallization time $t_{1/2}$ decreased, as exhibited in Table 3.2, indicating that the crystallization rate increased within the experimental range of branching degree. The introduction of a long-chain branch brought two opposite effects on crystallization. The long-chain branch, as a kind of defected of polymer chain when crystallizing, resulted in a longer nucleation time than is required to exclude the branching point from crystal nucleates. On the other hand, branching provided a larger crystal-growing space due to the mutual repulsion of chains,

TABLE 3.2

Parameters of Nonisothermal Crystallization under 0.1 MPa N_2 and 5 MPa CO_2 for VPET and Modified PETs

Gas Atmosphere	PET	R (°C/min)	Z_C (min^{-1})	$t_{1/2}$ (min)	n	ΔH_c (J/g)
0.1 MPa N_2	VPET	2	0.024	8.34	3.33	−40.96
		3	0.084	6.31	3.25	−39.86
		4	0.177	5.65	3.78	−38.91
		5	0.349	4.21	3.40	−38.22
	MPET2	2	0.039	8.42	3.07	−39.46
		3	0.163	5.12	3.11	−38.17
		4	0.216	4.38	3.90	−37.44
		5	0.375	4.20	3.16	−36.46
	MPET5	2	0.044	6.45	3.16	−34.05
		3	0.181	4.58	3.13	−33.84
		4	0.285	4.15	3.27	−33.07
		5	0.384	3.54	3.50	−32.09
	MPET8	2	0.044	6.17	3.23	−36.05
		3	0.123	5.14	3.62	−35.31
		4	0.241	4.31	3.64	−33.96
		5	0.408	3.39	3.37	−33.39
	MPET5	2	0.069	5.22	3.01	−32.45
		3	0.270	3.48	2.86	−32.29
		4	0.450	2.83	2.72	−32.01
5 MPa CO_2		5	0.603	2.08	2.27	−30.73
	MPET8	2	0.103	4.46	2.80	−34.00
		3	0.294	3.26	2.80	−29.23
		4	0.546	2.40	2.94	−25.06
		5	0.679	2.02	2.23	−20.56

Source: T. Xia et al., *Polymer Engineering & Science*, 55, 1528–1535, 2015.

which was in favor of the crystals' growth. For the crystallization of a modified PET, crystal growth was the rate-controlling step. Higher crystallization rates were obtained with CO_2 dissolution since Z_C enhanced whereas $t_{1/2}$ reduced with the increase of CO_2 pressure, as presented in Table 3.2. CO_2 dissolved into a PET matrix increased the free volume and weakened the interchain interaction, which facilitated the retraction and fold of molecular chains and meanwhile decreased the crystallization enthalpies. And during the cooling stage in the melt foaming process, the residual CO_2 in cells and polymer matrix could accelerate the crystallization process further to gain an ideal cell morphology. It was noteworthy that the values of the Avrami exponent n of all PET samples were between 3 and 4, which deduced that all PET samples were crystallized according to the three-dimensional growth of crystals with either a homogeneous or heterogeneous nucleation.

The highest foaming temperature (T_{f-H}) of a modified PET was determined by the melt strength, which reduced with the foaming temperature. The formed cells would collapse at a foaming temperature that is higher than T_{f-H}, and then no foams could be gained. The lowest foaming temperature (T_{f-L}) of melt foaming process was the onset temperature of nonisothermal crystallization from melts under compressed CO_2, below which the original densities of PET were measured as a result of extensive crystallization, and no PET foam could be gained. As shown in Figure 3.9, nearly no declined trend of crystallization onset temperature was observed with increased of CO_2 pressure for a modified PET, and the crystallization onset temperature of MPET5 was nearly equal to that of MPET8. The foaming temperature windows of MPET5 and MPET8 were 210°C–250°C and 210°C–280°C, respectively, as displayed in Figure 3.10, and no foaming temperature windows were explored for VPET and MPET2 due to lower melt strength and the crystallization rate.

With the increase of foaming temperature, the amount of CO_2 available for cell nucleation decreased, and the surface tension between the CO_2 and the melt, reducing with CO_2 dissolution, increased as a result of lower CO_2 solubility. Therefore, the nucleation rate decreased according to the classic nucleation theory, and a lower cell density was obtained, as shown in Figure 3.11. On the other hand, a higher foaming temperature decreased the viscosity of the polymer matrix, increased the diffusivity of CO_2 within the matrix, and extended the cell growth period. As a consequence, the average cell diameter and the expansion ratio increased. The characterization of modified PET foams is summarized in Table 3.3. More cells of MPET5 foams collapsed or coalesced due to the lower melt strength, as presented in Figure 3.10. Thereby, apparent lower cell density, especially at a higher foaming temperature, and a larger cell size along with higher expansion ratio

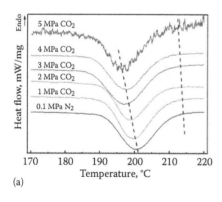
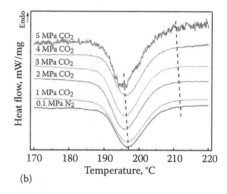

FIGURE 3.9
Differential scanning calorimetry cooling traces of modified PET melts at a cooling rate of 5°C/min under different gas atmospheres: (a) MPET5 and (b) MPET8. (From T. Xia et al., *Polymer Engineering & Science*, 55, 1528–1535, 2015.)

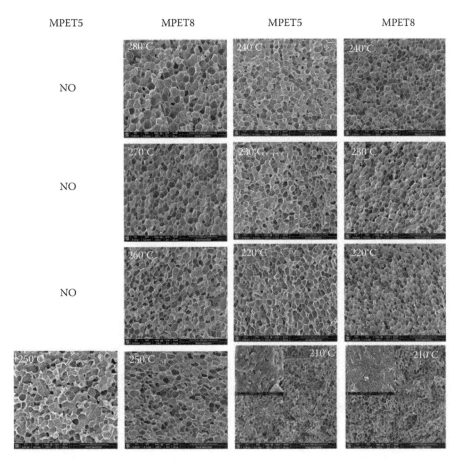

FIGURE 3.10
Scanning electron microscope (SEM) micrographs of modified PET foams obtained at different foaming temperatures. (From T. Xia et al., *Polymer Engineering & Science*, 55, 1528–1535, 2015.)

were obtained, compared to those of MPET8 foams. The cell size distribution was controlled by the cell growth period. A longer time was required to reach the crystallization temperature and to solidify the cell morphology with the increase of foaming temperature, and hence, the cell size distribution became broader, as demonstrated in Figure 3.12. In addition, the cell size distributions of MPET8 foams were more uniform than those of MPET5 foams at the same foaming temperature, since the cell growth was inhibited more intensively due to high melt strength.

In recent years, nanoclays, such as montmorillonite, have been widely used to improve the melt foamability of polymers because of their high aspect ratio, plate morphology, and natural abundance [14,15]. The polymer/clay nanocomposites, in which the clays were dispersed in nanosize by the intercalation of polymer chains into clay galleries, presented

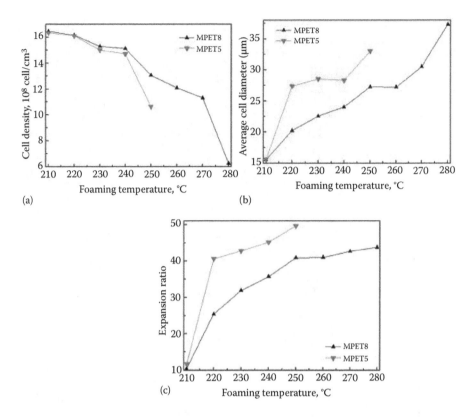

FIGURE 3.11
Characterization of MPET foams obtained at different foaming temperatures: (a) cell density,
(b) average cell diameter, and (c) expansion ratio. (From T. Xia et al., *Polymer Engineering &
Science*, 55, 1528–1535, 2015.)

attractive properties such as reduced gas permeability, flame retardance,
and enhanced mechanical properties. The nanocomposite foams were
expected to possess the combined advantages of polymer foams and
nanocomposites. The large interfacial areas between clay platelets and
the polymer matrix supplied many more heterogeneous nucleation sites,
which facilitated the cell nucleation process by reducing the energy barrier
of nucleation, resulting in a higher nucleation rate and ultimately a larger
number of cell nuclei. The inorganic platelets were nonpermeable to CO_2
and thus enhanced the tortuosity and path length of CO_2 diffusion in the
matrix, which reduced the rate of cell growth. Meanwhile, the amount of
CO_2 available for cell growth decreased while more cells were nucleated,
leading to a reduction of cell size. Furthermore, reorientation of clay platelets along the cell walls was observed by Okamoto et al. [16,17], which acted
as a second cell wall in favor of protecting the cells from being destroyed by
stretching force. The melt foamability of PET with relatively lower intrinsic
viscosity (IV) and melt elasticity could be significantly improved by the

TABLE 3.3

Examples of PET Foams Obtained by Different Foaming Processes with Supercritical CO$_2$

Parameters/ Properties	Li et al. [12]	Xia et al. [13]	Zhong et al. [18]	Fan et al. [19]	Zhong et al. [22]
Foaming method	Solid state/ temperature rising	Melt state/ pressure quench	Melt state/ pressure quench	Melt state/ extrusion foaming	Integrated process of CO$_2$-assisted melt polycondensation and foaming
Modifying method	N/M	Reactive extrusion	*in-situ* polymerization	*in-situ* polymerization +SSP	
Chain extender	N/M	PMDA	PMDA/PENTA	PENTA	PENTA
Intrinsic viscosity (dL/g)	1.0	0.88–1.36	0.865/0.860	1.5	0.860
Inorganic filler	N/M	N/M	N/M	Nano-SiO$_2$	N/M
Foaming temperature (°C)	235	210–280	265–280	232–240	280
Cell size (µm)	3.40–6.24(a)/ 0.19–2.57(c)	15–37	38–57/35–49	265–512	32–62
Cell density (cells/cm^3)	2.2×10^9–1.3×10^{10}(a)/ 4.2×10^{10}–3.4×10^{13}(c)	6.2×10^8–1.6×10^9	5.8×10^6–2.8×10^7/ 7.0×10^6–3.5×10^7	0.19×10^5–4.6×10^5	1.0×10^7–4.0×10^7
Expansion ratio	0.2–9.5	10–50	20–40	8.0–10.2	4.0–26*

Note: (a) amorphous layer, (c) crystalline layer.

* Assuming that the density of unfoamed PET was 1.3 g/cm^3.

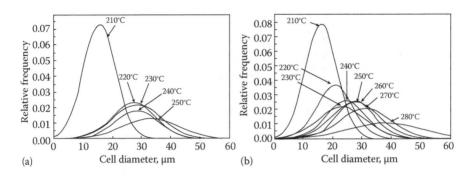

FIGURE 3.12
Cell size distribution of MPET foams obtained at different foaming temperatures: (a) MPET5 and (b) MPET8. (From T. Xia et al., *Polymer Engineering & Science*, 55, 1528–1535, 2015.)

well-dispersed clay, and the foaming temperature windows of PET/clay nanocomposites prepared by the melt blending method could be obviously broadened. It is worthwhile to note that the long-chain branch structure of PET molecular was still necessary to obtain a foamable PET, as demonstrated in Figure 3.13, although layered clays were added.

In the reactive extrusion process of recycled PET, the addition of heat stabilizer and antioxidant was necessary to avoid serious thermal and oxidation

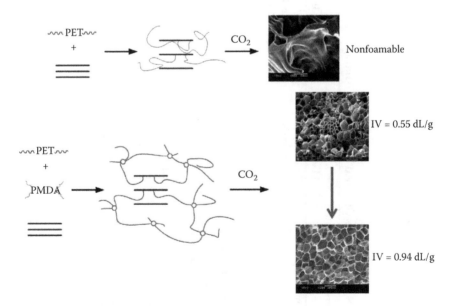

FIGURE 3.13
Melt foamability of PET/clay nanocomposites.

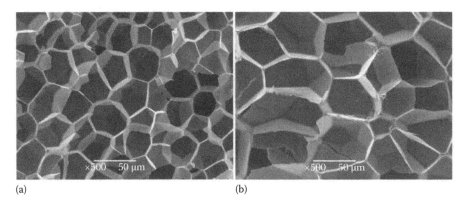

(a)   (b)

FIGURE 3.14
SEM micrographs of TGIC-modified bottle-recycled PET foams obtained at (a) 260°C and (b) 270°C under 16 MPa CO$_2$.

degradations in the extruder. It was discovered that triglycidyl isocyanurate (TGIC) was more effective than PMDA to increase the molecular weight and melt strength of recycled PET. And the TGIC-modified bottle-recycled PET could also be foamed in melt state, as displayed in Figure 3.14.

3.4 Melt-Foaming Behavior of *In-Situ* Modified PET in Batch and Continuous Processes

In *in-situ* polymerization–modification process, the third monomer, polyacid or polyalcohol with more than two functional groups, was added into the direct esterification–polycondensation process together with terephthalate acid and ethylene glycol (EG). The synthesis process and process conditions were similar to the traditional PET manufacture. PMDA could also be introduced into polymerization reactions to obtain a long-chain-branched PET [18]. The IV of *in-situ* modified PET increased with PMDA content, and the IV also reached a maximum value of 0.865 dL/g with a PMDA content of 0.8 wt%. With the continuous increase of PMDA content, gel formation was observed. Pentaerythritol (PENTA) was more effective in enhancing the IV and melt strength of PET as a chain extender, and the IV of PENTA-modified PET got up to 0.860 dL/g, as the PENTA content was only 0.35 wt%, because PENTA could react with the carboxyl group of terephthalate acid monomer, as well as the PET oligomer, and induce the formation of a branched structure in the esterification stage during the PET synthesis process. Gel also formed as the PENTA content exceeded 0.35 wt%. PENTA and PMDA *in-situ* modified

PET were foamed at different temperatures, as shown in Figures 3.15 and 3.16, both of which obtained uniform closed-cell structures. It was noted that PMDA-modified PET foams possessed larger cell sizes and lower cell densities than those of PENTA-modified PET foams, which could be explained by the lower carboxyl content of PENTA-modified PET, due to the consumption of a carboxyl-reactive chain extender PENTA, along with higher thermal stability. The characterization of *in-situ* modified PET foams is summarized in Table 3.3. Furthermore, via the prereaction between the monomer of EG and the phosphorus-based fire retardant of 2-carboxyethyl(phenyl)phosphinic acid (CEPPA), fire-resistant PET was prepared by *in-situ* polymerization and remained foamable due to the existence of polyalcohol as a multifunctional chain extender.

The melt-state foaming process of *in-situ* modified PET could also be implemented continuously, such as extrusion foaming. However, compared with the batch foaming process, a higher molecular weight and melt strength were required, which would deteriorate in the extruder owing to degradations. Therefore, PENTA-modified PET had to be polymerized in a solid state for another 24 h, and the IV reached 1.5 dL/g [19]. The rheological properties, including the shear and extension rheological behaviors of the polymer melt,

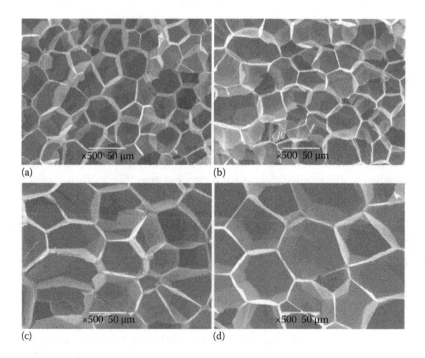

(a)   (b)

(c)   (d)

FIGURE 3.15
SEM micrographs of PENTA-modified PET foamed at a saturation pressure of 14 MPa and different temperatures of (a) 265°C, (b) 270°C, (c) 275°C, and (d) 280°C. (From H. Zhong et al., *Chinese Journal of Chemical Engineering*, 21, 1410–1418, 2013.)

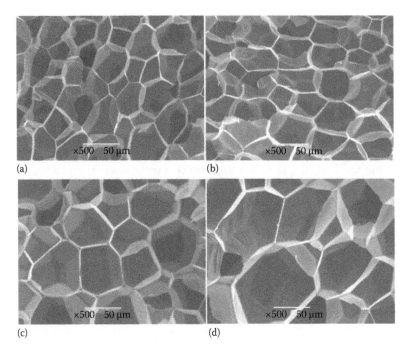

FIGURE 3.16
SEM micrographs of PMDA-modified PET foamed at a saturation pressure of 14 MPa and different temperatures of (a) 265°C, (b) 270°C, (c) 275°C, and (d) 280°C. (From H. Zhong et al., *Chinese Journal of Chemical Engineering*, 21, 1410–1418, 2013.)

played key roles in the extrusion foaming process. Relaxation time spectrum revealed the most basic function relation between viscoelasticity and time or frequency. All sorts of material functions obtained by rheology measurements were based on the same relaxation time spectrum, and rheological properties could be expressed by the relaxation time spectrum of different movement patterns and contributions. Four different PETs were investigated by Fan et al. [19]. PET1 was a regular spinning-grade PET with a linear molecular structure, and the IV was 0.65 dL/g. PET2 was obtained by solid-state polymerization (SSP) for 24 h at 220°C under vacuum with PET1 as a raw material, and the IV reached 0.90 dL/g. PENTA-modified PET was also polymerized in a solid state for 15 h to get PET3 with an IV up to 1.30 dL/g. Another 9-h SSP was carried out to obtain PET4 with an IV of 1.50 dL/g. Tschoegl equations were adopted to estimate the viscoelastic behavior of PET:

$$H(\tau) = \frac{dG'}{d\ln\omega} - \frac{dG'}{2\,d\ln\omega}\bigg|_{\frac{1}{\omega}=\frac{\tau}{\sqrt{2}}} \tag{3.3}$$

$$H(\tau) = \frac{2}{\pi}\left[G'' - \frac{4}{3}\left(\frac{dG''}{dln\omega}\right) + \frac{1}{3}\left(\frac{dG''}{dln\omega}\right)\right]_{\frac{1}{\omega}=\frac{\tau}{\sqrt{5}}} \tag{3.4}$$

where $H(\tau)$ is the relaxation spectra, and ω is the experimental frequency. For a better revelation of viscoelasticity in a long time, the relaxation time spectrum was calculated based on G' results. As shown in Figure 3.17, the linear chain (PET1 and PET2) was easier to slip to overcome macromolecular entanglement so that a shorter relaxation time was needed. When the relaxation time reached 10 s, the relaxation spectra were lower than 10 Pa, reflecting the liquid-like relaxation behavior. PET3 and PET4 had higher relaxation spectra due to the long-chain branch, especially at a long relaxation time. Macromolecular entanglements were more complicated with the existence of the long-chain branch, which made the relaxation process longer and the relax modulus higher. Modified PETs also exhibited a rubber-like behavior.

PET1, PET2, and PET3 all failed in the extensional rheology characterization with Sentmanat extensional rheometer (SER) unit at the temperature between 250°C and 280°C because the testing samples sagged during the heating, which as attributed to the lower melt elasticity. The extensional viscosity curves of PET4 are shown in Figure 3.18, and obvious strain-hardening behaviors were observed, which played an important role in stabilizing the cell structure and preventing cells from merging and collapsing. Moreover, the strain hardening was more obvious with the decrease of temperature, which made the strain-hardening behavior occur earlier and more significantly.

A single-screw extruder (screw diameter = 50 mm, L/D = 45) equipped with a rod die (2 mm in diameter) was used for the extrusion foaming

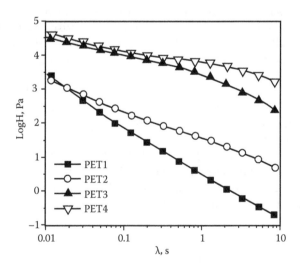

FIGURE 3.17
Relaxation spectra of PETs. (From C. Fan et al. *Journal of Cellular Plastics*, 52, 277–298, 2016.)

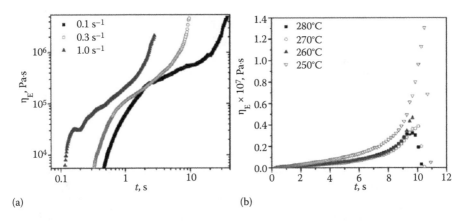

FIGURE 3.18
Extensional viscosities of PET4 (a) at 270°C under different deformation rates and (b) at 0.3 s^{-1} and different temperatures. (From C. Fan et al., *Journal of Cellular Plastics*, 52, 277–298, 2016.)

process by Fan et al. [19] A static mixer, with Sulzer mixture units aimed to facilitate gas dispersion and dissolution in a polymer matrix, was installed at the end of screw, as shown in Figure 3.19. Under proper operating conditions, the diameter of extrudates increased gradually with operating time, and PET foams were obtained, as demonstrated in Figure 3.20.

Die temperature was the key parameter in the extrusion foaming process. With the decrease of foaming temperature, the melt strength increased, and furthermore, higher CO$_2$ solubility was obtained. Therefore, the expansion ratio of PET foam got higher, and the cell morphology became better. And the highest expansion ratio of 9.37 was acquired at a die temperature of 240°C. As shown in Figure 3.21, idea cell morphologies were obtained with a uniform cell size distribution at foaming temperatures of 236°C and 240°C. With the continuous decrease of foaming temperature, the foam expansion would be hindered by the increasing stiffness of melt.

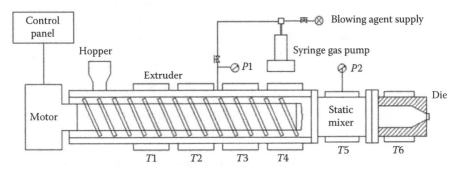

FIGURE 3.19
Schematic of the experimental apparatus for extrusion foaming. (From C. Fan et al., *Journal of Cellular Plastics*, 52, 277–298, 2016.)

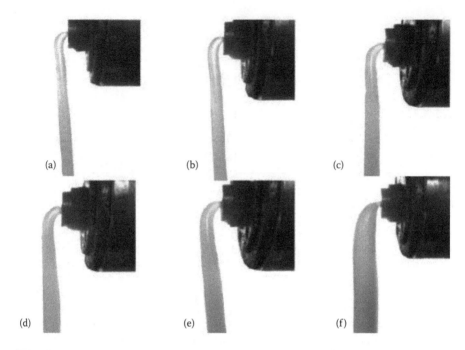

FIGURE 3.20
Schematic of extrusion foaming process: the extrudates expanded gradually from (a) the start of CO_2 injection to (f) the steady state of PET foaming.

The extrusion pressure increased with the screw speed and reached a maximum value of 11.6 MPa, and the expansion ratio of extruded PET foams got up to 9.0. Thermodynamic instability was triggered by the rapid pressure drop when the melt passed through the die. CO_2 solubility in the PET melt and the supersaturation of CO_2 increased with extrusion pressure, which directly facilitated the cell nucleation process. As a result, the cell size decreased from 512 to 449 μm, and the cell density, as well as the expansion ratio, got higher with the increase of screw speed from 37 to 46 rpm.

With the increase of gas input from 0.1 to 5 mL/min, the expansion ratio increased almost linearly and then leveled off as the gas input was up to 10.0 mL/min. More CO_2 dissolved in the PET melt with the increasing gas input, which could be used to explain the increase of cell density and the reduction of cell size. It was the noteworthy that the undissolved gas would cause the fluctuation of extrusion pressure, resulting in the instability and even the failure of extrusion foaming process, as the gas input exceeded the solubility.

The addition of nano-SiO_2 offered a large number of heterogeneous nucleation sites that improved the nucleation process via decreasing the nucleation energy barrier and had an obvious impact on the cell morphology. With the addition of only 0.1 wt% nano-SiO_2, the cell density increased by

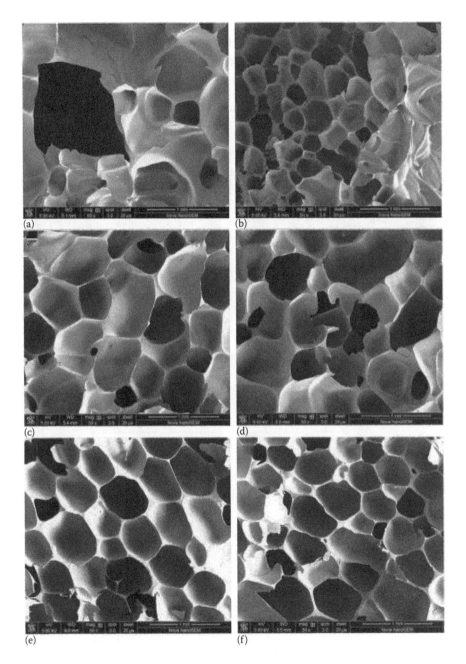

FIGURE 3.21

SEM images of the center (left) and border (right) regions of PET foams at different die temperatures: (a, b) 232°C, (c, d) 236°C, and (e, f) 240°C. (From C. Fan et al., *Journal of Cellular Plastics,* 52, 277–298, 2016.)

four times, whereas the cell size decreased from 499 to 265 μm, as displayed in Figure 3.22. Moreover, the cell size distribution was an important parameter to characterize the cell morphology in polymer foam, which also had a significant effect on the physical and mechanical properties of foams along with the cell size and cell density. Shortening the nucleation time interval was an effective method to obtain a narrow cell distribution, which could be

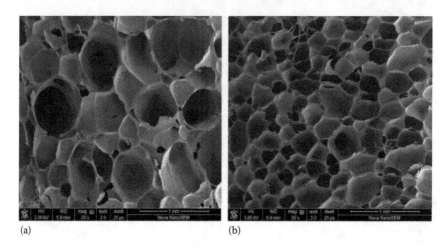

(a)   (b)

FIGURE 3.22
SEM images of PET foams: (a) without nano-SiO$_2$ and (b) with 0.2 wt% nano-SiO$_2$. (From C. Fan et al., *Journal of Cellular Plastics*, 52, 277–298, 2016.)

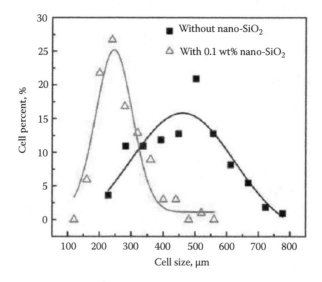

FIGURE 3.23
The effect of nano-SiO$_2$ on cell size distribution of PET foams. (Lines were fitted curves.) (From C. Fan et al., *Journal of Cellular Plastics*, 52, 277–298, 2016.)

reduced via accelerating the nucleation process. As shown in Figure 3.23, the fitted curves of the cell size distribution of PET foams with 0.1 wt% nano-SiO$_2$ demonstrated higher peak height and narrower peak width, meaning narrower cell size distribution.

3.5 Integrated Process of Supercritical CO$_2$-Assisted Modification and Foaming of PET

It was well known that the reversible reaction was usually controlled by the removal of volatile condensates in the final polycondensation stages, which was facilitated conventionally by high vacuum. However, there always existed mass-transfer limitations of volatiles in the polymer matrix. Supercritical CO$_2$, a kind of environmental alternative to traditional solvents, had been successfully applied to facilitate the devolatilization of the polymer [20,21]. The free volume of polymer swollen by CO$_2$ increased, which promoted the diffusion of volatiles in polymer matrix and the mobility of polymer chains along with their end groups, resulting in a higher polycondensation rate. Furthermore, many volatile condensates of polycondensation displayed significant solubility in liquid or supercritical CO$_2$, which may serve as a more effective sweeping fluid to bring the volatiles away and drive the polycondensation forward. EG, a volatile condensate generated in PET polycondensation, was soluble in CO$_2$ up to 2–3 wt%, which was much higher than the solubility of EG in the PET melt. The melt polycondensation–modification process was carried out with high-pressure CO$_2$ sweeping using PENTA-modified PET that is prepared via the *in-situ* polymerization method as a raw material, based on which an integrated process of CO$_2$-assisted PET further melt polycondensation–modification and foaming was carried out, as illustrated in Figure 3.24a [22].

The foam morphologies obtained under different CO$_2$ flow rates and pressures are displayed in Figure 3.25. The cell size and the foam density reduced, whereas the cell density increased, with CO$_2$ flow rates. Further, polycondensation promoted by CO$_2$ sweeping increased long-chain branches, which could improve the melt strength and was beneficial for the foaming process, and thus uniform cell a morphology could be expected. Although higher CO$_2$ flow rate was more effective in removing the condensates, it was found that the difference between foam densities obtained with CO$_2$ flow rates of 4 L/min and 5 L/min at each saturation pressure was undetectable, as shown in Figure 3.26. For each CO$_2$ flow rate, the foam density first decreased with saturation pressure (before 12 MPa) and then increased at 14 MPa. A higher saturation pressure had dual effects on foam morphology. The cell density increased, whereas the foam density decreased, with CO$_2$ pressure at first due to the increase of CO$_2$

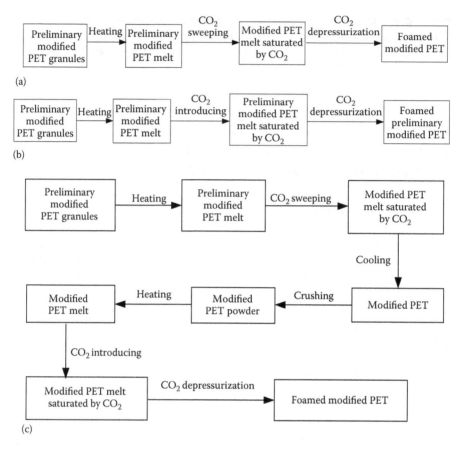

(a)

(b)

(c)

FIGURE 3.24
Flow chart of different PET foaming processes in melt state: (a) Integrated process, (b) one-step process, and (c) separated process.

solubility and depressurization rate. At the same time, a stronger plasticization effect caused by the continuous increase of CO_2 pressure would weaken the entanglements of the molecular chain and finally lead to more cell coalescence. Therefore, PET foams with relative lower expansion ratio and higher density were obtained.

The cell structures of PET foams obtained at different saturation pressures and treating times are exhibited in Figure 3.27. The foam density increased whereas the cell density decreased with treating time, as shown in Figure 3.28, indicating that the thermal degradation may dominate. The melt strength and foamability of PET deteriorated as a result of thermal degradation. Differential scanning calorimetry (DSC) and melt flow index (MFI) were measured for modified PET foams that are obtained under different treating times. The melt point of PET foams decreased slightly, whereas

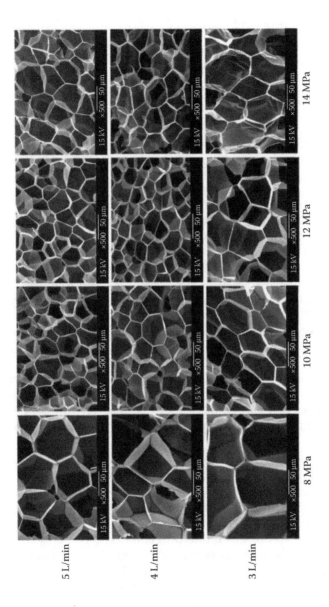

FIGURE 3.25
SEM micrographs of obtained PET foams from the integrated process under different saturation pressures and CO$_2$ flow rates at 280°C, 30 min. (From H. Zhong et al., *The Journal of Supercritical Fluids*, 74, 70–79, 2013.)

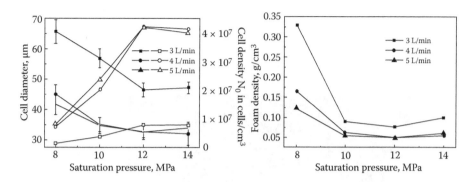

FIGURE 3.26

Characterization of PET foams obtained by the integrated process under different saturation pressures and CO_2 flow rates at 280°C, 30 min. (From H. Zhong et al., *The Journal of Supercritical Fluids*, 74, 70–79, 2013.)

the MFI increased with the treating time, which could be attributed to the branched structures and thermal degradations.

For comparison, another two foaming processes in the melt state were also carried out. One was the *one-step process*, as illustrated in Figure 3.24b, in which CO_2 was just introduced at the beginning to saturate polymer, i.e., no CO_2 sweeping, followed by rapid depressurization. A preliminary modified PET with an IV of 0.72 dL/g was foamed via the integrated process and the one-step process, as shown in Figure 3.29. It was obvious that no foam could be obtained with the one-step process due to the low melt strength of the preliminary PET. In regard to the integrated process, it was notable that cellular morphology could be easily distinguished, implying that the continuous sweeping of CO_2 led to further polycondensation and improved melt strength. Hence, it was deduced that the integrated process reduced the demand of the molecular weight of PET matrix in the melt foaming process.

The CO_2-assisted polycondensation–modification process of preliminary modified PET and the CO_2 foaming process were implemented separately in the so-called *separated process*, as illustrated in Figure 3.24c. It is clearly seen from Figure 3.30 that foams obtained by separated process were characterized by larger cell sizes, lower cell density, and thin cell walls. Cell collapse and cell-wall fractures could also be easily discerned. The cell morphologies were worse than those obtained by the integrated process. It was believed that degradation was the main reason, and DSC and MFI tests were carried out to investigate the decrease of the melting point (T_m) and the melt viscosity. There existed a 12°C–15°C reduction of T_m and a significant increase of MFI before and after foaming. It could be concluded that the degradation did occur seriously in the remelt process.

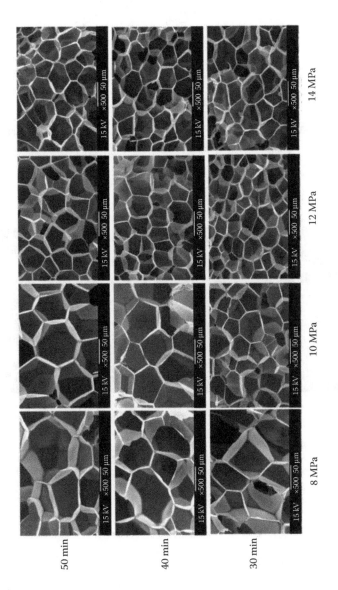

FIGURE 3.27
SEM micrographs of obtained PET foams from the integrated process under different saturation pressures and treating times at 280°C, 4 L/min. (From H. Zhong et al., *The Journal of Supercritical Fluids,* 74, 70–79, 2013.)

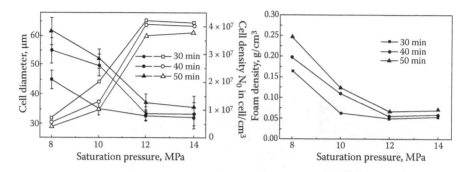

FIGURE 3.28
Characterization of PET foams obtained by the integrated process under different saturation pressures and treating times at 280°C, 4 L/min. (From H. Zhong et al., *The Journal of Supercritical Fluids*, 74, 70–79, 2013.)

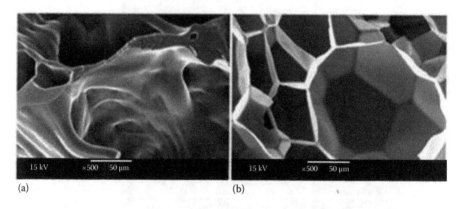

(a)   (b)

FIGURE 3.29
SEM micrographs of low-melt-strength PET foamed through (a) one-step process with saturation time 30 min and saturation pressure 12 MPa and (b) integrated process with sweeping time 30 min, saturation pressure 12 MPa, and CO_2 flow rate 4 L/min at 280°C. (From H. Zhong et al., *The Journal of Supercritical Fluids*, 74, 70–79, 2013.)

3.6 Summary and Outlook

Some industrially promising processes aided by CO_2 have been explored to tailor the structure of PET foams or simplify the manufacturing process of PET foams for high efficiency. For the melt-state foaming process of PET, a polymer matrix with high melt elasticity and strength was demanded, which could not be satisfied by commercial PET with linear molecular chain

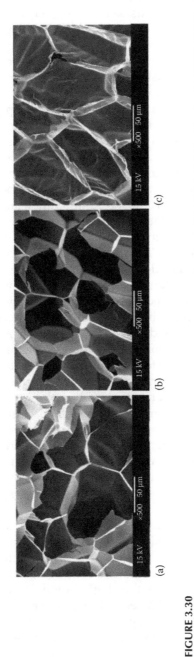

FIGURE 3.30
SEM micrographs of PET foams obtained by separated process with foaming conditions: (a) 280°C, 12 MPa, 30 min; (b) 275°C, 12 MPa, 30 min; and (c) 275°C, 10 MPa, 30 min. (From H. Zhong et al., *The Journal of Supercritical Fluids*, 74, 70–79, 2013.)

and relatively low molecular weight. Different modifying methods aiming to introduce the long-chain branch to the PET backbone were described in this chapter, including reactive extrusion, *in-situ* polymerization, and the so-called integrated process. It should be noted that a much-higher melt strength was essential in the extrusion foaming process because severe degradations were inevitable in the extruder, and the corresponding modifying process costs much more time and energy. Layered organoclay could also be employed to compound with the PET matrix to improve the melt foamability of PET with low melt elasticity. The cell collapse and coalescence were unavoidable in the melt foaming process, although the melt strength was improved a lot. PET foams prepared by the melt foaming process were always with a cell size of hundreds of microns and a cell density that is lower than 10^9 cells/cm^3. In other words, no microcellular PET foams could be acquired. Microcellular PET foams (cell size < 10 μm, cell density > 10^9 cells/cm^3) with excellent optical and mechanical properties were often produced by a solid-state foaming process. Crystallization, which could be facilitated due to the plasticization effect of CO_2, was the most important factor in the solid-state foaming process rather than the viscoelastic properties of polymer matrix, and thus the PET with a linear molecular chain could be used as a raw material without further modifying. The solid-state foaming process described in this chapter prepared a controllable sandwich structure of PET foams with two microcellular or even ultramicrocellular foamed crystalline layers outside and a microcellular foamed amorphous layer inside. The thickness of foamed crystalline layer could be regulated by the CO_2 saturation time and calculated by the proposed model coupling CO_2 diffusion and CO_2-induced crystallization of PET. In addition, the blowing agent supercritical CO_2 could also assist the regulation of the foaming process due to the strong plasticization effect on the polymer matrix, such as inducing the crystallization and facilitating the melt modification.

PET foams with better cell morphology and foam property should be prepared through a more simplified process in order to develop more applications. The long-chain branch structure of PET destroyed the ordered molecular structure and decreased the crystallinity of PET, as shown in Table 3.2, which reduced the mechanical and other properties of PET. The crystallinity enhancement of melt foaming products was necessary, which may be carried out via the addition of nucleation agents and/or increasing the melt foamability of PET with a linear molecular chain. Furthermore, the intensive scission of a PET chain in the extrusion foaming process decreased the melt strength again and thus weakened the foamability of PET. Hence, some new continuous foaming processes should be developed to avoid the degradation in the CO_2-dissolving stage. The *integrated process* involved in this chapter supplied an outstanding solution, which should be implemented in a continuous process, for example, in a falling film reactor with larger mass transfer interface that is operated under a high-pressure CO_2 atmosphere.

Acknowledgments

The authors are grateful to the National Natural Science Foundation of China (Grant No. 21176070 and 21306043), the National Programs for High Technology Research and Development of China (863 Project, 2012AA040211), the Research Fund for the Doctoral Program of Higher Education of China (Grant No. 20120074120019), the Fundamental Research Funds for the Central Universities, and the 111 Project (B08021).

References

1. V. Kumar, R.P. Juntunen, C. Barlow. 2000. Impact strength of high relative density solid state carbon dioxide blown crystallizable poly(ethylene terephthalate) microcellular foams. *Cellular Polymers*, 19 25–37.
2. S. Doroudiani, C.B. Park, M.T. Kortschot. 1996. Effect of the crystallinity and morphology on the microcellular foam structure of semicrystalline polymers. *Polymer Engineering & Science*, 36 2645–2662.
3. V. Kumar, P.J. Stolarczuk. 1999. Microcellular PET foams produced by the solid state process. *Imaging and Image Analysis Applications for Plastics*, 241–247.
4. D.F. Baldwin, M. Shimbo, N.P. Suh. 1995. The role of gas dissolution and induced crystallization during microcellular polymer processing: A study of poly(ethylene terephthalate) and carbon dioxide systems. *Journal of Engineering Materials and Technology*, 117 62.
5. V. Kumar, H.G. Schirmer. 1997. Semi-continuous production of solid state polymeric foams. In US Patent 5684055.
6. L. Di Maio, I. Coccorullo, S. Montesano, L. Incarnato. 2005. Chain extension and foaming of recycled PET in extrusion equipment. *Macromolecular Symposia*, 228 185–200.
7. M. Xanthos, C. Wan, R. Dhavalikar, G.P. Karayannidis, D.N. Bikiaris. 2004. Identification of rheological and structural characteristics of foamable poly(ethylene terephthalate) by reactive extrusion. *Polymer International*, 53 1161–1168.
8. R. Dhavalikar, M. Yamaguchi, M. Xanthos. 2003. Molecular and structural analysis of a triepoxide-modified poly(ethylene terephthalate) from rheological data. *Journal of Polymer Science Part A: Polymer Chemistry*, 41 958–969.
9. M.Y. Wang, Z.Q. Guo, B.Y. Lei, N.Q. Zhou. 2011. Rheological and thermal behavior of recycled PET modified by PMDA. *Advanced Materials Research*, 391–392 688–691.
10. L. Xiao, H. Wang, Q. Qian, X. Jiang, X. Liu, B. Huang, Q. Chen. 2012. Molecular and structural analysis of epoxide-modified recycled poly(ethylene terephthalate) from rheological data. *Polymer Engineering & Science*, 52 2127–2133.
11. S. Japon, Y. Leterrier, J.-A.E. Manson. 2000. Recycling of poly(ethylene terephthalate) into closed-cell foams. *Polymer Engineering & Science*, 40 1942–1952.

12. D. Li, T. Liu, L. Zhao, W. Yuan. 2012. Controlling sandwich-structure of PET microcellular foams using coupling of CO2 diffusion and induced crystallization. *Aiche Journal*, 58 2512–2523.
13. T. Xia, Z. Xi, T. Liu, X. Pan, C. Fan, L. Zhao. 2015. Melt foamability of reactive extrusion-modified poly(ethylene terephthalate) with pyromellitic dianhydride using supercritical carbon dioxide as blowing agent. *Polymer Engineering & Science*, 55 1528–1535.
14. B. Zhu, W. Zha, J. Yang, C. Zhang, L.J. Lee. 2010. Layered-silicate based polystyrene nanocomposite microcellular foam using supercritical carbon dioxide as blowing agent. *Polymer*, 51 2177–2184.
15. W. Zhai, T. Kuboki, L. Wang, C.B. Park, E.K. Lee, H.E. Naguib. 2010. Cell structure evolution and the crystallization behavior of polypropylene/clay nanocomposites foams blown in continuous extrusion. *Industrial & Engineering Chemistry Research*, 49 9834–9845.
16. P.H. Nam, P. Maiti, M. Okamoto, T. Kotaka, T. Nakayama, M. Takada, M. Ohshima, A. Usuki, N. Hasegawa, H. Okamoto. 2002. Foam processing and cellular structure of polypropylene/clay nanocomposites. *Polymer Engineering & Science*, 42 1907–1918.
17. M. Okamoto, P.H. Nam, P. Maiti, T. Kotaka, T. Nakayama, M. Takada, M. Ohshima, A. Usuki, N. Hasegawa, H. Okamoto. 2001. Biaxial flow-induced alignment of silicate layers in polypropylene/clay nanocomposite foam. *Nano Letters*, 1 503–505.
18. H. Zhong, Z. Xi, T. Liu, L. Zhao. 2013. In-situ polymerization-modification process and foaming of poly(ethylene terephthalate). *Chinese Journal of Chemical Engineering*, 21 1410–1418.
19. C. Fan, C. Wan, F. Gao, C. Huang, Z. Xi, Z. Xu, L. Zhao, T. Liu. 2016. Extrusion foaming of poly(ethylene terephthalate) with carbon dioxide based on rheology analysis. *Journal of Cellular Plastics*, 52 277–298.
20. S. Ye, K. Jiang, C. Jiang, Q. Pan. 2005. Dynamic supercritical fluid devolatilization of polymers. *Chinese Journal of Chemical Engineering*, 13 732–735.
21. S. Alsoy, J.L. Duda. 1998. Supercritical devolatilization of polymers. *Aiche Journal*, 44 582–590.
22. H. Zhong, Z. Xi, T. Liu, Z. Xu, L. Zhao. 2013. Integrated process of supercritical CO_2-assisted melt polycondensation modification and foaming of poly(ethylene terephthalate). *The Journal of Supercritical Fluids*, 74 70–79.

4

Formation Mechanism and Tuning for Bimodal Cell Structure in Foams by Synergistic Effect of Temperature Rising and Depressurization with Supercritical CO_2*

Han-Xiong Huang and Lin-Qiong Xu

CONTENTS

4.1 Introduction

Recently, polymer foams with a bimodal cell structure (BMCS) have attracted a great deal of interest in industry and academia because they combine the advantages of both small cells that exhibit outstanding mechanical properties and large cells that lower bulk density [1–4]. Compared with foams with a monomodal cell structure (MMCS), foams with a BMCS have a higher degree of cellular interconnectivity, which helps to expand their applications

* Financial support provided by the National Natural Science Foundation of China (11172105) is gratefully acknowledged.

in biomedicine for tissue engineering [5–11]. More interestingly, the present authors [12] found for the first time that the fractured surface of the foam with a BMCS exhibits a superhydrophobic wetting property like a lotus leaf, which is similar to that of biomimetic superhydrophobic surfaces [13–16]. Although most publications are out of research laboratories, some features are of high commercial value. The development route may take time, and this chapter lays the foundational elements on BMCS.

Foams with a BMCS may be fabricated by manipulating the foaming conditions using supercritical carbon dioxide (Sc-CO_2) as a physical foaming agent in a batch foaming process [3,4,7,8,17–21]. Arora et al. [17] firstly prepared polystyrene (PS) foams with a BMCS using a two-step depressurization process, which generates two nucleation stages sequentially. The holding stage between the two depressurization stages is the key to controlling the BMCS. The depressurization rate and temperature in the depressurization stages also influence the cellular structure to a certain extent [18,20]. Li et al. [21] prepared isotactic polypropylene (iPP) foams with a BMCS in a batch process by the synergistic effect of temperature decreasing and depressurization, which produces two nucleation stages sequentially. This method, however, is invalid for poly(lactic acid) (PLA) because the solubility variation of CO_2 in it from saturation to foaming temperatures is not as obvious as that of CO_2 in iPP [21,22]. Yokoyama and Sugiyama [19] found that a BMCS is formed in poly[styrene-block-4-(perfluorooctylpropyloxy) styrene] (PS-PFS) foams when the temperature of depressurization is in the vicinity of the glass transition temperature (T_g) of PS using a batch foaming process. In PS-PFS foams with a BMCS, nanocells and microcells are formed in the fluorinated spherical block copolymer domains and rubbery matrix, respectively, because of different surrounding resistance for cell growth in the two domains. So, the method proposed by Yokoyama and Sugiyama for the formation of the BMCS is limited to materials with a heterogeneous structure. Another type of approach for preparing foams with a BMCS combines two different foaming agents to induce two different nucleating mechanisms in an extrusion foaming process [1,2]. For example, Zhang et al. [2] prepared PS foams with a BMCS using CO_2 and water as cofoaming agents and using nanoclay and activated carbon as the nucleation agents.

Microcellular foams can be prepared via a temperature rising or depressurization manner using Sc-CO_2 as a physical foaming agent in a batch process [23–38]. Huang et al. [25,32] proposed a foaming process involving temperature rising and depressurization stages. This work pursues the previous studies in this laboratory but puts emphasis on the formation of the BMCS in both PS and PLA foams by the synergistic effect of temperature rising and depressurization. The formation mechanism of the BMCS is revealed, and the influences of foaming parameters on the BMCS are investigated.

4.2 Experimental

4.2.1 Materials

Two polymers were used in this work. One was PS (grade 951F, Taita Chemical Company, Ltd., Taiwan), an amorphous polymer. The PS had a melt flow index (MFI) of 32.2 g/10 min (5 kg, 200°C), a density of 1.05 g/cm^3, and a T_g of 90.5°C. The other was PLA (grade 2002D, NatureWorks LLC, USA), a semi-crystalline polymer. The PLA had a D-lactic acid monomer content of approximately 4%, an MFI of 8.8 g/10 min (2.16 kg, 210°C), and a density of 1.24 g/cm^3. Industrial CO_2 with a purity of 99.5% was used as a foaming agent.

4.2.2 Sample Preparation

A batch foaming apparatus consisting of a high-pressure vessel and a syringe pump (ISCO 500D) [25,32] was used.

The PS was dried at 80°C for 8 h in a vacuum oven and then extruded into strands with a diameter of approximately 1.5 mm using a single-screw extruder at a temperature of 180°C. The strands were cut into rods with a length of approximately 10 mm.

The as-prepared rods were foamed with the batch foaming apparatus via a pressure-quenching manner by using Sc-CO_2 as a physical foaming agent. The foaming procedure was as follows: (1) heat the vessel to saturation temperature (T_S) within approximately 15 min; (2) place the aforementioned samples in the vessel; (3) slowly flush the vessel using CO_2 gas; (4) pressurize the vessel to foaming pressure (P) using high-pressure CO_2; (5) use two different temperature modes, that is, constant- and varying-temperature modes (CTM and VTM, as shown in Figure 4.1); (6) depressurize the vessel to atmospheric pressure in less than 0.5 s; and (7) inject the foamed samples out of the vessel and cool them to the room temperature in the air. In the CTM, the saturation and foaming temperatures (T_S and T_F) were both 150°C, and the saturation time was 220 min. In the VTM, considering the effect of T_S on the diffusion rate for the CO_2 in the samples, the samples were saturated within the vessel for 180 and 300 min (the first saturation time, t_{S1}) when the T_S was equal to and lower than 100°C, respectively; then, the vessel was heated to the second saturation temperature (i.e., T_F) within approximately 15 min, followed by keeping at this temperature for some time (the second saturation time, t_{S2}), that is, there exist two sequential saturation stages in the VTM. The saturation pressure is the same as the P in both the CTM and the VTM. The foaming parameters used in the VTM are listed in Table 4.1.

The PLA was dried at 90°C for 2 h in a vacuum oven, and then compression molded into sheets with a thickness of 4 mm at 180°C and 15 MPa for 10 min, followed by cooling the mold by water. The sheets were cut into

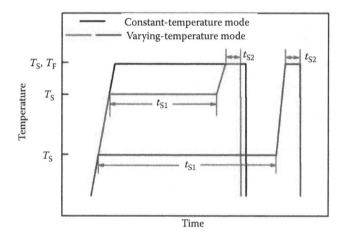

FIGURE 4.1
Schematics of temperature evolution versus time in foaming process used in this work.
T_F: foaming temperature; T_S: saturation temperature; t_{S1}: first saturation time; t_{S2}: second saturation time.

TABLE 4.1

Foaming Parameters Used for PS in VTM

Variable	Values
Saturation temperature, T_S (°C)	30, 40, 60, 80, 90 ($T_F = 100$°C); 40, 60, 100 ($T_F = 150$°C)
Foaming temperature, T_F (°C)	60, 80, 100, 150 ($T_S = 40$°C); 110, 120, 130, 140, 150, 160 ($T_S = 100$°C)
First saturation time, t_{S1} (min)	180 ($T_S = 100$°C); 300 ($T_S < 100$°C)
Second saturation time, t_{S2} (min)	25, 45, 105
Foaming pressure, P (MPa)	14, 16, 18

cubes with a length of approximately 10 mm and a width of approximately 4 mm. The cubic PLA samples were foamed with a batch foaming apparatus via both CTM and VTM. In the CTM, the T_S and T_F were both 140°C, and the saturation time was 40 min; in the VTM, after saturating the samples in the vessel for 40 min (t_{S1}), the vessel temperature was raised to the T_F within approximately 15 min and then kept at the T_F for 25 min (t_{S2}).

4.2.3 Characterization

The foamed samples were immersed in liquid nitrogen for approximately 10 min and fractured. Then they were gold sputtered, and their cryofractured surfaces were examined using a scanning electron microscope (SEM; Quanta 200, FEI Company, Eindhoven, Holland) at an accelerating voltage of

15 kV. The mean cell diameter (D) and cell density (ρ_c) of the foamed samples were estimated by

$$D = \frac{\Sigma d_i n_i}{\Sigma n_i}$$
(4.1)

$$\rho_c = \left(\frac{NM^2}{A} \right)^{3/2}$$
(4.2)

where d_i is the single cell diameter, n_i is the number of cells with d_i, A is the area of an SEM micrograph, N is the number of cells in area A, and M is the magnification factor of an SEM micrograph. The mean cell diameters and cell densities of large (D_1 and ρ_{cl}) and small (D_s and ρ_{cs}) cells were calculated in this work.

The density (ρ_f) of the foamed samples was measured using a specific gravity bottle with a volume of 25 mL by the water displacement method in accordance with ASTM D792-00 [39] and calculated by

$$\rho_f = \frac{W_0}{W_0 + W_1 - W_2} \rho_w$$
(4.3)

where W_0 is the weight of the foamed sample, W_1 is the weight of the specific gravity bottle that is filled with water, W_2 is the weight of the specific gravity bottle containing both water and the sample, and ρ_w is the density of water, which is 1 g/cm³.

The volume expansion ratio (V_r) of the foamed samples was defined as the ratio of the density of the unfoamed sample (ρ_0) to the ρ_f:

$$V_r = \frac{\rho_0}{\rho_f}$$
(4.4)

4.3 Results and Discussion

4.3.1 Formation Mechanism Analyses on BMCS

Figure 4.2 shows the typical SEM micrographs of the fractured surfaces for the foamed PS samples that were prepared at T_ss of 150°C using the CTM and 40°C using the VTM. It can be seen that an MMCS is obtained using the CTM, and the values of D and ρ_c for the cells are 29 μm and 7.2×10^7 cells/cm³, respectively. It is interesting to find from Figure 4.2b that the foamed sample

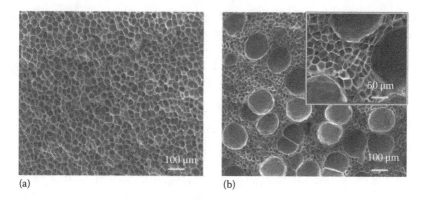

(a)   (b)

FIGURE 4.2

Typical SEM micrographs of foamed PS samples prepared at saturation temperatures of (a) 150°C using CTM and (b) 40°C using the VTM. Foaming temperature: 150°C; foaming pressure: 14 MPa; second saturation time: 25 min.

prepared using the VTM presents a BMCS, and the D_l, D_s, ρ_{cl} and ρ_{cs} are 139 µm, 18 µm, 1.4×10^5 cells/cm^3, and 1.5×10^8 cells/cm^3, respectively. The D and ρ_c for the former sample fall in between those of the large and small cells for the latter one, respectively.

The foamed PLA samples prepared at T_Ss of 140°C using the CTM and 120°C using the VTM display high V_rs and a more uniform cellular structure across their fractured surfaces, and thus the micrographs were taken from the regions that are close to their outer surfaces, as illustrated in Figure 4.3. As can be seen, an MMCS is obtained using the CTM, and a higher magnification allows observing a stamen-like cell structure with entities. A BMCS is achieved using the VTM, and the D_l and D_s are 1400 µm and 36 µm, respectively. A higher magnification allows observing the stamen-like cell structure with small entities in the small cell region (as indicated by the arrow).

To reveal the formation mechanism of the aforementioned BMCS, a rapid cooling experiment was carried out. Specifically, during the VTM foaming under the same foaming parameters as those that are used in Figure 4.2b, the vessel with PS samples was quenched in iced water to 0°C for approximately 20 min following the second gas saturation, and then CO_2 was released at a very low depressurization rate to avoid PS foaming. The SEM micrograph of the fractured surface for the as-prepared sample is displayed in Figure 4.4. It is clearly visible from Figure 4.4 that a number of cells with nonuniform size distribution and thick cell walls are formed. Compared to the large cells in the sample foamed using the VTM (Figure 4.2b), the cells exhibit a much smaller D (46 µm) and one order of magnitude larger ρ_c (2.1×10^6 cells/cm^3). Combining Figures 4.2b and 4.4, a development mechanism of the BMCS for the amorphous polymer (such as PS in this work) prepared using the VTM is proposed, as schematically shown in Figure 4.5. At a giving temperature and pressure in the first gas saturation stage, the Sc-CO_2 is dissolved into the

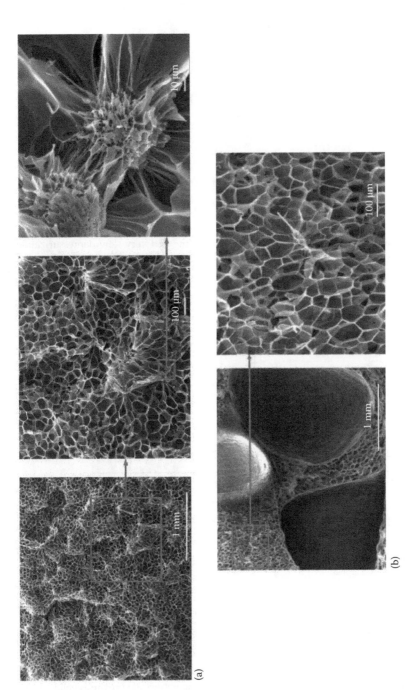

FIGURE 4.3
SEM micrographs taken close to outer surfaces of foamed PLA samples prepared at saturation temperatures of (a) 140°C using CTM and (b) 120°C using the VTM. Foaming temperature: 140°C; foaming pressure: 14 MPa.

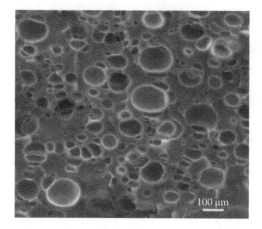

FIGURE 4.4

SEM micrograph of PS sample prepared by rapid cooling and very slow depressurization using the VTM. The same foaming parameters as those used in Figure 4.2b are used.

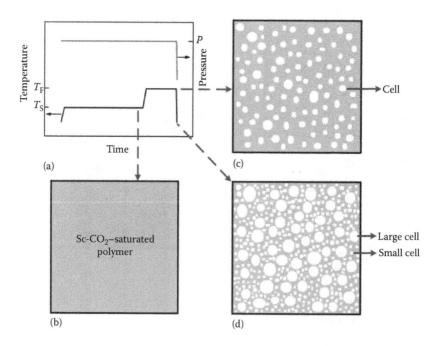

FIGURE 4.5

Schematics of development mechanism for BMCS: (a) temperature and pressure evolution versus time in foaming process, (b) initial state, (c) large cell nucleation and growth, and (d) small cell nucleation and growth and large cell further growth.

polymer to form a Sc-CO_2–saturated polymer (Figure 4.5b). When raising the temperature from the T_S to T_F, the CO_2 solubility in the polymer decreases, resulting in gas supersaturation and thus cell nucleation. In the second gas saturation stage, as the cell starts to grow, the increase in the volume of the cell induces a gas concentration to decrease in the cell, further encouraging the diffusion of the gas from the polymer to the cell [40]. Moreover, some neighboring cells may collapse into larger ones to reduce the total gas–polymer interfacial energy. Consequently, the cells grow and evolve into relatively large cells (Figure 4.5c). In the rapid depressurization stage, the relatively large cells grow and develop into large cells; meanwhile, a large number of cell nuclei are formed in unfoamed regions, which evolve into small cells due to restricted growth. Moreover, the supersaturated CO_2 gas is easier to diffuse into the growing large cells due to lower pressure within them than small cells [40], which promotes the further growth and even coalescence of the large cells. Finally, the BMCS is developed (Figure 4.5d). Certainly, some CO_2 gas would diffuse outside the sample in the temperature rising, second gas saturation, and rapid depressurization stages [20].

The entities appearing in the foamed PLA samples (shown in Figure 4.3) are supposed to be the corresponding spherulites. The stamen-like cell structure with small entities is observed for the foamed PLA sample that was prepared using the VTM (shown in Figure 4.3b). This means that a small amount of small spherulites are formed within the sample in the gas saturation stage at T_S of 120°C using the VTM. These small spherulites evolve into the stamen-like cell structure after depressurization because of the exclusion effect of CO_2 from the crystal growth front. So, the development of the BMCS for the PLA foam prepared using the VTM can also be explained by the mechanism that is shown in Figure 4.5.

From the aforementioned formation mechanism of the BMCS, it can be speculated that the polymers with high CO_2 solubility and a great dependence of CO_2 solubility on temperature are easier to form a BMCS using the VTM that is proposed in this work.

4.3.2 BMCS Tuning via Changing Foaming Parameters in VTM Foaming

Four main foaming parameters, that is, T_F, T_S, t_{S2}, and P, are involved during the VTM foaming. In the following, the four parameters are changed to investigate their tuning effects on the cellular structure and foam densities of the foamed PS samples, and the results are explained based on the aforementioned formation mechanism of the BMCS. For the foamed PLA samples, the T_F and P are changed to investigate their effects on the cellular structure.

4.3.2.1 Foaming Temperature

Figures 4.6 and 4.7 display the effect of the T_F on the cellular structure of the PS samples that were foamed at the T_Ss of 40°C and 100°C, respectively. From the two figures, it can be seen that the sample prepared at 100°C T_S and 110°C T_F

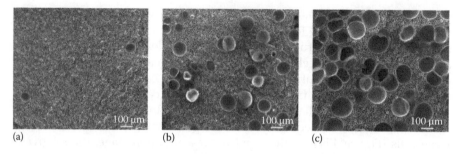

FIGURE 4.6
SEM micrographs of foamed PS samples prepared at foaming temperatures of (a) 60°C, (b) 80°C, and (c) 100°C using the VTM. Saturation temperature: 40°C; foaming pressure: 14 MPa; second saturation time: 25 min.

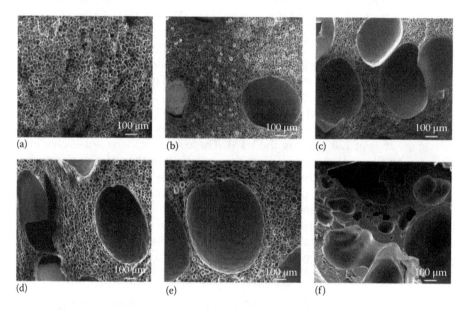

FIGURE 4.7
SEM micrographs of foamed PS samples prepared at foaming temperatures of (a) 110°C, (b) 120°C, (c) 130°C, (d) 140°C, (e) 150°C, and (f) 160°C using the VTM. Saturation temperature: 100°C; foaming pressure: 14 MPa; second saturation time: 25 min.

presents an MMCS with a D of 25 μm and ρ_c of 6.1 × 10^7 cells/cm^3; the sample prepared at 100°C T_S and 160°C T_F displays a nonuniform size distribution and thick cell wall; all other samples present a BMCS, and their D_l, D_s, ρ_{cl}, ρ_{cs}, mean diameter ratio (D_l/D_s), and density ratio (ρ_{cl}/ρ_{cs}) of large to small cells are shown in Figure 4.8. It is clear from Figure 4.8 that for the samples prepared at T_Ss of 40°C and 100°C, the D_l and D_s gradually increases, the ρ_{cl} increases first and then decreases, the ρ_{cs} gradually decreases, the D_l/D_s ratio

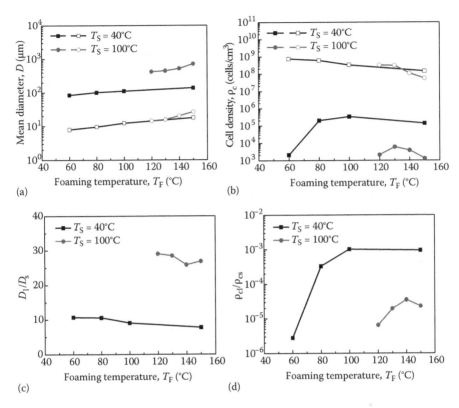

FIGURE 4.8
Parameters of cells versus foaming temperature curves for foamed PS samples prepared under different saturation temperatures (T_S) using VTM. (a) Mean diameter and (b) density of large (closed symbols) and small (open symbols) cells, ratios of (c) mean diameters, D_l/D_s, and (d) densities, ρ_{cl}/ρ_{cs}, of large to small cells. Foaming pressure: 14 MPa; second saturation time: 25 min.

exhibits a slight change, and the ρ_{cl}/ρ_{cs} ratio increases sharply first and then decreases slightly with raising T_F.

The aforementioned results can be explained as follows. Raising the temperature from 100°C (T_S) to 110°C (T_F) lowers the CO_2 concentration in the sample slightly, resulting in the formation of low supersaturation degree and thus very low cell nucleation [40]. Consequently, almost no cells are formed in the temperature rising stage. In the rapid depressurization stage, a large number of cell nuclei are formed, which evolve into small cells due to restricted growth. Finally, an MMCS with small cells (Figure 4.7a) is formed in the sample foamed at the T_F of 110°C. At the T_F of 160°C, some CO_2 gas may diffuse outside from the sample; meanwhile, low melt strength leads to the collapse of some cells. As a result, a nonuniform cell structure appears in the foamed sample (Figure 4.7f). For the foaming of the samples with the BMCS, raising the T_F at a given T_S results in an increase in both the degree of supersaturation and the difference between the saturation pressure and the surrounding pressure at which nucleation occurs. This increases the nucleation rate and thus the ρ_{cl} when raising the temperature from the T_S to T_F. On the other hand, raising the T_F lowers the resistance for cell growth and even induces the coalescence of large cells in the second gas saturation and depressurization stages, which increases the D_l and D_s and decreases the ρ_{cl} and ρ_{cs}. Combining the two aspects, the ρ_{cl} and ρ_{cl}/ρ_{cs} increase followed by a decrease with raising T_F. The D_l and D_s exhibit a similar increasing trend with raising T_F, and so the D_l/D_s changes slightly with raising T_F. The BMCS formed in the samples foamed at temperatures (60°C and 80°C) lower than the T_g of the unsaturated PS is attributed to the fact that dissolution of CO_2 into polymer plasticizes it, depressing its T_g [41–43].

The cellular structure of the foamed PS samples exerts a significant impact on their densities (ρ_fs). As shown in Figure 4.9a, for the samples prepared at the T_S of 40°C, the ρ_f decreases from 0.73 to 0.21 g/cm³ with raising the T_F

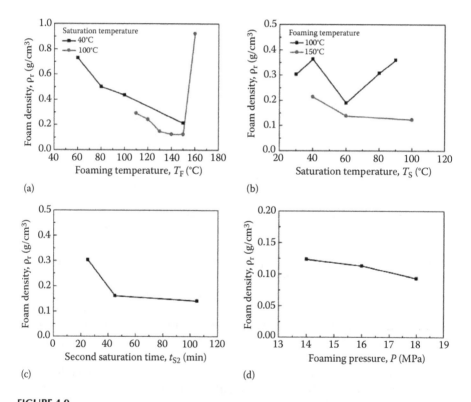

FIGURE 4.9
Foam densities versus (a) foaming temperature, (b) saturation temperature, (c) second saturation time, and (d) foaming pressure curves for foamed PS samples prepared using the VTM. Foaming pressure is 14 MPa, and second saturation time is 25 min in (a) and (b); saturation temperature is 30°C, foaming temperature is 100°C, and foaming pressure is 14 MPa in (c); and saturation temperature is 100°C, foaming temperature is 150°C, and second saturation time is 25 min in (d).

from 60°C to 150°C; for the samples prepared at a T_S of 100°C, the ρ_f decreases from 0.29 to 0.12 g/cm^3 with raising the T_F from 110°C to 150°C, and increases rapidly to 0.92 g/cm^3 with further raising of the T_F to 160°C. The results can be explained as follows. When the T_F is lower than 150°C, the Sc-CO$_2$–saturated PS samples have a relatively high melt strength; thus, raising the T_F is beneficial to their expansion. On the other hand, as raising the T_F, although the ρ_{cl} increases followed by a decrease, the D_l gradually increases, which increases the volume of the large cells (as shown in Figures 4.6 and 4.7). So the ρ_f of the samples gradually decreases with raising T_F. At a T_F of 160°C, the aforementioned CO$_2$ gas diffusion from the sample to the outside and the collapse of some cells resulting from low melt strength lead to very high ρ_f.

Further observing Figures 4.6 through 4.8 demonstrates that the T_S plays a large role on the cellular structure, D_l, and ρ_{cl} for the foamed PS samples. The ρ_{cl} values for the samples prepared at the T_Ss of 40°C and 100°C are in the range of $2.1 \times 10^3 - 3.3 \times 10^5$ and $2.1 \times 10^3 - 5.9 \times 10^3$ cells/cm^3, respectively. The former samples have D_l values of 85–140 µm and the D_l/D_s values of 7.7–10.7, much smaller than the corresponding values (420–715 µm, 25.9–29.0) for the latter ones. The T_S exerts a slight effect on the D_s and ρ_{cs} for the samples. The effect of the T_S on the cellular structure and size is further investigated in Section 4.3.2.2.

Using the VTM, the foamed PLA samples were prepared at two T_Fs (140°C and 170°C) when fixing T_S (100°C) and P (14 MPa). The sample foamed at 140°C T_F exhibits a low V_r (1.9), and so the micrograph was taken across its overall fractured surface, as shown in Figure 4.10. It is clearly visible that two regions—skin and core regions—can be distinguished across the overall fractured surface. In the skin region, there exist two different cellular structures: (1) elongated cells existing around the circular entities (Figure 4.10b) and (2) submicron-sized cells in the interlamellar regions of the large spherulites (Figure 4.10c). In the core region, a relatively uniform cellular structure with a D of approximately 80 µm is formed. A higher magnification allows observing a stamen-like cell structure (Figure 4.10d). The formed cellular structure is explained as follows. Large spherulites are formed in the skin region and small spherulites in the core region in the first saturation stage (40 min). Raising the temperature from 100°C (T_S) to 140°C (T_F) and keeping the T_F in the second saturation stage (25 min) leads to more perfect crystals, which induces higher crystallinity in the skin region. As a consequence, microcells are formed in the skin region of the foamed sample after rapid depressurization. Entities surrounded by the elongated cells are evolved from some large spherulites, and submicron-sized cells are formed in the interlamellar zones of other large spherulites. In the core region, small spherulites and a lower crystallinity result in the formation of a relatively uniform cellular structure with some stamen-like cells after rapid depressurization. No BMCS formation is attributed to higher stiffness of a sample at 100°C T_S and 140°C T_F, which restrains the growth of the cell nuclei that are formed in the temperature rising stage.

The PLA sample foamed at 170°C T_F presents a more uniform cellular structure and higher V_r (42.2), and the micrographs shown in Figure 4.11

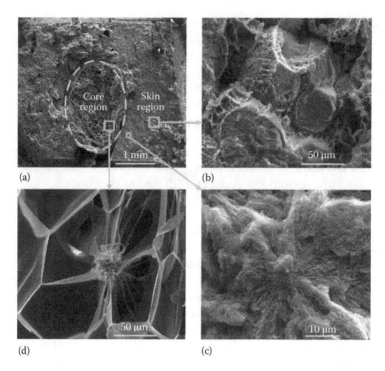

FIGURE 4.10
SEM micrographs across overall fractured surface of foamed PLA sample prepared at 140°C foaming temperature using the VTM. Saturation temperature: 100°C; foaming pressure: 14 MPa.

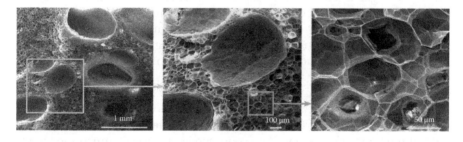

FIGURE 4.11
SEM micrographs taken close to outer surface of foamed PLA sample prepared at 170°C foaming temperature using the VTM. Saturation temperature: 100°C; foaming pressure: 14 MPa.

were taken from the region close to its outer surface. It is clearly visible that the foamed sample displays a BMCS in which the size distribution of the cells is nonuniform. The D_l and D_s are 592 and 16 µm, respectively. This can be explained briefly as follows. When increasing the temperature from 100°C (T_S) to 170°C (T_F), the spherulites formed in the first saturation stage melt, leading to low stiffness. The BMCS formation is attributed to the aforementioned synergistic effect of rising temperature and depressurization.

4.3.2.2 Saturation Temperature

Figures 4.12 and 4.13 illustrate the SEM micrographs of the fractured surfaces for the foamed PS samples that were prepared at four T_Ss (30°C, 60°C, 80°C, and 90°C)/100°C T_F and 60°C T_S/150°C T_F, respectively. As can be seen, the sample prepared at 90°C T_S and 100°C T_F presents an MMCS with a D of 20 µm and ρ_c of 1.4×10^8 cells/cm³; all other samples present a BMCS, and their D_l, D_s, ρ_{cl}, ρ_{cs}, D_l/D_s, and ρ_{cl}/ρ_{cs} are shown in Figure 4.14, in which the relevant parameters of the cells for the samples shown in Figures 4.2b, 4.6c, and 4.7e are also given. It is clear from Figure 4.14 that the samples foamed at 100°C T_F exhibit slightly smaller D (D_l and D_s) and slightly larger ρ_c (ρ_{cl} and ρ_{cs}) than the ones foamed at 150°C T_F under the same T_S. For the samples prepared at 100°C T_F, by raising the T_S from 30°C to 80°C, the D_l increases from 50 to 325 µm, the ρ_{cl} decreases from 3.6×10^6 to 1.2×10^4 cells/cm³, the D_s increases from 10 to 16 µm, the ρ_{cs} increases from 5.6×10^7 to 3.3×10^8 cells/cm³ first and then decreases to 2.2×10^8 cells/cm³, the D_l/D_s increases from 4.8 to 20.3, and the ρ_{cl}/ρ_{cs} decreases from 6.5×10^{-2} to 5.6×10^{-5}. The relevant parameters of the cells for the samples prepared at 150°C

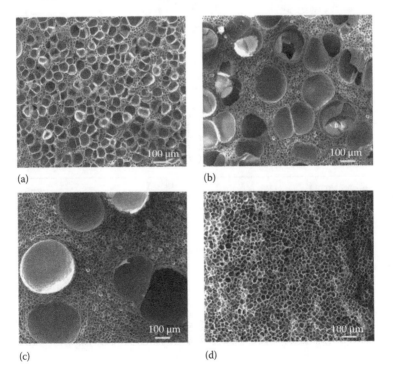

(a) (b) (c) (d)

FIGURE 4.12
SEM micrographs of foamed PS samples prepared at saturation temperatures of (a) 30°C, (b) 60°C, (c) 80°C, and (d) 90°C using the VTM. Foaming temperature: 100°C; foaming pressure: 14 MPa; second saturation time: 25 min.

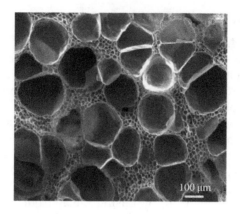

FIGURE 4.13
SEM micrograph of foamed PS sample prepared at saturation temperature of 60°C using the VTM. Foaming temperature: 150°C; foaming pressure: 14 MPa; second saturation time: 25 min.

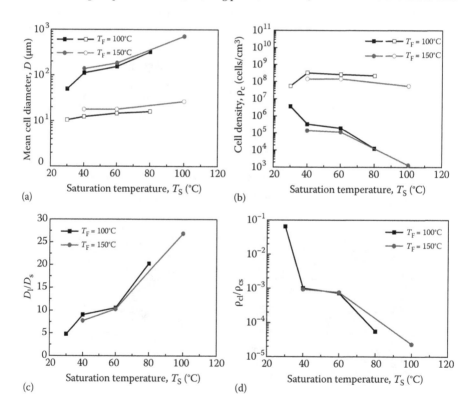

FIGURE 4.14
Parameters of cells versus saturation temperature curves for foamed PS samples prepared under different foaming temperatures (T_F) using the VTM. (a) Mean diameter and (b) density of large (closed symbols) and small (open symbols) cells, (c) ratios of mean diameters, D_l/D_s, and (d) densities, ρ_{cl}/ρ_{cs}, of large to small cells. Foaming pressure: 14 MPa; second saturation time: 25 min.

T_F exhibit similar varying trends as those for the ones prepared at 100°C T_F. The samples foamed at 100°C and 150°C T_Fs have ρ_f values of 0.19–0.36 g/cm^3 and 0.12–0.21 g/cm^3, respectively, as shown in Figure 4.9b.

The aforementioned results can be briefly explained as follows. The MMCS formed in the sample prepared at 90°C T_S and 100°C T_F is attributed to the aforementioned low supersaturation degree and thus very low cell nucleation. Raising the T_S at a given T_F reduces the degree of supersaturation resulting from the decrease of CO_2 solubility in the samples, which decreases the nucleation rate in the temperature rising stage and thus the ρ_{cl} [44]; meanwhile, raising the T_S, although it would weaken the plasticizing effect of CO_2 due to the decrease of its solubility in the samples [41–43], lowers the resistance for cell growth in the temperature rising and second saturation stages, which increases the D_l and thus decreases the ρ_{cl}. For the sample prepared under 100°C T_F, a very large ρ_{cl} is obtained at 30°C T_S (Figure 4.14b). This means that unfoamed regions used for the second nucleation in the depressurization stage are small, which results in a small ρ_{cs}. When raising the T_S to 40°C, the ρ_{cl} decreases by one order of magnitude, which increases the unfoamed regions and thus the ρ_{cs}. On the other hand, because the Sc-CO_2–saturated sample is in a state of supersaturation before depressurization owing to short t_{S2} (25 min), raising the T_S decreases the CO_2 concentration in the samples and thus the nucleation rate of small cells in the depressurization stage, which decreases the ρ_{cs} and increases the D_s. Combining the two aspects, the ρ_{cs} increases first and then decreases with raising T_S.

4.3.2.3 Second Saturation Time

As mentioned in Section 4.3.1, some neighboring cells formed when raising the temperature from the T_S to T_F may collapse into larger ones in the second saturation stage; on the other hand, the supersaturated CO_2 gas diffuses from the unfoamed regions into the cell and promotes their growth. In this way, the D_l increases, and the ρ_{cl} decreases with increasing t_{S2}. This is corroborated by Figure 4.15, which illustrates the SEM micrographs of the fractured surfaces for the foamed PS samples obtained at t_{S2}s of 45 and 105 min. The D_l, D_s, ρ_{cl}, ρ_{cs}, D_l/D_s, and ρ_{cl}/ρ_{cs} for the two samples and the sample foamed at the t_{S2} of 25 min (Figure 4.12a) are shown in Figure 4.16. It can be seen that the D_l increases from 50 to 370 μm, and the ρ_{cl} decreases from 3.6 × 10^6 to 5.9 × 10^3 cells/cm^3 with increasing t_{S2} from 25 to 105 min. As shown in Figure 4.9c, the ρ_f for the foamed samples decreases rapidly initially and then gently with increasing t_{S2}.

Figure 4.16 also displays that the D_s and D_l/D_s increase, the ρ_{cl}/ρ_{cs} decrease, and the ρ_{cs} increases first and then decreases slightly with increasing t_{S2}. This can be briefly explained as follows. The solubility of the CO_2 in the samples at T_F (100°C) is lower than that at T_S (30°C). Therefore, extending the t_{S2} decreases the CO_2 concentration in the samples and thus the nucleation rate in the depressurization stage, which decreases the ρ_{cs} and increases the D_s.

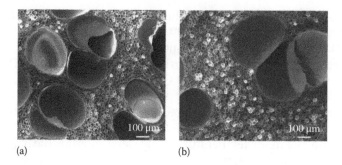

(a)   (b)

FIGURE 4.15
SEM micrographs of foamed PS samples prepared at second gas saturation times of (a) 45 min and (b) 105 min using the VTM. Saturation temperature: 30°C; foaming temperature: 100°C; foaming pressure: 14 MPa.

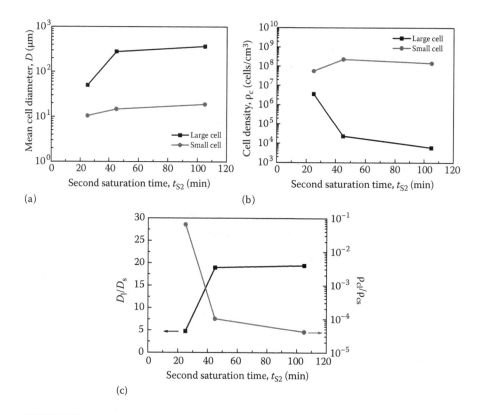

(a)   (b)

(c)

FIGURE 4.16
Parameters of cells versus second saturation time curves for foamed PS samples prepared using the VTM. (a) Mean diameter and (b) density of large and small cells, and (c) ratios of mean diameters, D_l/D_s, and densities, ρ_{cl}/ρ_{cs}, of large to small cells. Saturation temperature: 30°C; foaming temperature: 100°C; foaming pressure: 14 MPa.

However, at the t_{S2} of 25 min, high ρ_{cl} (Figures 4.12a and 4.16b) leads to fewer unfoamed regions for the second nucleation, resulting in a small ρ_{cs}.

4.3.2.4 Foaming Pressure

In the traditional batch foaming process, a higher depressurization rate can increase nucleation rate and thus cell density. In the batch foaming process with the synergistic effect of rising temperature and depressurization in this work, however, the effect of the P on the cellular structure is somewhat complicated. Figure 4.17 displays the SEM micrographs of the fractured surfaces for the foamed PS samples that were prepared at Ps of 16 and 18 MPa. The ρ_f for the foamed samples decreases from 0.12 to 0.09 g/cm³ with increasing the P from 14 to 18 MPa (as shown in Figure 4.9d). The D_1, D_s, ρ_{cl}, ρ_{cs}, D_1/D_s, and ρ_{cl}/ρ_{cs} for the two samples and the sample foamed at the P of 14 MPa (Figure 4.7e) are shown in Figure 4.18. Increasing the P leads to a decrease in the D_1, D_s, and D_1/D_s and an increase in the ρ_{cl}, ρ_{cs}, and ρ_{cl}/ρ_{cs}. These can be explained as follows. Raising the P, on the one hand, decreases the amount of CO_2 gas diffusing from the unfoamed regions into the growing cells [44] and weakens the cell collapse in the second gas saturation stage; on the other hand, it increases the nucleation rate for the small cells, which decreases the amount of the CO_2 gas for cell growth in the depressurization stage. Consequently, the D_1 decreases and the ρ_{cl} increases with increasing P. Moreover, the P plays a larger effect on the D_1 and ρ_{cl} than on the D_s and ρ_{cs}. This is attributed to the fact that the D_1 and ρ_{cl} are affected by the P in the temperature rising, second gas saturation and depressurization stages, whereas the D_s and ρ_{cs} are affected by the P only in the depressurization stage.

No BMCS is formed in the foamed PLA sample that was prepared at 100°C T_S, 140°C T_F, and 14 MPa P (as shown in Figure 4.10). Fixing the T_S (100°C)

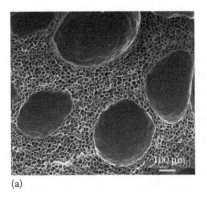
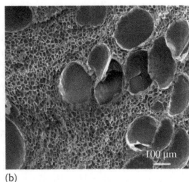

(a)   (b)

FIGURE 4.17
SEM micrographs of foamed PS prepared at foaming pressures of (a) 16 MPa and (b) 18 MPa using the VTM. Saturation temperature: 100°C; foaming temperature: 150°C; second saturation time: 25 min.

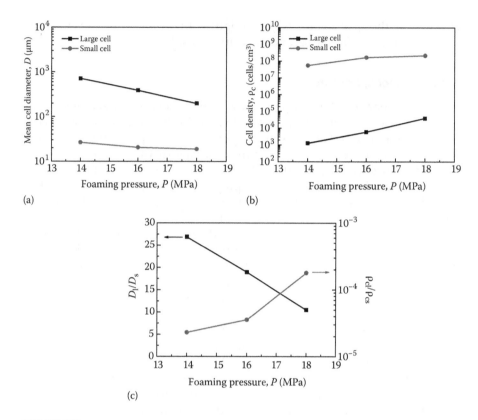

FIGURE 4.18
Parameters of cells versus foaming pressure curves for foamed PS samples prepared using the VTM. (a) Mean diameter and (b) density of large and small cells, and (c) ratios of mean diameters, D_l/D_s, and densities, ρ_{cl}/ρ_{cs}, of large to small cells. Saturation temperature: 100°C; foaming temperature: 150°C; second saturation time: 25 min.

and T_F (140°C), the foamed PLA samples were prepared using the VTM under 18 and 22 MPa Ps. The samples display high V_rs (13.7 and 76.2, respectively) and more uniform cellular structure across their fractured surfaces, and so the micrographs were taken from the regions close to their outer surfaces, as shown in Figure 4.19. As can be seen from Figure 4.19a, a BMCS is obtained under 18 MPa P, and the D_l and D_s are 437 and 21 μm, respectively. A higher magnification allows observing a stamen-like cell structure with entities in the small cell region. Interestingly, as for the sample prepared at 22 MPa P, a trimodal cellular structure appears, and some cells display an open-cell structure (shown in Figure 4.19b). The mean cell diameters of large, middle, and small cells are 635, 24, and 1 μm, respectively. The test demonstrates that the sample exhibits a porosity of approximately 31%.

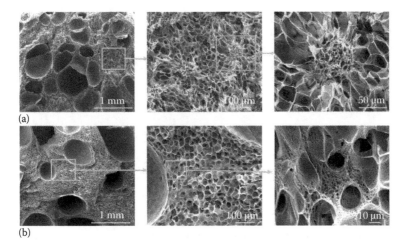

FIGURE 4.19
SEM micrographs taken close to outer surfaces of foamed PLA prepared at foaming pressures of (a) 18 MPa and (b) 22 MPa using the VTM. Saturation temperature: 100°C; foaming temperature: 140°C.

4.4 Conclusions

In this work, both PS and PLA foams were prepared in a batch process via CTM and VTM. An MMCS is obtained through depressurization in the CTM, whereas a BMCS is achieved through the synergistic effect of temperature rising and depressurization in the VTM. The development mechanism for the BMCS is proposed based on a rapid cooling experiment: cell nucleation occurs in the supercritical CO_2–saturated polymer both when raising temperature and depressurizing; the cells nucleating in the temperature rising stage develop into large cells owing to long growth time and the entrance of CO_2 gas, whereas the cells nucleating in the depressurization stage evolve into small cells due to restricted growth.

The BMCS and foam densities of the foamed PS samples can be tuned by changing four foaming parameters: (1) foaming temperature, (2) saturation temperature, (3) second saturation time, and (4) foaming pressure. The higher degree of supersaturation resulting from larger difference between foaming and saturation temperatures, higher resistance for cell growth resulting from lower saturation temperature, shorter second saturation time, and higher foaming pressure are favorable for the formation of smaller diameters and larger densities of large cells. As for the PLA foams, the low saturation temperature (100°C) in the first saturation stage is favorable to form large spherulites in the skin region, which generally evolve into the circular entities that are surrounded by elongated cells or the submicron-sized cells in the interlamellar regions after foaming at 140°C. Increasing the foaming temperature to 170°C leads to a BMCS. Small spherulites evolve into a stamen-like cell structure. Interestingly, increasing the foaming pressure to 22 MPa results in a trimodal cellular structure and a very high expansion ratio (76.2) for the foamed sample.

References

1. K.M. Lee, E.K. Lee, S.G. Kim, C.B. Park, H.E. Naguib. 2009. Bi-cellular foam structure of polystyrene from extrusion foaming process. *Journal of Cellular Plastics* 45 539–553.
2. C.L. Zhang, B. Zhu, D.C. Li, L.J. Lee. 2012. Extruded polystyrene foams with bimodal cell morphology. *Polymer* 53 2435–2442.
3. Z.L. Ma, G.C. Zhang, Q. Yang, X.T. Shi, A.H. Shi. 2014. Fabrication of micro-cellular polycarbonate foams with unimodal or bimodal cell-size distributions using supercritical carbon dioxide as a blowing agent. *Journal of Cellular Plastics* 50 56–79.
4. J.B. Bao, G.S. Weng, L. Zhao, Z.F. Liu, Z.R. Chen. 2014. Tensile and impact behavior of polystyrene microcellular foams with bi-modal cell morphology. *Journal of Cellular Plastics* 50 381–393.
5. A. Salerno, D. Guarnieri, M. Iannone, S. Zeppetelli, P.A. Netti. 2010. Effect of micro- and macroporosity of bone tissue three-dimensional-poly(ε-caprolactone) scaffold on human mesenchymal stem cells invasion, proliferation, and differentiation in vitro. *Tissue Engineering: Part A* 16 2661–2673.
6. A. Kramschuster, L.S. Turng. 2010. An injection molding process for manufacturing highly porous and interconnected biodegradable polymer matrices for use as tissue engineering scaffolds. *Journal of Biomedical Materials Research Part B: Applied Biomaterials* 92B 366–376.
7. A. Salerno, E. Di Maio, S. Iannace, P.A. Netti. 2011. Solid-state supercritical CO_2 foaming of PCL and PCL-HA nano-composite: Effect of composition, thermal history and foaming process on foam pore structure. *Journal of Supercritical Fluids* 58 158–167.
8. A. Salerno, S. Zeppetelli, E Di Maio, S. Iannace, P.A. Netti. 2011. Design of bimodal PCL and PCL-HA nanocomposite scaffolds by two step depressurization during solid-state supercritical CO_2 foaming. *Macromolecular Rapid Communications* 32 1150–1156.
9. A. Salerno, S. Zeppetelli, E. Di Maio, S. Iannace, P.A. Nettia. 2012. Architecture and properties of bi-modal porous scaffolds for bone regeneration prepared via supercritical CO_2 foaming and porogen leaching combined process. *Journal of Supercritical Fluids* 67 114–122.
10. A. Salerno, E. Di Maio, S. Iannace, P.A. Nettia. 2012. Tailoring the pore structure of PCL scaffolds for tissue engineering prepared via gas foaming of multi-phase blends. *Journal of Porous Materials* 19 181–188.
11. E. Di Maio, A. Salerno, S. Iannace. 2013. Scaffolds with tubular/isotropic bimodal pore structures by gas foaming and fiber templating. *Materials Letters* 93 157–160.

12. L.Q. Xu, H.X. Huang. 2013. Preparation of polystyrene foams with bi-modal cells. *Acta Polymerica Sinica* 1357–1362.
13. J. Lin, Y. Cai, X. Wang, B. Ding, J. Yu, M. Wang. 2011. Fabrication of biomimetic superhydrophobic surfaces inspired by lotus leaf and silver ragwort leaf. *Nanoscale* 3 1258–1262.
14. E. Huovinen, J. Hirvi, M. Suvanto, T.A. Pakkanen. 2012. Micro-micro hierarchy replacing micro-nano hierarchy: A precisely controlled way to produce wear-resistant superhydrophobic polymer surfaces. *Langmuir* 28 14747–14755.
15. W.S. Guan, H.X. Huang, B. Wang. 2013. Topographic design and application of hierarchical polymer surfaces replicated by microinjection compression molding. *Journal of Micromechanics and Microengineering* 23 105010.
16. W.S. Guan, H.X. Huang, A.F. Chen. 2015. Tuning 3D topography on biomimetic surface for efficient self-cleaning and microfluidic manipulation. *Journal of Micromechanics and Microengineering* 25 035001.
17. K.A. Arora, A.J. Lesser, T.J. McCarthy. 1998. Preparation and characterization of microcellular polystyrene foams processed in supercritical carbon dioxide. *Macromolecules* 31 4614–4620.
18. X.H. Sun, G. Li, X. Liao, J.S. He. 2004. An investigation on the cell nucleation and cell growth of microcellular foaming by means of dual depressurization. *Acta Polymerica Sinica* 93–97.
19. H. Yokoyama, K. Sugiyama. 2005. Nanocellular structures in block copolymers with CO_2-philic blocks using CO_2 as a blowing agent: Crossover from micro- to nanocellular structures with depressurization temperature. *Macromolecules* 38 10516–10522.
20. J.B. Bao, T. Liu, L. Zhao, G.H. Hu. 2011. A two-step depressurization batch process for the formation of bi-modal cell structure polystyrene foams using $scCO_2$. *Journal of Supercritical Fluids* 55 1104–1114.
21. D.C. Li, T. Liu, L. Zhao, W.K. Yuan. 2011. Foaming of linear isotactic polypropylene based on its non-isothermal crystallization behaviors under compressed CO_2. *Journal of Supercritical Fluids* 60 89–97.
22. D.C. Li, T. Liu, L. Zhao, X.S. Lian, W.K. Yuan. 2011. Foaming of poly(lactic acid) based on its nonisothermal crystallization behavior under compressed carbon dioxide. *Industrial and Engineering Chemistry Research* 50 1997–2007.
23. M.R. Holl, J.L. Garbini, W.R. Murray, V. Kumar. 2001. A steady-state mass balance model of the polycarbonate-CO_2 system reveals a self-regulating cell growth mechanism in the solid-state microcellular process. *Journal of Polymer Science Part B-Polymer Physics* 39 868–880.
24. L. Singh, V. Kumar, B.D. Ratner. 2004. Generation of porous microcellular 85/15 poly (DL-lactide-co-glycolide) foams for biomedical applications. *Biomaterials* 25 2611–2617.
25. H.X. Huang, J.K. Wang. 2007. Improving polypropylene microcellular foaming through blending and the addition of nano-calcium carbonate. *Journal of Applied Polymer Science* 106 505–513.
26. D. Miller, P. Chatchaisucha, V. Kumar. 2009. Microcellular and nanocellular solid-state polyetherimide (PEI) foams using sub-critical carbon dioxide: I. Processing and structure. *Polymer* 50 5576–5584.
27. L. Urbanczyk, C. Calberg, C. Detrembleur, C. Jérôme, M. Alexandre. 2010. Batch foaming of SAN/clay nanocomposites with $scCO_2$: A very tunable way of controlling the cellular morphology. *Polymer* 51 3520–3531.

28. J.E. Weller, V. Kumar. 2010. Solid-state microcellular polycarbonate foams: I. The steady-state process space using subcritical carbon dioxide. *Polymer Engineering and Science* 50 2160–2169.
29. D. Miller, V. Kumar. 2011. Microcellular and nanocellular solid-state poly-etherimide (PEI) foams using sub-critical carbon dioxide: II. Tensile and impact properties. *Polymer* 52 2910–2919.
30. A.B. Martínez, V. Realinho, M. Antunes, M.L. Maspoch, J.I. Velasco. 2011. Microcellular foaming of layered double hydroxide-polymer nanocomposites. *Industrial and Engineering Chemistry Research* 50 5239–5247.
31. J.A.R. Ruiz, J. Marc-Tallon, M. Pedros, M. Dumon. 2011. Two-step micro cellular foaming of amorphous polymers in supercritical CO_2. *Journal of Supercritical Fluids* 57 87–94.
32. H.X. Huang, H.F. Xu. 2011. Preparation of microcellular polypropylene/polystyrene blend foams with tunable cell structure. *Polymers Advanced Technologies* 22 822–829.
33. Y.M. Corre, A. Maazouz, J. Duchet, J. Reignier. 2011. Batch foaming of chain extended PLA with supercritical CO_2: Influence of the rheological properties and the process parameters on the cellular structure. *Journal of Supercritical Fluids* 58 177–188.
34. K. Taki, D. Kitano, M. Ohshima. 2011. Effect of growing crystalline phase on bubble nucleation in poly(L-Lactide)/CO_2 batch foaming. *Industrial and Engineering Chemistry Research* 50 3247–3252.
35. S. Costeux, L. Zhu. 2013. Low density thermoplastic nanofoams nucleated by nanoparticles. *Polymer* 54 2785–2795.
36. N. Najafi, M.C. Heuzey, P.J. Carreau, D. Therriault, C.B. Park. 2014. Rheological and foaming behavior of linear and branched polylactides. *Rheologica Acta* 53 779–790.
37. D. Kohlhoff, M. Ohshima. 2014. Influence of polyethylene disperse domain on cell morphology of polystyrene-based blend foams. *Journal of Cellular Plastics* 50 241–261.
38. J. Pinto, J.A. Reglero-Ruiz, M. Dumon, M.A. Rodriguez-Perez. 2014. Temperature influence and CO_2 transport in foaming processes of poly(methyl methacrylate)–block copolymer nanocellular and microcellular foams. *Journal of Supercritical Fluids* 94 198–205.
39. Designation: D 792-00. 2001. Standard test methods for density and specific gravity (relative density) of plastics by displacement. ASTM Committee D20.
40. S.T. Lee, C.B. Park, N.S. Ramesh. 2007. *Polymeric Foams: Science and Technology.* Boca Raton: CRC Press LLC 40–80.
41. D. Condo, I.C. Sanchez, C.G. Panayiotou, K.P. Johnston. 1992. Glass transition behavior including retrograde vitrification of polymers with compressed fluid diluents. *Macromolecules* 25 6119–6127.
42. T. Otsuka, K. Taki, M. Ohshima. 2008. Nanocellular foams of PS/PMMA polymer blends. *Macromolecular Materials and Engineering* 293 78–82.
43. H. Guo, V. Kumar. 2015. Solid-state poly(methyl methacrylate) (PMMA) nanofoams. Part I: Low-temperature CO_2 sorption, diffusion, and the depression in PMMA glass transition. *Polymer* 57 157–163.
44. S.T. Lee, N.S. Ramesh. 2004. *Polymeric Foams: Mechanisms and Materials.* Boca Raton: CRC Press LLC 82–90.

5

Extrusion Foaming of Polylactide

Richard Gendron and Mihaela Mihai

CONTENTS

5.1 Introduction

Over the past two decades, considerable efforts have been devoted to the development of so-called biopolymers, i.e., produced from renewable agricultural resources. The drive to gradually replace oil-derived plastics such as polystyrene (PS) or polyethylene relies on several claimed benefits: reduction of the dependency upon foreign and domestic oil supplies, and thus reduced fossil fuel use and therefore decreased greenhouse gas emissions, resin cost stability, etc.

Among those biopolymers, polylactide (PLA) is probably one of the most popular. Being bioresorbable and biocompatible with the human body, its use was originally focused on biomedical applications, such as porous biodegradable scaffolds for tissue regeneration [1].

PLA was also restricted to high-value packaging applications due to its initial high cost. Nowadays, it competes fairly on a price basis with commodity polymers and has found its way in day-to-day applications. Innovations in biotechnology have enabled the transformation of plant sugar coming from field corn into a high-quality polymer. PLA was introduced in 2003 on a commercial scale by NatureWorks LLC, with this production having no impact on international or local food chains [2,3].

PLA is rigid and has an excellent gloss, transparency, and clarity, as well as good flavor and aroma barrier properties, which make it as a good candidate for packaging applications [4]. PLA can be processed similarly to petroleum-based plastics into trays, drinking cups, shrink films, and various other applications. Additionally, PLA bears the appreciable reputation of being biodegradable; composted under appropriate temperature and humidity conditions, PLA can be degraded down to lactic acid (LA), its basic natural component, which is readily assimilated by bioorganisms that are commonly found in nature.

Despite its usefulness in terms of thermal insulation and cost-effectiveness, and even in light of recent studies underlining the benefits of PS in terms of energy consumption, water use, solid waste, and greenhouse gas emission [5], packaging based on PS foams have been at the heart of decades-long debates that are fed by environmental advocates, which have led to the progressive ban of the PS foam food containers in many communities and US states, starting as early as 1990.

Businesses having to find alternatives have seen in PLA foams an interesting solution to this dilemma and have started to develop products based on foamed PLA. As early as 2005, the first foamed trays, christened *Naturalbox*, were being launched by an Italian packaging company, Coopbox Europe S.p.A. (Figure 5.1). The foamed sheets were produced on a standard tandem extrusion line using carbon dioxide [6]. The density reduction, however, was modest, from 1.25 g/cm^3 for the neat PLA down to 0.30 g/cm^3 for the foam [7]. Commercialization of PLA foam trays in North America followed

FIGURE 5.1
Naturalbox, produced by Coopbox, was the first biodegradable foamed PLA tray commercialized in 2005 in Europe, using NatureWorks' Ingeo PLA resin.

shortly in 2007 by Cryovac Inc. (Sealed Air), with products under the name of NatureTray [7]. The third company to have developed foamed PLA meat trays was Pactiv with their Ingeo EarthChoice trays in 2010 [8]. Still for food tray application, a Canadian company, Dyne-A-Pak Inc., also manufactured PLA foams using a PS foam line that was equipped with a special cooling screw. The density of their foam was as low as 50–60 kg/m^3 [7].

However, some inherent characteristics of PLA and related processing issues make it difficult to fully implement the conversion from PS to PLA. Despite some commercially available manufacturing lines, foaming PLA has not yet met the economic conditions to be fully competitive with its styrenic predecessor.

In order to fully grasp the level of difficulty encountered in the extrusion of low-density PLA foams, a thorough understanding of the chemical nature of PLA, as well as its interaction with required additives such as the physical foaming agent, is required. Such physicochemical issues will be addressed first in Sections 5.2 through 5.5 in order to appreciate the processing developments that were undertaken over the last decade, as detailed in Section 5.6.

5.2 What Is Polylactide?

5.2.1 Chemistry

Polylactide is an aliphatic polymer that is constituted of LA (2-hydroxy propionic acid) units. Because of this basic composition, the polymer is also frequently referred to as poly(lactic acid), although the term polylactide would be more appropriate given that the commercial PLA used for commodity applications is produced from lactide units, which is a dimer of LA, through ring-opening polymerization.

PLA has a carboxyl (acid radical—COOH) and a hydroxyl end group. LA is obtained from natural sources through bacterial carbohydrate fermentation, or it can be synthetized chemically. Although corn is usually identified as the principal natural source for manufacturing PLA, potato (glucose and maltose), cane and beet sugar (sucrose), or cheese whey (lactose) can also provide the simple sugars that are required for the fermentation process.

Since LA has a chiral nature, the stereochemistry of PLA is rather complex. LA has two stereoisomers due to an asymmetric carbon atom, D and L isomers (D-LA and L-LA), which makes PLA much more than a single material; it encompasses a large family of materials with different properties. One main difference between the different grades of PLA relies on the ratio between the L and D components. This has a huge impact on the material properties, through the percent of crystallization achievable, as semicrystalline PLA can only be obtained with a content of L-lactic acid greater than 93%. A L-LA content ranging from 50% to 93% leads to totally amorphous PLA since the D-LA units disrupt the crystallization of the L-LA chains. Crystallinity can be as high as 45% with pure poly(L-lactide). Commercially available PLA resins are constituted mainly of the L isomer, this form being the most common in nature.

5.2.2 Polymerization and Production

Although different methods can be used to polymerize the high-molecular-weight PLA required for commercial applications, the polymerization through lactide formation, known as the Cargill process [9], is by far the most current method being used for PLA production. Condensation of two LA molecules forms a lactide (LA dimer), which is followed by their ring-opening polymerization to produce high-molecular-weight PLA, as illustrated in Figure 5.2.

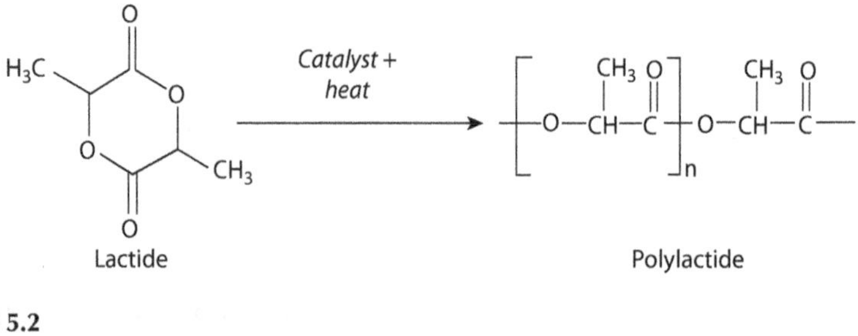

FIGURE 5.2
Transformation of the lactide dimer into PLA.

Obviously, three combinations are possible for the lactide unit: (1) LL-lactide (two L-lactic acids), (2) DD-lactide (two D-lactic acids), and (3) LD-lactide (combination of one L- and one D-lactic unit). The PLAs resulting from pure L- or pure D-units bear the denomination of PLLA and PDLA, respectively. However, commercial PLA grades consist predominantly of L-LA with a minimum of 1%–2% D units.

PLA's production is the world's largest in terms of the niche bioplastic market. Some 25 companies are involved in its production, but NatureWorks LLC dominates this market with its 150,000 t of Ingeo PLA polymer produced per year from its manufacturing plant located in Blair, NE. The typical properties of the Ingeo resin are listed in Table 5.1. Building another production plant having a capacity of 75,000 t in Thailand to supply the Asian market, which was expected in 2015, has been delayed due to the collapse in the global oil price. Other producers are Synbra/Purac Biomaterials in the Netherlands, Futerro (Galactic–Total Petrochemicals joint venture) in Belgium, Pyramid Bioplastics in Germany, Zhejiang Hisun Biomaterials in China, and Toyobo in Japan [10–12]. It is estimated that the PLA market could reach 800,000 t in 2020 [12].

5.2.3 Thermal Transitions

The glass transition temperature, T_g, has a great impact on the heat deflection temperature (HDT) which is related to the higher temperature of service for most of the applications. The T_g for PLA is around 57°C for polymers having a molecular weight of over 80–100 kg/mol, which is the case for commercial PLA. This T_g value is significantly lower than that of PS, which is approximately 100°C. Thus, the low PLA's T_g severely limits its utilization in many applications unless crystallization has reached an adequate level in the final product.

TABLE 5.1

Typical Properties of NatureWorks PLA Compared to General-Purpose PS

Property	PLA	PS
Density (g/cm³)	1.24–1.25	1.05
Glass transition temperature T_g (°C)	55–60	100
Melting temperature T_m (°C)	145–170	None
Crystallization temperature T_c (°C)	100–105	None
Thermal conductivity at 25°C (J/m-K-s)		0.140
Amorphous sheet	0.130	
25% crystallinity	0.160	
Heat capacity C_P at 25°C (J/g°C)	1.20	1.30
Tensile strength (MPa)	53.1	45.5
Elongation at break (%)	4.1	1.4
Tensile modulus (GPa)	3.45	3.03
Izod impact (J/m)	16	21.4

The melting temperature, T_m, is very sensitive to the D/L ratio. The highest melting temperature $T_m \sim 175°C–180°C$ is obtained for the PLA based solely on L-LA units. Increasing the D-LA content lowers the T_m value by approximately 5°C for each percent of D-LA unit, down to 120°C for a 10:90 D/L ratio (Figure 5.3). However, a much higher melting temperature and thus better temperature resistance can be obtained through the cocrystallization of one PLLA and one PDLA chain, leading to stereocomplex crystals having a T_m of approximately 230°C [13].

Many properties of the PLA will be a function of the level of crystallinity: (a) mechanical properties, (b) the HDT, (c) the permeability, and (d) the time of degradation in a composting environment. However, contrarily to many other semicrystalline polymers like polyethylene terephthalate (PET), the rate of crystallization of PLA is very slow because of the high rigidity of the chains, which hampers their mobility.

An excellent review of the abundant literature on PLA crystallization has been done by Saeidlou et al. [13]. Crystallization kinetics is obviously related to the main molecular structure characteristics of PLA, especially the ratio of the two enantiomeric forms of LA.

The crystallization rates of different PLAs can be easily compared with respect to their crystallization half-time, $t_{1/2}$, which is the time that is required to reach 50% of the final crystallinity. This indicator of the crystallization rate is usually reported as a function of the temperature, and follows a parabolic curve corresponding to the optimal temperature window, as illustrated in Figure 5.4 for PLA with varying D-LA contents and molecular weights in the 100–150 kg/mol range. This parabolic shape results from two competing

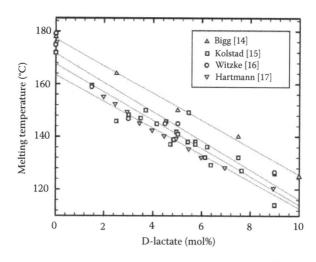

FIGURE 5.3

Melting temperature as a function of the D-LA content. (From Saeidlou, S. et al., *Prog. Polym. Sci.*, 37: 1657–1677, 2012.)

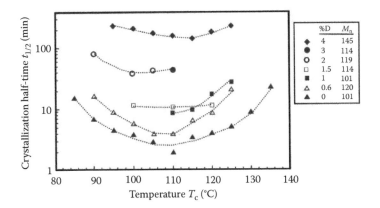

FIGURE 5.4
Crystallization half-time as a function of crystallization temperature for PLAs having different D-LA contents and molecular weights. (From Saeidlou, S. et al., *Prog. Polym. Sci.*, 37: 1657–1677, 2012.)

mechanisms: (1) the low-crystallization-temperature region is associated with a reduced chain mobility, which controls the crystal growth, and (2) for the high-crystallization-temperature region, where the chains have high mobility, the crystal nucleation rate is the limiting factor. This competition results in a minimum $t_{1/2}$ in the 105°C–110°C range, and even for the fastest PLA (PLLA of low molecular weight), the crystallization process takes several minutes to be initiated and completed. Increasing the D-LA content tremendously delays the occurrence of crystallization and it is not unusual to see reported values for the crystallization half-time in the tens of minutes, if not an hour in a few cases.

Nevertheless, this major drawback could be circumvented through the use of adequate nucleation agents that increase the heterogeneous nucleation density and plasticizers that increase the chain mobility, then widening the crystallization window. An efficient plasticizer is carbon dioxide and its impact on both plasticization and crystallization kinetics will be detailed in Section 5.5.

Crystallinity development in PLA can be amplified and accelerated through stretching, as reported in studies performed under uniaxial and biaxial stretching [18,19]. Such biaxial stretching occurs during the expansion phase of the foaming process. Crystallization can then be obtained under much lower temperatures than those usually encountered for quiescent crystallization. For example, the crystallinity development induced during biaxial stretching for two different grades of PLA having 2% and 4% of D-LA (Ingeo 4032D and 2002D, respectively) is illustrated in Figure 5.5 [19]. The 0.5 mm PLA amorphous cast sheets were deformed biaxially at 100°C at different stretch ratios from 2 × 2 to 9 × 9. Since stretching was done at a relatively high speed, 1 m/min, the crystalline phase developed over a pretty short time, e.g., 15 sec for a 4 × 4 deformation, which is estimated to be the minimum stretch ratio corresponding to the crystallization onset. The PLA

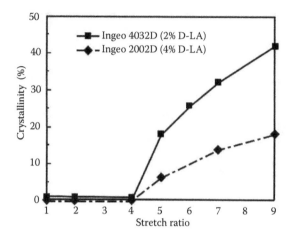

FIGURE 5.5
Crystallinity resulting from biaxial stretching performed at 100°C on two semicrystalline PLAs, for different stretch ratios. (From Mihai, M. et al., *J. Appl. Polym. Sci.*, 113: 2920–2932, 2009.)

containing more L-LA crystallized more rapidly and reached a much higher final crystalline content at a deformation of 9 × 9, i.e., 42% versus 18% for the other PLA grade containing more D-LA.

5.3 Solubility and Diffusivity of Physical Foaming Agents in PLA

The solubility of the chosen physical foaming agent (PFA) into the polymer matrix is a key parameter in foam extrusion processing. It dictates the pressure conditions that have to be met and maintained in the extruder and the die to prevent the occurrence of a two-phase system, detrimental to the expansion process and the quality of the resulting foam.

The solubilities and diffusivities of some potential foaming agents in PLA have been reported in the scientific literature: carbon dioxide [20–25], nitrogen [23], and butane [26]. CO_2 is obviously the one that has been the most thoroughly investigated, over wide temperature (0°C–200°C) and pressure (0.30 MPa) ranges. The gravimetric technique using a microbalance (CAHN or Rubotherm magnetic suspension balance) is usually chosen, and the reported results were corrected for a buoyancy effect through estimation of the density of PLA using various equations of state (EoSs), the Simha–Somcynsky and Sanchez–Lacombe EoSs being the most popular ones. The diffusion coefficients of the gases were extracted from the sorption uptake curves, i.e., the weight gain that is monitored as a function of time.

Below the melting point, the solubility is restricted to the amorphous zones only, and the overall solubility is then a function of the crystalline content. Moreover, as it will be detailed in Section 5.5.2, the dissolution of carbon dioxide can induce premature crystallization, since the gas acts as an efficient plasticizer and increases the mobility of the macromolecules, which eases and speeds up the crystallization process. Typical sorption curves are displayed in Figure 5.6 [27] where the second pressure step shows an overshoot of the gas that is dissolved followed by the solubility decrease due to the formation of the crystals that expel the CO_2 from the polymer–gas solution.

In the molten state, the effect of the D-LA content on the solubility of CO_2 has been investigated using three different commercial PLAs (Ingeo 3001D, 8051D, and 4006D). Interestingly, no significant difference was observed in this study [25].

The graphs displayed in Figure 5.7 summarize most of the solubility results that are reported for carbon dioxide, nitrogen, n-butane, and isobutane. An interesting abacus can be constructed to facilitate a comparison between various gases and polymers [28], with master curves being a function of the Henry's constants that are obtained from the solubility experiments, as a function of a dimensionless number that is based on the critical temperature of the gas (Figure 5.8). Results for PS have also been included in this last graph [29]. The solubilities of butane and isobutane in PLA are lower than those in PS. On the opposite, PLA can dissolve twice more CO_2 than PS under the same pressure and temperature conditions.

The higher solubility of CO_2 in PLA is obviously linked to the interactions between the oxygen of the carbonyl group and the carbon of the CO_2. This has also been reported for other polymers containing carbonyl groups such as poly(methyl methacrylate) (PMMA) and polycarbonate (PC). This CO_2-philicity

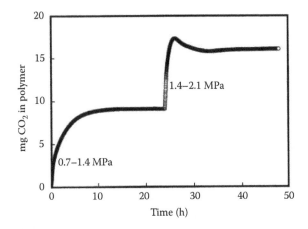

FIGURE 5.6
Sorption curves of carbon dioxide in PLLA performed at 25°C for two pressure steps. (From Liao, X. et al., *Biomacromolecules*, 7: 2937–2941, 2006.)

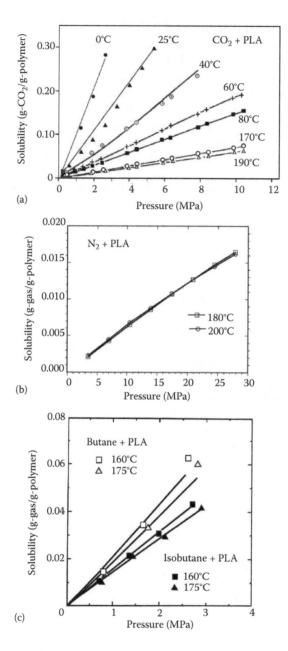

FIGURE 5.7
Solubility of various PFAs in PLA: (a) carbon dioxide, (b) nitrogen, and (c) n-butane and iso-butane. (From Liao, X. et al., *Polym. Int.*, 59: 1709–1718, 2010; Sato, Y. et al., Solubility and diffusion coefficient of carbon dioxide in polylactide, *Proceedings of the 21st Japan Symposium on Thermophysical Properties*, Nagoya, Japan, pp. 196–198, 2000; Li, G. et al., *Fluid Phase Equilib.*, 246: 158–166, 2006; Sato, Y. et al., Solubility of butane and isobutane in molten poly(lactide) and poly(butylene succinate), *PPS 2004*, Akron, OH, Paper 114, 2004.)

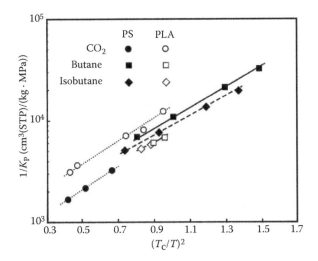

FIGURE 5.8
Comparison of Henry's constants K_P of various PFAs between PLA and PS. T_C is the critical temperature of the PFA in Kelvin. (From Sato Y. et al., Solubility of carbon dioxide in polymers, *Proceedings of Polymer-Supercritical Fluid Systems and Foams* (P-(SF)²), Tokyo, Japan, Dec. 4–5, 139–140, 2003; Sato, Y. et al., *J. Supercrit. Fluids*, 187–198, 2001.)

would be related to the Lewis acid–Lewis base interactions since the electron-donating carbonyl group of the rather rigid PLA chain can be easily accessed by the CO_2 molecules [30]. Partitioning of the CO_2 molecules has been proposed by Kazarian et al. [31] where the CO_2 molecules dissolved under low pressure in PMMA interact with the carbonyl groups, whereas those dissolved under higher pressures do not exhibit such specific interactions. Moreover, entrapment of the CO_2 molecules by such interactions has been proposed as the explanation for the lack of desorption issues in PMMA and PC with some evidence of residual plasticization in PMMA films after a six-month ageing [32].

Diffusion coefficients have also been reported for carbon dioxide over a wide range of temperature and under various CO_2 conditions, as displayed in Figure 5.9 [21]. Values for nitrogen lie in the 3×10^{-9} to 4×10^{-9} m²/s range for temperatures between 180°C and 200°C [23]. As expected, the gas diffuses much faster in PLA under molten conditions, as reported for CO_2. In fact, a rapid increase in diffusivity occurs when the polymer gets into the rubbery state, i.e., when the temperature under which the measurements are conducted exceeds the glass transition of the plasticized polymer. This is illustrated in Figure 5.10 [33].

Comparing the diffusivity of CO_2 in PS to that in PLA, interesting observations can be made. In the molten state, i.e., 190°C–200°C, the two polymers have roughly similar diffusion coefficients (10^{-9} m²/s). However, at temperatures close to their respective glass transition temperatures (100°C for PS, 60°C for PLA), CO_2 diffuses 10 times faster in PS than in PLA (10^{-10} m²/s for

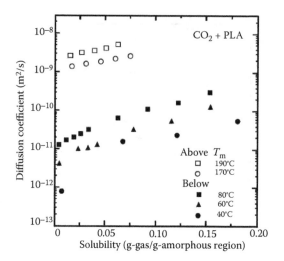

FIGURE 5.9
Coefficient of diffusivity of CO_2 in PLA, as a function of CO_2 concentration. (From Sato, Y. et al., Solubility and diffusion coefficient of carbon dioxide in polylactide, *Proceedings of the 21st Japan Symposium on Thermophysical Properties*, Nagoya, Japan, pp. 196–198, 2000.)

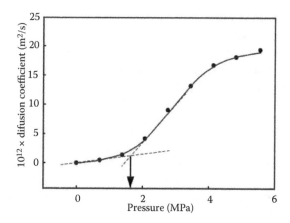

FIGURE 5.10
Coefficient of diffusivity of CO_2 in PLLA measured at 25°C. The arrow corresponds to the CO_2 pressure when the plasticized polymer undergoes transition from glassy to a rubbery state at 25°C. (From Liao, X. and Nawaby, A.V., *J. Polym. Res.*, 19: 9827–9835, 2012.)

PS versus 10^{-11} m²/s for PLA). It should be underlined, however, that the PLA chosen for such measurements was semicrystalline with a crystallinity content in the 30%–45% range. Since the transport of a gas in the polymer can occur only in the amorphous phase, the crystals act as a barrier and make the diffusion path rather tortuous [34]. The diffusion coefficient is then a function on the amorphous fraction available.

Although PLA is known for its good barrier properties, it shows poor results when it comes to permeability to water vapor and CO_2. Bao et al. [35] have characterized the gas-permeation properties of three PLAs with different L-LA/D-LA ratios with respect to carbon dioxide, oxygen, and nitrogen. Despite the fact that their film samples were solution casted and amorphous, the permeability values were higher for PLAs that were richer in L-LA. For films having a L/D ratio of 98.7:1.3, they obtained values of 1.10, 0.26, and 0.05 Barrer, respectively, for CO_2, O_2, and N_2. Thus the permeability of CO_2 is largely greater than that of air composed essentially of N_2 (78%) and O_2 (21%).

5.4 Rheology of PLA

5.4.1 Rheology of Neat PLA

The shear viscosity of typical molten PLA resins (Ingeo 2002D and 8302D) is displayed in Figure 5.11, from a combination of measurements made under dynamic (plate–plate oscillatory) and steady-state (capillary) conditions. The results from the capillary experiments were corrected according to standard practices (Bagley and Rabinovitch corrections). Both sets of data, complex and steady-state viscosities, overlap very well in the midrange of the

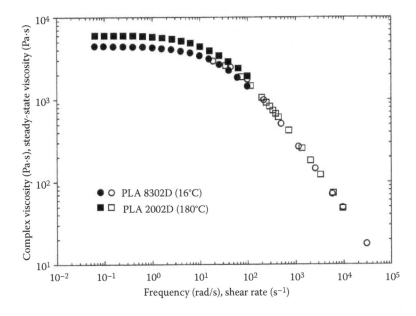

FIGURE 5.11
Dynamic (closed symbols) and steady-state (open symbols) shear viscosity of two different PLAs.

frequency-shear rate, making the Cox–Merz rule ($\eta^* \approx \eta_{ss}$ when $\omega = \dot{\gamma}$) verified for PLA.

Under low rates of deformation, PLA displays a nice plateau corresponding to Newtonian behavior. As the molten polymer is exposed to higher shear rates, i.e., typical of those encountered during extrusion, the viscosity of PLA starts to undergo shear-thinning due to the alignment of the macromolecules, until the viscosity at very high shear rates follows a straight line that can be described by a power-law equation, with the power-law exponent $n = d \log \eta / d \log \dot{\gamma} = 0.14$.

The melt-flow activation energy, E_a, was determined from the dynamic shear experiments that were performed at temperatures ranging from 160°C to 200°C, i.e., in a temperature window at least 100°C above the T_g of the neat PLA, which allows the use of the Arrhenius equation. Shift factors were obtained from superposition of the modulus curves, as displayed in Figure 5.12. The excellent agreement between the curves obtained at different temperature suggests that PLA is a rheologically simple system due to its linear structure. The resulting E_a for Ingeo 8302D is 74.4 kJ/mol. Other investigated PLAs having different viscosities and D-LA contents lead to values in a very close range: (a) 70.9 kJ/mol for 4032D, (b) 73.0 kJ/mol for 3051D, (c) 77.4 kJ/mol for 4033D, (d) 80.5 kJ/mol for 4044D, and (e) 84.6 kJ/mol for 2002D. Such numbers indicate that the viscosity of PLA is rather sensitive to temperature and that modest temperature variation can significantly impact its viscosity.

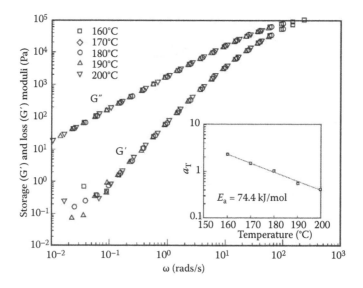

FIGURE 5.12
Time–temperature superposition of storage G' and loss G'' moduli of Ingeo 8302D at 180°C (reference temperature); insert relates the shift factor a_T to the temperature for the computation of the energy of activation E_a.

Decreasing the temperature may induce the crystallization process and the crystals can be probed through a viscosity increase. The time sweeps displayed in Figure 5.13 illustrate this phenomenon, with the tested PLA having a D-LA content of 2% with a melting point of 170°C [36]. An indication of the crystallization induction time can be extracted from the delay observed before the rise of the viscosity. The magnitude of time delays observed is typical of the crystallization process of neat PLA. Another observation that can be made from this figure is the thermal degradation of PLA that translates into a slight but significant viscosity reduction.

5.4.2 Rheology of PLA Modified with Chain Extender

In order to alleviate the processing issues that are associated with the low melt strength inherent to commercial PLA, an epoxy-functionalized chain extender (CE) is frequently utilized. In polyester/epoxide reactions, the epoxide groups can react with carboxyl- or hydroxyl-chain end groups of the PLA, with a preference for the carboxyl groups.

The most frequently used multifunctional epoxide for PLA foaming is a styrenic-glycidyl acrylate copolymer, manufactured by BASF under the trade name of Joncryl® 4368 [37]. Its formula is given in Figure 5.14.

A thorough investigation of the conditions amenable to the desired chain extension and its impact on the rheology have been performed with various grades of PLA: 3051D [38], 4032D, and 8302D [39]. The epoxide used was the Joncryl 4368 under a masterbach form, the CESA Extend OMAN 698493 (Clariant). From the study conducted by Corre et al. [38], the CE was found thermally stable below 250°C, a temperature above that recommended for PLA to avoid thermal degradation. Since residence time and temperature

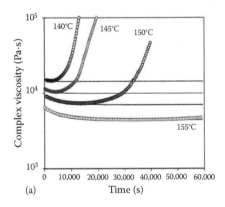

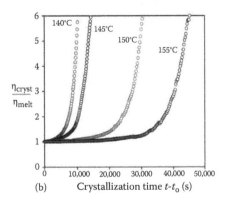

FIGURE 5.13

(a) Time sweeps showing the variation of the complex viscosity of PLA during various isothermal crystallization conditions. (b) The same data displayed using reduced variables. (From Yuryev, Y. and Wood-Adams, P., *J. Polym. Sci. Part B. Polym. Phys.*, 48: 812–822, 2010.)

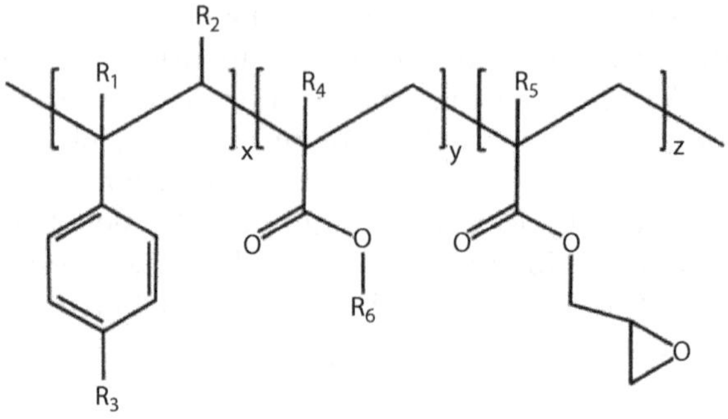

FIGURE 5.14
Formula of Joncryl® (BASF).

are two critical variables for the completeness of the reaction and the extent of the chain branching, reactions were performed at 210°C with various contents of CESA-Extend ranging from 1 to 3 wt%. The reaction time was found to be proportional to the content of CE, from 4 min with 1 wt% CE to 12 min for 3 wt% CE. Results obtained from size exclusion chromatography indicated that the reaction involved preferably low-molecular-weight species, due to their greater chain diffusivity. With the decrease of the low-molecular-weight population and the creation of a high-molecular-weight shoulder, the resulting molecular weight distribution tends to be bimodal.

The rheology of modified PLA was found to be highly sensitive to the modification of the chain structure that is imparted by the CE. In a study of Mihai et al. [39] comparing an amorphous and a semicrystalline grade, the viscosity of the two grades rose approximately to the same extent with CE content as shown in Figure 5.15a. Figure 5.15b illustrates that the increase in the complex viscosity is largely attributable to the elastic contribution to the rheological response; such elasticity is absent in the neat PLA (no melt strength).

It should be noted that the two grades underwent different levels of degradation with their passage in the extruder. The semicrystalline grade 4032D that exhibited a much higher viscosity prior to its extrusion suffered from greater chain scission due to the aggressive shearing environment in the twin-screw extruder. Time sweeps performed on the same modified PLA, shown in Figure 5.16, suggested that this increased degradation was compensated by a greater usage of the CE. Even with a residence time less than 3 min, the reaction was completed for PLA 4032D and the samples rapidly underwent further degradation during their stay in the rheometer. On the other hand, the initially lower-viscosity 8302D grade is still reactive during the rheological testing for CE content above 1 wt%. With residual amounts of unreacted CE, chain extension could still occur and compensates for the shear and thermal degradation leading to chain scission. Finally, this study suggested that the maximum relative elasticity would be obtained around

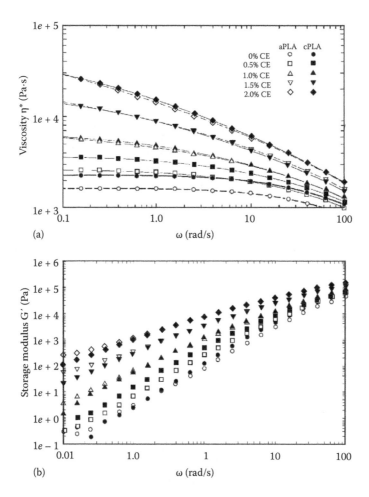

FIGURE 5.15
(a) Complex viscosity of amorphous (open symbols) and semicrystalline (close symbols) PLA for various degree of branching, measured at 180°C. (b) Storage modulus G' for the same sets of PLA at 180°C. (From Mihai, M. et al., *Polym. Eng. Sci.*, 50: 629–642, 2010.)

1.5–2 wt% CE, as the *tan δ*, i.e., the ratio G"/G', tended to level off as such CE concentration levels (Figure 5.17).

The presence of branching induced by the reactive CE, as suspected by the results shown in Figure 5.17, was also verified through elongational rheology. The response of polymers having such branched structures like low-density polyethylene (LDPE) is characterized by the presence of a so-called strain-hardening signature, i.e., a large increase of the viscosity with the extent of deformation. The presence of such a feature has been frequently reported to be highly beneficial, if not essential, for the stability of polymer processes where the deformations are mainly extensional, such as film blowing and

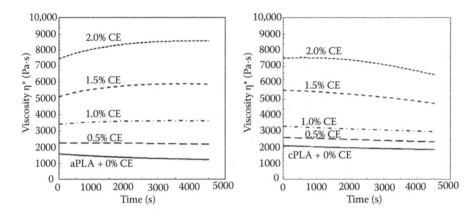

FIGURE 5.16
Thermal stability of linear (0% CE) and branched (with CE) PLA at 180°C illustrated through the relative variation of the complex viscosity at 10 rad/s during time sweeps.

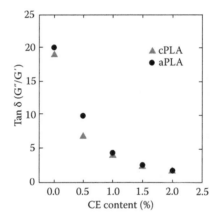

FIGURE 5.17
Loss tangent as a function of CE content determined at 180°C and frequency of 1 rad/s for two PLAs (amorphous and semicrystalline). (From Mihai, M. et al., *Polym. Eng. Sci.*, 50: 629–642, 2010.)

foaming. It has a self-healing effect such that during expansion, the stress associated with the thinning of the membrane does not concentrate on any local point, but it is evenly distributed. The striking difference between the elongational response of neat PLA and that of the one modified with 2% CE definitely confirmed the presence of a branched structure. The magnitude of the strain-hardening effect was found to be proportional to the CE content over the range of the CE that was investigated (Figure 5.18).

Lehermeier and Dorgan [40] have reported the activation energy of the viscosity for different PLA blends based on various ratios between a linear grade and a peroxide-initiated cross-linked one. The values for E_a were almost similar in the 72–78 kJ/mol range. The activation energy results of

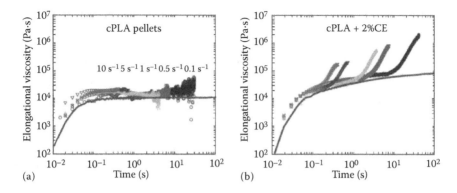

FIGURE 5.18

Transient elongational viscosity of neat PLA (a) and chain-branched PLA (b) at 180°C, measured for different strain rates. (From Mihai, M. et al., *Polym. Eng. Sci.*, 50: 629–642, 2010.)

Corre et al. [38] for the CE-modified PLA have been calculated at a constant rate of deformation. Since E_a computed at constant shear stress is a better estimate of the zero-rate-of-shear E_a, these results, once corrected according to Utracki and Schlund [41], the yield also a nearly constant value lying in the same range, i.e., 70–80 kJ/mol. Thus, contrarily to polyolefins where E_a increases with the degree of branching, e.g., HDPE versus LDPE, modified PLA exhibits a constant dependence of viscosity with respect to temperature.

Corre et al. [42] have also investigated the influence of the chain extension on the crystallization behavior of PLA. Their main findings are summarized in Figure 5.19, which illustrates that the kinetics of crystallization of a branched PLA is accelerated, whereas its crystallinity ratio is decreased.

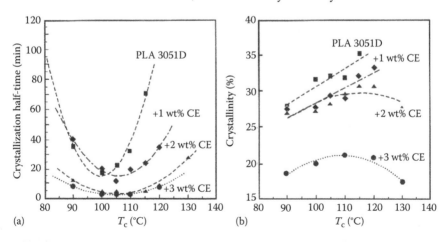

FIGURE 5.19

Evolution of the crystallization behavior as a function of the crystallization temperature. (a) Crystallization kinetics (half-times); (b) crystalline ratio. (From Corre, Y.-M. et al., *Polym. Eng. Sci.*, 54(3): 616–625, 2014.)

5.5 Effect of PFA on PLA

5.5.1 Plasticization of PLA by PFA and Its Impact on Shear Viscosity

The glass transition temperature (T_g) is associated with the transition from the glassy to the rubbery state. Plasticization increases the chain's mobility and thus enables the glass-to-rubber transition to occur at a lower temperature. Plasticization induced by volatile chemical species modifies tremendously the conditions under which the polymer can be melt processed. For foam extrusion, it allows the melt temperature to be reduced to close to the T_g of the neat polymer, which contributes to the stabilization of the cell structure. Plasticization also shifts the crystallization temperature window to lower temperatures for semicrystalline polymers. In addition, due to the enhanced molecular mobility, crystallization occurs much faster which might be highly beneficial for slow-crystallization polymers such as PLA.

The determination of T_g can be performed in different ways, although special care and specific apparatus might be needed if a volatile chemical is used as a plasticizer. One conventional determination technique, dilatometry, consists in locating the kink in the specific volume as observed during cooling. This has been done with an amorphous PLA (Ingeo 8302D) in which various amounts of carbon dioxide were dissolved [43]. A PVT measuring device complemented with ultrasonic sensors was used and the results are

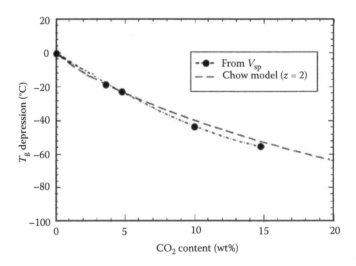

FIGURE 5.20
Depression in glass transition temperature (ΔT_g) determined by volumetric measurements (break-point in specific volume curve) as a function of CO_2 content for an amorphous PLA. Experimental points were obtained by specific volume measurements. Chow model was used for comparison purpose with a value of $z = 2$. The lines are only guides for the eyes. (From Reignier, J. et al., *J. Appl. Polym. Sci.*, 112: 1345–1355, 2009.)

displayed in Figure 5.20. The experimental data are compared to estimates obtained with a theoretical model developed by Chow [44], the function of the molecular weight of the plasticizer and that of the repeating unit of the polymer [44]. In this model, the lattice coordinate number z was set equal to 2, and despite the fact that the Chow model does not take into account any chemical interaction that may exist between the polymer and the plasticizer, a fair agreement can be observed.

Another technique relies on the significant diffusivity increase observed during sorption experiments, when the plasticized polymer undergoes transition from a glassy to a rubbery state, as previously illustrated in Figure 5.10. The threshold pressure determined from this method refers to the minimum pressure of CO_2 that is needed to plasticize the PLA, such as the resulting T_g is equal to the temperature at which the measurements have been performed. Surprisingly, due to the higher solubility observed at low temperatures (see Figure 5.7a), the needed CO_2 pressure shifts to lower values as illustrated in Figure 5.21 [33]. This means that such a polymer–gas system has two T_gs for a given gas pressure: the higher one corresponds to the expected glass-to-rubber transition as the temperature is increased at a given CO_2 pressure and an unusual one called *retrograde vitrification* since this transition is a rubber-to-glass transition. Such behavior has been previously reported for other CO_2-philic systems like PMMA/CO_2 and poly(ethyl methacrylate)/CO_2 [34].

One frequent method for determining the T_g under gas-plasticized conditions uses a pressurized differential scanning calorimeter (DSC). Unfortunately, the results are usually reported in terms of CO_2 pressure and not gas concentration. Since the CO_2 concentration is a function of both pressure and temperature, the plasticized conditions are those of the polymer sample before the DSC temperature sweep. This could explain the huge difference reported

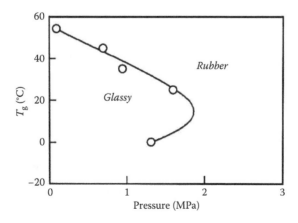

FIGURE 5.21
Variation of T_g as a function of the CO_2 pressure for the PLA/CO_2 system. (From Liao, X. and Nawaby, A.V., *J. Polym. Res.*, 19: 9827–9835, 2012.)

when results from different sources are compared as in Figure 5.13b from Nofar and Park [45].

Lastly, the rheology of a polymer is function of the difference between the temperature and its T_g. Thus, reduction of T_g translates into viscosity decrease at constant temperature. This dependence clearly manifests itself within the range of $T - T_g$ less than 100°C. So, measurement of the viscosity of gas-swollen polymer can provide a fair estimate of the decrease of T_g. The difficulty for such a method relies on the availability of a close-pressurized rheometer capable of preventing phase separation between the gas and the polymer during the measurements. The use of an online rheometer fulfills these conditions, and such a method has been used by Reignier et al. [43] to measure the effect of foaming agent concentration on the rheological properties of the PLA/CO_2 mixtures. Results for the shear stress as a function of apparent shear rate are displayed in Figure 5.22 for measurements conducted at two temperatures, 140°C and 160°C, for various concentrations of carbon dioxide. Comparing these results on a constant stress basis, one can see that increasing concentrations of CO_2 shifts the stress–rate curve to the right, i.e., to lower viscosity values. For example, comparing the curves of neat PLA 8302D at 160°C with that using 4.9 wt% CO_2, viscosity is reduced by a factor of 5. Moreover, using curves of similar viscosity, for example 1.7 wt% CO_2 at 160°C and 4.7 wt% at 140°C, one can infer that a 20°C decrease can be compensated for by a CO_2 increase of 3 wt%, meaning a shift in the glass transition temperature of –20°C/3 wt% = –6.7°C/wt% CO_2. Unpublished data based on the same kind of experiments but using n-pentane instead of CO_2 lead to a lesser plasticization efficiency, approximately –6°C/wt% of n-pentane [46]. For comparison purposes, Gendron et al. [47,48] reported the same level of plasticization in PS for both CO_2 and n-pentane, which is approximately –8°C/wt% PFA.

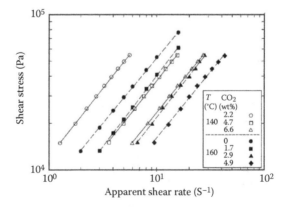

FIGURE 5.22
Apparent viscosity measurements for PLA charged with various concentrations of CO_2 at two different temperatures, 140°C and 160°C. (From Reignier, J. et al., *Cell. Polym.*, 26(2): 83–115, 2007.)

5.5.2 CO₂-Induced Crystallization

As displayed in Figure 5.6, the presence of CO_2 can induce unexpected premature crystallization. As was illustrated in Section 5.5.1, dissolution of CO_2 into PLA lowers significantly the T_g, which translates into an increase of the segmental mobility of the chains. This enables the crystallization process to occur at much lower temperatures and with faster rates. However, the time frame under which such a phenomenon occurs is not only dependent upon the crystallization kinetics but also relies on the diffusion of the gas within the polymer. Until the moment that the ambient temperature remains below the T_g the diffusion process is slow (see Figure 5.10). However, PLA crystallization can occur as soon as the T_g gets below the set temperature due to CO_2-plasticization accompanied by a speed up of CO_2 diffusion within the PLA under such adequate conditions.

The following example will help to appreciate the length of time that is needed for diffusion and crystallization from solid-state PLA. Two semicrystalline grades of PLA having 2% and 4% of D-LA, in the shape of sheets having a thickness of 0.5 mm, were conditioned with CO_2 for different exposure times ranging from 1 min up to 2 h. The gas pressure was set at 6 MPa and the soaking was performed at room temperature.

Figure 5.23 displays the crystallinity results as determined by X-ray diffraction, where the crystalline content increases progressively for the first 10 min of soak time and stabilizes at a crystallinity level of 25%–30% [19]. The effect of exposure time on the level of crystallinity achieved is very similar for the two PLA grades. Under such conditions, the crystallization process still remains relatively slow; the maximum crystallinity is achieved only after several minutes.

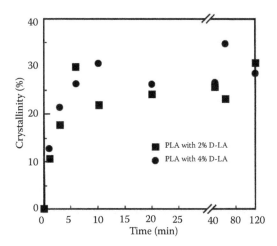

FIGURE 5.23
Crystallinity developed in PLA sheets exposed to 6 MPa of CO_2 as a function of exposure time. (From Mihai, M. et al., *J. Appl. Polym. Sci.*, 113: 2920–2932, 2009.)

Since CO_2 can plasticize the glassy state of PLA into its rubbery state and provide enough mobility to the molecules to rearrange and form crystals, one would expect that such enhanced mobility due to the presence of CO_2 would tremendously affect the crystallization kinetics as measured under isothermal conditions. In fact, it is well recognized that carbon dioxide can even contribute to forming a crystalline phase in polymers that are commonly reported as amorphous ones, such as polycarbonate [49].

Various studies have been conducted on the CO_2-modified kinetics of PLA. Takada et al. [50] have used an enthalpic method based on high-pressure differential scanning calorimeter (HP-DSC) measurements, whereas Reignier et al. [51] utilized a novel technique relying on the variation of the ultrasonic velocity, with the velocity being very sensitive to the amorphous or semicrystalline state. While experiments performed with an HP-DSC do not allow the uncoupling of the effect of pressure from that of the CO_2 molecules dissolved into the polymer sample, as the CO_2 concentration is pressure dependent in such an experimental setup, the ultrasonic technique relies on a given specific concentration of CO_2 dissolved in the polymer sample, so, in addition, it can provide direct information on the effect of pressure on crystallization kinetics.

Figure 5.24 clearly illustrates the difference on the half-crystallization time $t_{1/2}$-curves for a neat semicrystalline PLA (Ingeo 8302D) and when 3.9 wt%

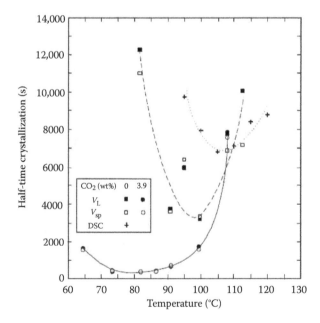

FIGURE 5.24
Effect of CO_2 content on the crystallization half-time ($t_{1/2}$) of cPLA as a function of temperature, with pressure set at 20 MPa, except for the DSC measurements (atmospheric pressure). Values of $t_{1/2}$ were determined either by DSC (cross symbol), by ultrasonic velocity measurements (solid symbol), or by specific volume measurements (open symbol).

of CO_2 was dissolved in it, both samples submitted to the same hydrostatic pressure (20 MPa). In addition, this graph also displays how the crystallization under isothermal conditions can be accelerated by high pressures, comparing the neat PLA results that were obtained from more classical DSC measurement at atmospheric pressure and those that have resulted from increasing the pressure at 20 MPa, which became two times faster. Nevertheless, without carbon dioxide, crystallization practically only occurs in the hour range.

Findings from Takada and Reignier on the effect of CO_2 are very similar, giving the increased mobility of the PLA molecules at lower temperatures. The crystallization rate is accelerated in the low-temperature region, usually termed the self-diffusion-controlled region, but exhibits longer times at higher temperatures, i.e., in the nucleation-controlled region. The crystallization process, characterized by its parabola shape, is not only shifted to lower temperatures in the presence of CO_2, but its kinetics is also significantly modified: crystallization now occurs in the tens of minutes, which is a drastic improvement. In addition, as mentioned in Takada's study, the CO_2 content also impacts the final crystallinity level. For instance, at a saturation pressure of 2 MPa and a temperature of 140°C, crystallinity increases from 26% to more than 46% due to the presence of CO_2.

The addition of CO_2 also has a tendency to modify the parabolic V-shape of the half-crystallization time into a flatter and less temperature-dependent U-shape. The following mechanism can be suggested: when crystallization occurs, the crystals expel carbon dioxide in the surrounding amorphous region, thus enhancing the mobility of the chains, which, in turn, accelerate the crystallization process. This may contribute to a more abrupt decrease of the crystallization half-times in the high-temperature zone, and a flatter minimum half-time results over a wider range of temperature.

Nonisothermal melt crystallization, evaluated under various CO_2 pressures and at different cooling rates, also contributes to illustrate the strong modification in the crystallization process that is imparted by the dissolved gas molecules that facilitate the molecular mobility of the PLA matrix (Figure 5.25a). Such mobility, in turn, will affect the crystalline structure and the perfection that is achieved for the crystals under the prevailing CO_2 conditions. This will be reflected in the crystal melting temperature T_m, with a decreasing value with increasing CO_2 pressure because of the presence of less perfect crystals [52]. Such dependence is illustrated in Figure 5.25b.

5.6 Extrusion Foaming

Foaming PLA can be performed in several ways. Batch processes such as solid-state foaming and the autoclave method have their utilities when it

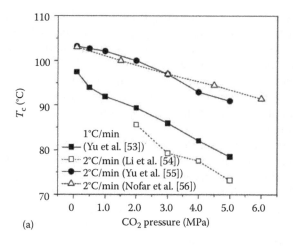

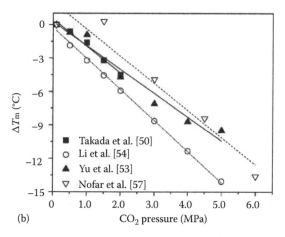

FIGURE 5.25

(a) T_c and (b) T_m variation as a function of CO_2 pressure. (From Nofar, M. and Park, C.B., *Prog. Polym. Sci.*, 39: 1721–1741, 2014.)

comes to producing small-scale foam samples or studying the influence of different process variables. Many scientific publications have addressed these topics [1,45]. However, mass-production methods are the only processing routes to supply goods for industrial applications and the foam extrusion process is one of two preferred choices, the other one being expandable beads. The technologies for expandable PLA beads has been the object of several studies and patents [58–61] and commercialization is currently underway in Europe [62–64]. The development of the extrusion foaming of PLA has occurred over the past decade. Literature on extrusion foaming PLA is unfortunately relatively scarce due to the many challenges that need to be taken care of: (a) the inherent low melt strength of PLA, (b) its high shear sensitivity, and finally, (c) its high heat retention. A list of the studies published can be found in Table 5.2, along with details on the experimental conditions.

Table 5.2 can be divided into two groups of experiments. The first includes small- or laboratory-scale experiments that have been essentially focused on the use of carbon dioxide as the PFA. The flow rate is usually less than 15 kg/h and the resulting foam is obtained from a strand (or capillary) die. In this group, it should be noted that many of the first trials were based on amorphous PLA, as amorphous resins are well-known to be easier to foam with the extrusion process.

The second group deals with pilot- or industrial-scale equipment for production of foam sheets (annular die), with throughput in the few hundreds of kilograms per hour. Published experiments from this second group may address problems that are not targeted by the former laboratory studies such as efficiency compared with the production of PS foam sheets. Transition from the current PS process was attempted using the common PFAs that are utilized in this business, i.e., hydrocarbons such as butane and pentane. Sections 5.6.1 and 5.6.2 will summarize the findings for each group, respectively.

5.6.1 Laboratory-Scale Experiments (CO_2-Based Foams)

Most of the studies available in the open literature are based on carbon dioxide as the foaming agent. Since one of the benefits of using PLA instead of PS is its reduced impact on environmental issues, carbon dioxide became an obvious choice as this foaming agent is considered as a green solvent. This also probably explains why most of the research published on extruded PLA foams have used essentially carbon dioxide as PFA.

As early as 2005, NatureWorks filed for a patent on extruded PLA foams blown with carbon dioxide [65]. The examples provided in the patent were obtained with amorphous PLA grades, with the percentage of D-LA greater than 9.5%. The concentration of CO_2 was set at 7.9% and talc at 0.5%. The temperature of the melt was approximately 100°C and the resulting foams exhibited low-density values, like 32.5 kg/m³ for the first provided example. The claimed carbon dioxide concentration could be within 5%–15% by weight, more preferably between 7 and 11 wt%.

The PLA resins used in the reported studies are commercial ones that belong to the NatureWorks catalog. As illustrated in Figure 5.26, they are of viscosity compatible with extrusion and vary from semicrystalline to completely amorphous grades.

Amorphous PLA grades were the first candidates chosen for extrusion foaming, as listed in Table 5.2. The following experiments were made also on semicrystalline PLA, and different attempts to modify their rheological behaviors were subsequently performed using a chemical chain extender (CE) and nanoclays.

TABLE 5.2

List of Published Experiments on PLA Extrusion Foaming

Authors–Publication Date	Extruder	Flow Rate (kg/h)	Blowing Agent (Content in wt%)	PLA Grade	Rheology Modifier	Talc (Additives)	Lowest Foam Density (kg/m³)
Laboratory-Scale Experiments (Extrudate: Foam Strand)							
Cink–2011 [65]	TSE 27 mm	n/a	CO_2 (7.9%)	6300D	n/a	n/a	32.5
Lee, S.T.–2006 [66]	Tandem	n/a	CO_2 (2%–9%)	6300D	None	n/a	30
Reignier–2007 [43]	TSE (cn) 50 mm	15	CO_2 (2%–9%)	8302D	None	0, 0.5, and 1 wt% talc	21
Mihai–2007 [67]	TSE (co) 34 mm	10	CO_2 (2%–9%)	2002D	MA	0.5% talc	25
Mihai–2009 [19]	TSE(co) 34 mm	5, 7.5, 10	CO_2 (5%, 7%, and 9%)	2002D, 4032D, 8302D	None	No talc	32 (4032D–7% CO_2) 35 (2002D–9% CO_2)
Pilla–2009 [68]	SSE 19.1 mm	n/a	CO_2 (4%)	3001D	CE (0–1.3 wt%)	0 and 0.5 wt% talc	310 (0.5%talc – 1–1.3%CE)
Mihai–2010 [39]	TSE (co) 34 mm	10	CO_2 (3%, 5%, 7%, and 9%)	4032D, 8302D	CE (0–2 wt%)	n/a	19 (8302D, 1–2%CE 9%CO_2) 28 (4032D, no CE, 7–9%CO_2)
Ramesh–2011 [69]	Tandem (TSE)	5.7	4.5%–6.5% (80 n-pentane/20 isobutane)	3051D, 4032D	CE (2.35 wt%)	0.4% talc or 1% NAS516 (Sukano)	38
Larsen–2013 [70]	TSE (co) 25 mm	2.4	CO_2 (5%)	2002D, 3251D, 8052D	CE (2 wt%)	n/a	23 (2002D)
Wang–2012 [71]	Tandem (SSE 19.1 mm / SSE 38.1 mm)	0.66–0.84	CO_2 (5%, 7%, and 9%)	2002D, 8051D	CE (0–2.3 wt%)	0.4 wt% (with CE)	30 (8051D, 2.3%CE, 9%CO_2)

(Continued)

TABLE 5.2 (CONTINUED)

List of Published Experiments on PLA Extrusion Foaming

Authors–Publication Date	Extruder	Flow Rate (kg/h)	Blowing Agent (Content in wt%)	PLA Grade	Rheology Modifier	Talc (Additives)	Lowest Foam Density (kg/m³)
Matuana-2010 [72]	SSE 19.1 mm	2.3	CO_2 (5%)	2002D, 8302D	NC (5 wt%)	n/a	No mention
Keshtkar-2014 [73]	Tandem (SSE 19.1 mm/ SSE 38.1 mm)	n/a	CO_2 (5% and 9%)	2002D	NC (0.5–5 wt%)	n/a	28 (9%CO_2–5% Cloisite 30B)
Industrial-Scale Experiments (Foam Sheet)							
Corre-2010 [74]	Tandem (SSE115 mm/ SSE150 mm)	n/a	Butane (5%) (20 wt% isobutene/80% n-butane)	3051D	CE (0.5–2.3 wt%)	3 wt% talc 0.5 wt% Biomax 120	50 (with 1.3% CE)
Lee, E.K.–2015 [75]	Tandem (150 mm/ 180 mm)	n/a	Isobutane	8052D	CE	Talc	78
Fogarty-2007 [76]	Tandem (4.5 in./ 6 in.—Turbo-Screw)	325–400	Isobutane + isopentane	n/a	n/a	n/a	
Nangeroni-2014 [77]	Tandem (3.5 in./4.5 in.)	60	HFC-152a (5.8%)	3051D	CE	Talc	53

Note: CE: CESA-Extend Oman 698493; cn: counterrotating; co: corotating; MA: PLA-g-MA; NC: nanoclay Cloisite 30B; SSE: single-screw extruder; TSE: twin-screw extruder.

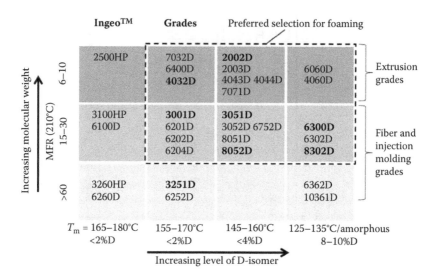

FIGURE 5.26
A selection of Ingeo PLA grades from NatureWorks classified according to their viscosity and
D-LA content. The group within the dotted rectangle includes preferentially the grades chosen
for the studies based on carbon dioxide (numbers in bold). Ingeo grades developed for foam
application have grade numbers starting with 8.

5.6.1.1 Foaming Amorphous PLA

5.6.1.1.1 Effect of CO₂ Content on Foam Density

The very first trials on foam extrusion of PLA with carbon dioxide, published
between 2007–2009 [43,66], were performed using amorphous grades of PLA
(Ingeo 6300D and 8302D). The main finding was that the concentration of the
PFA was critical, and that an unusually high threshold concentration of CO_2
was required to lead to low-density foams. In comparison with PS foam that
required roughly 2.5 wt% of CO_2 to get to densities lower than 100 kg/m³, at least
5 wt% of CO_2 was needed with PLA. The best results in terms of density reduction have been achieved in the CO_2 concentration range of 7–8 wt% (Figure 5.27).

Figure 5.28 summarizes the findings from different experimental sources
and for different grades of PLA. It displays the foam density as a function of
carbon dioxide concentration for a nominal processing temperature of 100°C
in most of the cases. All curves follow a similar shape, with a sharp transition toward lower foam densities at intermediate CO_2 concentrations. Thus,
three distinct zones can be distinguished in Figure 5.28, corresponding to
low, moderate, and high PFA concentrations [43]:

Region I: Low concentrations of CO_2 have very little impact on the density
reduction, with densities remaining above 500 kg/m³. These unsuc-
cessful foams have been associated with a rather poor cell nucleation
(less than 10⁷ cell/cm³ based on unfoamed material).

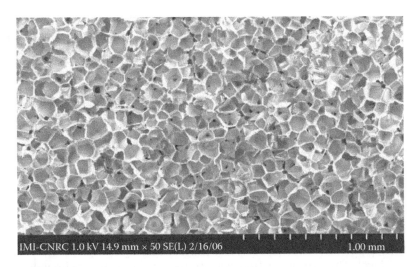

FIGURE 5.27
SEM photograph of a PLA foam with density of 33 kg/m^3 extruded at 100°C using 7.3 wt% of CO_2 and 0.5 wt% of talc. (From Reignier, J. et al., *Cell. Polym.*, 26(2): 83–115, 2007.)

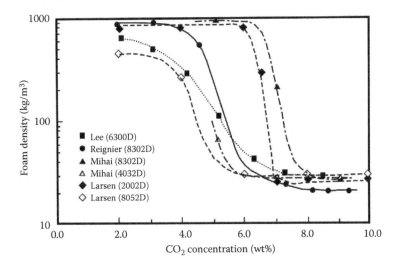

FIGURE 5.28
Extruded PLA foam density reduction with increased CO_2 concentration, excerpted from various studies. (From Mihai, M. et al., *J. Appl. Polym. Sci.*, 113: 2920–2932, 2009; Reignier, J. et al., *Cell. Polym.*, 26(2): 83–115, 2007; Lee, S.T. et al., *J Cell. Plast.*, 44(4): 293–305, 2008; Larsen, A. and Neldin, C., *Polym. Eng. Sci.*, 53(5): 941–949, 2013.)

Region II: This is the transition region, where a gradual increase in the CO_2 content has a huge impact on the density reduction. The onset of this transition varies from one study to another, since different polymer systems, as well as different processing equipments, have been used.

Region III: At higher PFA concentrations, i.e., above 7 wt% CO_2 for most of the studies reported in Figure 5.28, low-density foams in the 30 kg/m³-range have been easily obtained, with no further significant change in the density with an increase in the blowing agent content. It appears from the results above that a critical gas concentration of roughly 7 wt% is thus required for a favorable nucleation and expansion process leading to low-density foams.

5.6.1.1.2 Nucleation

Another phenomenon that occurred when the CO_2 concentration reached the 7% level. The cell nucleation density increased by approximately three decades, from 10^5 to nearly 10^8 cells/cm³. For a low-density foam with such nucleation density, cell sizes lie around 100 μm. Reignier [43] also studied the influence of talc on the foaming process and found out that improvement of the nucleation process due to talc was effective at low contents of CO_2, but the difference with foams free of nucleating agents vanished as the CO_2 content was increased over the said 7 wt% level (Figure 5.29). Thus nucleation in PLA foams could be triggered by three different mechanisms: CO_2 clusters formation when CO_2 reaches high content level, crystallite formation enhanced by

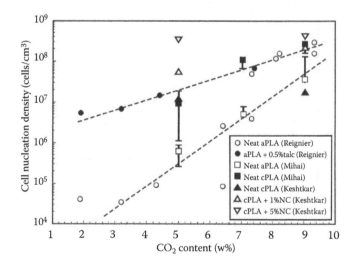

FIGURE 5.29

Cell nucleation density for various systems as a function of CO_2 content. (From Mihai, M. et al., *J. Appl. Polym. Sci.*, 113: 2920–2932, 2009; Mihai, M. et al., *Polym. Eng. Sci.*, 50: 629–642, 2010; Reignier, J. et al., *Cell. Polym.*, 26(2): 83–115, 2007; Keshtkar, M. et al., *Polymer*, 55: 4077–4090, 2014.)

the presence of CO_2, and the standard external nucleation agent utilization. One could suspect, however, that these three mechanisms could at some point be linked: for instance, the formation of a crystalline phase expelled its CO_2 content toward the amorphous phase, then increased this local CO_2 concentration up to possibly its self-nucleating threshold.

5.6.1.1.3 Dimensional Stability versus Open-Cell Content

Unfortunately, the study from Reignier et al. [43] also reported that the foams blown at CO_2 contents in the 8–10 wt% range may exhibit severe deformations, shrinkage, and densification upon ageing, as shown in Figure 5.30. This dimensional instability issue has been associated with higher closed-cell contents. As mentioned in Section 5.3, carbon dioxide permeates much faster than oxygen and nitrogen, so the CO_2 exiting from the foam cells toward ambient atmosphere would not be compensated fast enough by the inward air diffusion, resulting in a vacuum taking place within the closed cells. As mechanical compression test results will indicate (Section 5.6.1.5), foams in the low-density range exhibit a very low modulus and a yield stress close to 0.1 MPa, i.e., a level similar to atmospheric pressure. So the pressure differential across the foam is sufficient to generate a shrunken structure. This behavior is typical of low-density polyolefin foaming that requires a permeability modifier to circumvent the occurrence of dimensional instability [78].

However, in the presence of opened cells, the mode of transport of the gas molecules is convection and the pressure equilibrium is achieved rapidly with little if no dimensional changes. Surprisingly, foams with high

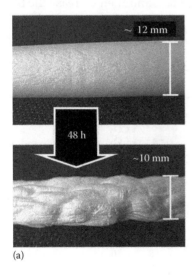

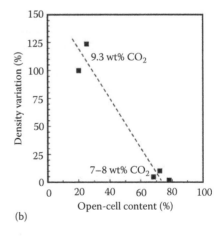

(a)

(b)

FIGURE 5.30
(a) Shrinkage of a PLA foam strand extruded with 9.3 wt% CO_2 over a 48 h ageing period following extrusion. (b) Dependence of the magnitude of shrinkage (relative density variation) as a function of open-cell content. (From Reignier, J. et al., *Cell. Polym.*, 26(2): 83–115, 2007.)

closed-cell content were those showing the thinnest cell walls. In addition, the presence of talc had no impact on the open-cell content results.

5.6.1.2 Foaming Semicrystalline PLA

Since the level of crystallinity can be tuned in PLA according to its D-LA content, Mihai et al. [19] have chosen to investigate the impact of crystallization on the foaming process using different grades of PLA (2%, 4%, and 10% D-LA). DSC examination performed on the extruded foams highlighted the progressive increase of the crystallinity levels with the CO_2 content. In comparison with a pristine PLA that can achieve its full crystallization only after a few hours, crystallinity levels as high as 45% were obtained with the grade containing the lowest percentage of D-LA (4032D) and this, within the time frame of the extrusion process, i.e., in the minute range (Figure 5.31). Obviously, the crystallization process was tremendously facilitated by the presence of CO_2. In addition, the authors have suggested that "it is obvious that only a small amount of crystal nuclei was present in the flowing material," which makes the crystallization process occur in two steps: the first one taking place in the extruder and die, and the second one during the expansion phase, which implied a strain-induced crystal nucleation during biaxial stretching of the cell walls.

Moreover, this enhanced crystallization prior to foaming might have triggered the foam expansion at lower levels of CO_2 and thus eased the foam density reduction (Figure 5.28). Gas molecules expelled from the crystallites enrich the CO_2 content of the amorphous phase, and then increase the blowing power.

The presence of a small and rigid crystalline phase during expansion might be at the origin of a particular morphology that is observed at high magnification in the cell walls (Figure 5.32). Those observations point out

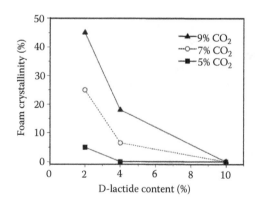

FIGURE 5.31

Foam crystallinity resulting from foaming PLAs having varying D-LA contents. (From Mihai, M. et al., *J. Appl. Polym. Sci.*, 113: 2920–2932, 2009.)

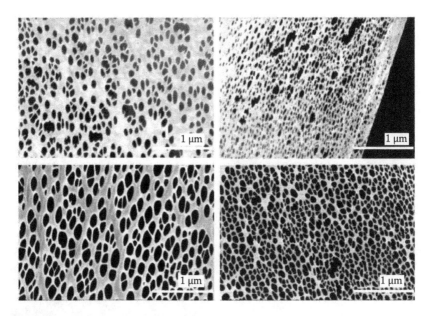

FIGURE 5.32
Lace structures observed in the cell walls of a foamed semicrystalline PLA. (From Mihai, M. et al., *J. Appl. Polym. Sci.*, 113: 2920–2932, 2009.)

the presence of *loose networks of thick and fine fibrils*, termed *lace structures*, for PLAs having high crystallinity content and good dimensional stability [19]. Small cavities may have developed in the amorphous regions with the nods being rich in crystals. Cavity and fibril sizes were obviously a function of the crystallinity content: in the case of foams made with 9 wt% of CO_2, cavities were as small as 75 nm. Such porous membranes may thus facilitate gas exchange from and to the foam cells through convection while maintaining the pressure at the atmospheric level, which may contribute to keeping the cellular structure with adequate dimensional stability and free of shrinkage.

The presence of a crystalline phase formed inside the extruder was validated through in-line visualization using sapphire windows that were located before the extrusion die. The gas–polymer system investigated was PLA Ingeo 2003D charged with 9 wt% of CO_2 [79,80]. Varying the temperature on the cooling extruder of the tandem line impacted the crystallization kinetics and, consequently, the amount of heterogeneities that are suspected to be PLA crystals, as illustrated in Figure 5.33. The resulting foam expansion was proportional to the crystalline content that is observed in the image. The presence of premature crystals might be beneficial to the rheology of molten PLA for the expansion phase, preventing cell rupturing and coalescence. The crystalline phase might also decrease the gas loss by lowering the diffusion process and thus enhancing the expansion process, as reported in this study.

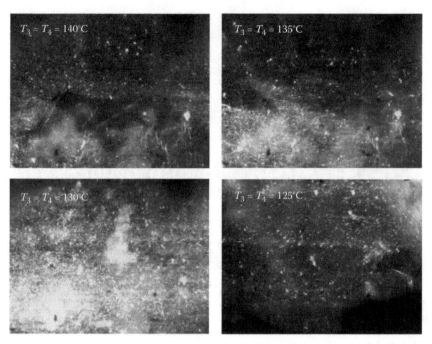

FIGURE 5.33
Images taken from a visualization chamber located prior to the extrusion die for various temperature profiles in the cooling extruder. White heterogeneities on the images were associated with PLA crystals. (From Tabatabaei, A., *Visualization of the Crystallization in Foam Extrusion Process*. M. Applied Sci. Thesis, University of Toronto, Canada, 2012, p. 118.)

5.6.1.3 Foaming Chain-Branched PLA

One of the main drawbacks of PLA is its lack of melt strength and the absence of strain hardening, features that are known to be highly desirable in foaming semicrystalline polymers. Foaming polyester resin such as PET was only possible through the chain branching of its initial linear structure [81], this chemical modification being performed at an early stage during the extrusion foaming process using what is generally referred to as a chain extender (CE). However, similarities between PLA and PET cannot be extended to the full process since their processing temperature windows are very dissimilar due to the very high melting point of PET, which, in return, requires a PFA with adequate properties, such as vapor pressure and solubility, in this high temperature range (close to 300°C).

As for the CE, epoxy-based oligomers are preferred as they react preferentially with the carbonyl end groups of the PLA chain, resulting in a branched structure. This approach was investigated by several research groups in the early 2010. Experiments were performed using a commercial masterbatch from Clariant, CESA Extend OMAN 698493, which is a blend of PLA and 30 wt% of a multifunctional epoxide manufactured by BASF, a styrenic-glycidyl acrylate copolymer commercialized under the trade name of Joncryl® 4368.

Mihai et al. [39] tested this branching modification on both amorphous and semicrystalline PLA with the concentration of the CE varying between 0 and 2 wt%. Surprisingly, for the amorphous PLA, the presence of a CE had no impact on the resulting foam characteristics, namely, density and morphology. A simple explanation was obtained through viscosity measurements that are performed on the foams that showed that chain branching was hindered by the low-temperature profile that is chosen for the extrusion of the amorphous grade, which limited the reactivity of the CE (see Figure 5.34).

This was not the case for the semicrystalline grade as it required a higher mechanical energy input, limiting the range of accessible low temperatures. Increasing the CE content systematically raised the achievable lowest temperature, due to its higher viscosity. In all cases, this limiting temperature was associated with the premature crystallite formation. Such early crystallization increased the viscosity, as shown in Figure 5.13, which then generated more shear heating, maintaining the temperature to relatively high levels, i.e., in the 130°C–160°C range.

While the concentration of carbon dioxide clearly had a positive impact on the cell nucleation density (Figure 5.29), the presence of CE at any concentration investigated did not affect the cell nucleation and had little if no impact on the foam density. A comparison between amorphous and semicrystalline grades also confirmed the significant role that is played by the premature crystallites on the cell nucleation during foaming, which led to lower foam densities.

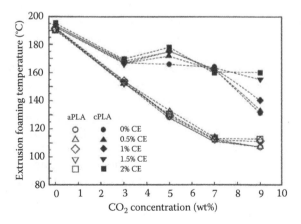

FIGURE 5.34
Minimal extrusion temperature as a function of CO_2 content during the extrusion foaming of various branched PLA. (aPLA is the amorphous grade 8302D, and cPLA is the semicrystalline 4032D.) (From Mihai, M. et al., *Polym. Eng. Sci.*, 50: 629–642, 2010.)

Nevertheless, despite the fact that the CE modified the rheology of the PLA in the expected way with respect to elasticity and strain hardening, as verified through rheological experiments (Figure 5.18), these modifications did not translate into any significant improvement in the properties of the PLA that is foamed with carbon dioxide.

5.6.1.4 Foaming PLA/Clay Nanocomposite

The use of nanoparticles in polymer foaming is often associated with enhanced cell nucleation, with the objective of achieving a microcellular and even a nanocellular structure claimed to present several advantages when compared to conventional polymer foam morphology. Polymer nanocomposite based on nanoclay has been a very popular field of investigation and a few studies on the batch foaming of PLA/nanoclay systems have been published, as reviewed by Nofar and Park [45]. The use of nanoclay in PLA extrusion foaming has also been attempted by Keshtkar et al. [73] using Ingeo 2002D.

A major impact due to nanoclays was obtained when these nanoparticles were at least intercalated or, even better, when they were fully exfoliated. A few percent in the composite was sufficient to tremendously modify both rheological behavior and crystallization kinetics. Those changes increased cell nucleation, and thus the overall morphology of the cellular structure, with cell diameters that were less than 25 μm.

The foaming experiments performed by Keshtkar et al. used a commercial nanoclay, Cloisite 30B (Southern Clay Products), that displayed better foaming performance than another commercial nanoclay, Cloisite 20A. The two types of nanoclays differ by the chemical nature of the surfactant that is utilized, as well as the mean interlayer spacing, 1.85 nm for 30B and 2.42 nm for 20A. Foaming, expressed in terms of expansion ratio, increased rapidly with very small amounts of nanoclay, such as a low content of 1 wt%, as illustrated in Figure 5.35a. Expansion even increased two-fold with 5 wt% of Cloisite 30B. The lowest foam density achieved with 9 wt% of CO_2 was approximately 28 kg/m^3, which is, however, in the same range as many other results that were obtained using different systems and reviewed previously. As expected, an additional impact of nanoclay was the tremendous increase of the cell nucleation density by almost two decades (Figure 5.29). However, it was not clear from this study if the increase of the heterogeneous nucleation sites was directly related to the population of the nanoparticles, usually claimed to be a very efficient nucleating agent. As mentioned in the study, the presence of nanoclays might affect the solubility of CO_2 [82]. The adsorption of the macromolecules of PLA to the surface of the clay platelets created a rigid-polymer volume in which the gas molecules could not dissolve, a similar situation to that of a crystalline phase. Thus, for a given CO_2 input in the extrusion trials that would correspond to a mean gas concentration, the effective concentration of CO_2 in the amorphous phase would increase

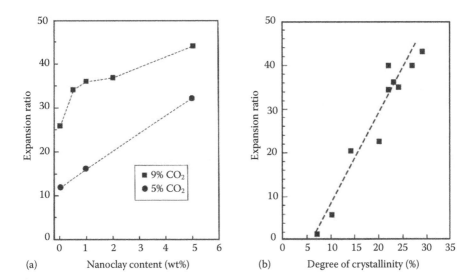

FIGURE 5.35
Expansion ratio as a function of (a) nanoclay concentration and (b) the degree of crystallinity for PLA nanocomposite foams. (From Keshtkar, M. et al., *Polymer*, 55: 4077–4090, 2014.)

proportionally to the nanoclay content. As seen in Section 5.6.1.1.2, a higher carbon dioxide concentration leads to higher cell nucleation.

Moreover, the nanoclays may have promoted in the first place an increase of the degree of crystallinity, which then enhanced both cell nucleation and expansion (Figure 5.35b). Tatibouët et al. [83] studied the effects of the combination of CO_2 and nanoclay on the crystallization of PLA. Nanoclay was found as a very effective crystal nucleating agent. When combined with CO_2, even if each contribution of both CO_2 and nanoclays did not sum up, the kinetics of crystallization was still enhanced.

5.6.1.5 Mechanical Properties

Gibson and Ashby [84] have highlighted in their extensive textbook on the structure–property relationship of cellular solids that foam mechanical properties can be related to density through power–law relationships. Typically, reduced mechanical properties of the foam with respect to its unfoamed polymer counterpart should follow a single master curve proportional to the reduced density. The exponent of the power–law dependency could vary according to the property that is tested, a value of 2 being considered for compression modulus and 1.5 for compression yield [85].

This immediately implies that polyesters such as PET and PLA, due to their relatively high density in the solid state ($\rho = 1.38$ g/cm^3 for PET and 1.24 g/cm^3 for PLA), will always be less efficient than PS ($\rho = 1.05$ g/cm^3) for a given foam density. This is illustrated in Figure 5.36 in which a comparison

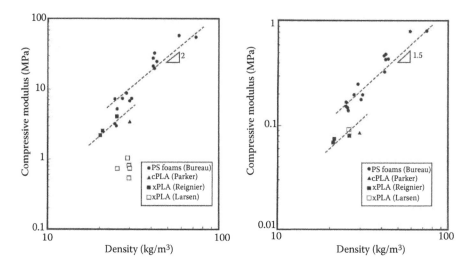

FIGURE 5.36
Comparison between PS and PLA low-density foams for the compressive modulus and the compressive strength. (From Reignier, J. et al., *Cell. Polym.*, 26(2): 83–115, 2007; Larsen, A. and Neldin, C., *Polym. Eng. Sci.*, 53(5): 941–949, 2013; Parker, K. et al., *J. Cell. Plas.*, 47(3): 233–243, 2011; Bureau, M.N. and Gendron, R., *J. Cell. Plast.*, 39(5): 353–367, 2003.)

is being made between the few published results on the compression testing of PLA foams [43,70,86] and PS foam data [87]. The differences observed between the PS and PLA groups of data are directly associated to their resin density, since the mechanical properties in compression for the two different resins are roughly similar.

5.6.1.6 Foaming PLA with CO_2: Conclusive Remarks

Extrusion foaming using carbon dioxide has always been a considerable challenge usually due to its low solubility in most of the commodity polymers. Although PLA exhibits enhanced CO_2 solubility attributable to its carbonyl groups, adequate foaming to low density can only be accomplished at CO_2 concentration exceeding 5 wt% in the best cases. This results in very high pressures that are required to be maintained in the extrusion equipment to prevent premature phase separation.

Crystallization occurring prior the die exit has appeared to be beneficial to the foaming process; it provides enhanced elasticity to the molten PLA and increases the local CO_2 concentration in the amorphous phase. Most of the experiments have reported that the higher the CO_2 content is, the better the foam will be with respect to a low-density target and a fine morphology. Crystallization kinetics can be improved through the proper choice of the PLA grade (low viscosity and low D-LA content), high CO_2 concentration, and, obviously, the use of a nucleating agent (talc or nanoclays).

For the foams to be dimensionally stable, a high open-cell content is a prerequisite. Surprisingly, the semicrystalline foams obtained have shown microporosity in their cell walls, which contributes to a fast equilibrium between the cells and atmosphere through convection.

5.6.2 Large-Scale Manufacturing

Despite the fact that the first PLA foam trays introduced in 2005 were produced with carbon dioxide, the industrial practice relied more on the use of hydrocarbons for low-density foam sheet production, such as isobutane [75], a mixture of isopentane and isobutane [76], a mixture of n-pentane and isobutane with an 80:20 wt% ratio [69], and isobutane and n-butane with a 20:80 ratio [74]. NatureWorks also reported some results using HFC-152a [77]. Obviously, these blowing agent formulations are those that are known to have been successful for PS foaming, the two polymers being processed in a very similar temperature window. To some extent, one may expect that blowing agents suitable for PS might be good candidates for PLA, including the fluorocarbon chemical families that are used for extruded PS foam insulation boards. However, lack of solubility data and absence of reported processing trials may restrain the validity of such assertion.

Nevertheless, for foam sheet manufacturing, it still looks like some inherent characteristics of PLA and related processing issues have made difficult the full implementation of the conversion from PS to PLA. Literature dealing with industrial-scale practice, with throughput over 300 kg/h, mentioned issues that are not necessarily grasped at the laboratory scale such as high shear sensitivity and high heat retention, thus limiting the efficiency of an extrusion process based on PLA rather than PS. Even if some commercial solutions are available and were implemented, foaming PLA has not met yet the economic conditions to be fully competitive with its styrenic predecessor.

5.6.2.1 NatureWorks Formulation

NatureWorks commercializes a specific PLA grade, Ingeo 8052D, that is dedicated to extrusion foaming for low-density foam sheet production (see Table 5.3 [88] for resin properties). It is a semicrystalline PLA that has a D-LA content of 4.7%. The processing guide supplied by NatureWorks specifies that this grade must be modified with the branching agent Joncryl 4368C (BASF) or its masterbatch CESA Extend OMAN 698493 (Clariant). This masterbatch should be precisely added at a rate of 2.3 wt% to provide an efficient branched structure to the PLA. Most of the conventional foaming agents traditionally used for manufacturing foamed PS are also claimed to be compatible for foaming the Ingeo 8052D, such as butane, pentane, and HFC-152a [89]. Hydrocarbons are nonpolar and are less soluble in PLA than CO_2 and polar molecules such as HFC-152a.

TABLE 5.3

Ingeo 8052D Foam Grade

Physical Properties	Ingeo Resin	ASTM Method
Specific gravity	1.24	D792
Melt index, g/10 min (210°C, 2.16 kg)	14	D1238
Relative viscosity	3.3	D5225
Crystalline melt temperature (°C)	145–160	D3418
Glass transition temperature (°C)	55–60	D3418
Clarity	Transparent	
Mechanical Properties		
Tensile yield strength, MPa	62	D638
Tensile elongation, %	3.5	D638
Notched Izod impact, J/m	16.0	D256
Flexural strength, MPa	108	D790
Flexural modulus, GPa	3.6	D790

Source: NatureWorks LLC. Ingeo™ Biopolymer 8052D Technical Data Sheet,
NW8052D_120213V1.

Since hydrocarbons do not possess the self-nucleating effect of CO_2 when used at high dosages, it is recommended to work with a typical concentration of talc of 1.25–2.5 wt% in order to get cell sizes in the 250–500 μm range.

5.6.2.2 Extruder: Tandem Line

Although a conventional tandem foam extrusion line can be used, several important modifications should be implemented to take into account the specific properties of the PLA, especially its rheology, which differs significantly from PS. For instance, the higher pressures encountered in the primary extruder due to the viscosity of PLA that has a lower tendency to shear thin compared to PS may necessitate such modifications as to increase the gas pressure at the injection zone, in order to deliver an adequate flow rate of the foaming agent. The branching reaction should also be completed in the primary extruder, with adequate residence time and processing temperatures that are maintained in the 200°C–225°C range prior to the crossover piping leading to the cooling extruder.

The cooling capabilities of the second extruder are nevertheless the critical parameter that requires greater modifications. In order to solve this issue, increasing the mandrel cooling capacity is not sufficient, and the high temperatures resulting from the shear heating due to the screw rotation should be addressed through a novel screw design.

Corre [74] investigated the use of an industrial direct gassing tandem extrusion line for the production of PLA foamed sheet using 5 wt% of a mixture of isobutane and n-butane (20:80) along with the Ingeo 3051D PLA grade. The extrusion trials were more specifically focused on the impact of the concentration of the CE [74]. The optimal CE content was found near 1.3 wt% for a foam density of approximately 50 kg/m³ and foamed sheet thicknesses in the range of 1.5–3.5 mm.

Turn-on key processing lines for PLA foam sheet are available on the market. Such equipment is manufactured, for instance, by Macro Engineering and Technology Inc. [90] and can produce a PLA foam sheet with a density of 70 kg/m³ at a throughput rate of 300 kg/h (see Figure 5.37). Processing is done according to the NatureWorks formulation (Ingeo 8052D and CESA Extend) using isobutane as the foaming agent. In a recent published study, the characteristics of the extruded PLA foam were reported as a foam density of 78 kg/m³, a cell density of 4.5×10^5 cells/cm³, and a crystallinity of 27% [75]. With the heat treatment occurring during the thermoforming stage,

FIGURE 5.37
PLA foam sheet extrusion line of Macro Engineering and Technology Inc. (From Macro Engineering and Technology Inc. PLA Foam Sheet Extrusion Line Specification Sheet.)

the crystallinity of the PLA foam sheet increased to approximately 46%. Interestingly, this suggests that postprocessing can be envisaged to further boost the crystallinity of the extruded foam, i.e., maintain a viable crystalline content at the extrusion step and raise the crystallinity up to its ultimate value through subsequent operations such as thermoforming.

5.6.2.3 Screw Design

An energy balance conducted over the second extruder from a tandem extrusion line implies that the rate of cooling provided by the barrels should largely exceed the shear heating coming from the viscous melt. With limited sources of cooling, it became imperative to reduce the shear heating through an innovative screw design.

Plastic Engineering Associates Licensing Inc. has designed the Turbo-Screw for increasing the efficiency of extrusion foaming through the addition of circulation channels [91,92]. Later, its design was modified specifically for PLA foaming [76]. The novel screw design relied on gentle mixing and cross channel flow that is achieved by numerous shaped openings, as can be seen in Figure 5.38.

Trials performed on amorphous and semicrystalline PLAs validated the efficiency of this innovative screw design, which enabled access to commercially viable extrusion rates, i.e., close to 400 kg/h. The foam manufactured from the amorphous PLA grade could match low-heat applications under 50°C, but the semicrystalline grade was said to enable the higher temperature range of 50°C–125°C [93]. This technology has been licensed to a few

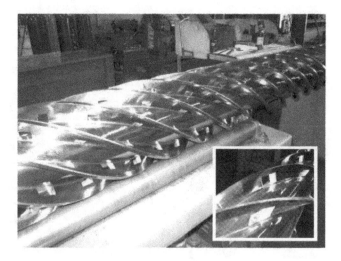

FIGURE 5.38
Turbo-Cool screw designed for PLA extrusion foaming, showing details on screw flight venting. (From Fogarty, D. and Nangeroni, J., Processing NatureWorks foams using new screw technology, *Polymer Foam* 2007, Newark, NJ, October 2–3, 2007.)

companies in North America, such as Sealed Air Corp and Dyna-A-Pak [94], and in Europe [95,96].

5.6.2.4 Industrial Foaming of PLA: Conclusive Remarks

Considerable progress has been made over the past decade to improve the economics of the foam sheet extrusion of PLA. Retrofitting the existing processing equipment to take into account some particularities of the new resin has been inevitable, and reaching the viability of this system necessitated new tooling. The driving force to switch from PS to PLA foam for food packaging applications relies essentially on environmental concerns, as PS foam remains cheaper on one hand and provides better performances on the other hand. Nevertheless, the PLA-based technology exists and is commercially available, but it is still mainly devoted to the niche markets originating from the ban of PS foam in specifically localized areas.

In order to access to a more competing level with its styrenic counterpart, PLA still requires technological advancements to be achieved, with goals that are set for increasing the efficiency of the foam extrusion process and improving the thermal resistance of the final foam. As manufacturers might be reluctant to implement costly major modifications to their existing equipment, material formulation is still the best route to tackle down the remaining problematic issues.

5.7 Conclusions

Extrusion foaming of PLA is only one facet of the roles that PLA can play in the diversified market of thermoplastic foams. As for more traditional polymers, blending may result in new materials with combined or unexpected properties. Foaming PLA-based blends has been the subject of several investigations [45,97], most of them targeted on fully bio-based blends such as PLA/starch [98], PLA/thermoplastic starch [67], and PLA/polybutylene succinate [99]. With similar objectives, foam composites made of PLA and biofibers have also been envisioned [100,101].

However, the biased perception that PLA-based products are essentially envisaged for composting has restrained at some point its development into long-term applications. Such opportunities should be rationally considered and foam-related companies such as JSP has foreseen automotive applications for the expandable PLA beads that they have developed [102]. Other long-term applications usually associated to PS can also be considered in various domains such as sports and leisure (Figure 5.39), as well as in the construction area, e.g., insulation panels. Although composting is obviously not an immediate target for this last application, sustainable development

FIGURE 5.39
Tecniq, a company based in California, has commercialized surfboards made with Synbra's PLA Biofoam.

now requires that materials and products have a low environmental impact in terms of the carbon footprint and global warming potential.

A life-cycle assessment conducted on various thermal insulating products has identified PLA as an interesting alternative to fossil resource-dependent resins [103]. Clearly, with the forthcoming increase of the PLA resin production capacity, PLA will find new niches in unexpected domains.

References

1. Shah Mohammadi, M., Bureau, M.N. and Nazhat, S.N. 2014. Ch. 11—Polylactic acid (PLA) biomedical foams for tissue engineering. In *Biomedical Foams for Tissue Engineering Applications*, Ed. P. Netti, Woodhead Publishing Limited, Elsevier Science Technology, Cambridge, United Kingdom, pp. 313–334.
2. Gruber, P. and O'Brien, M. 2002. Polylactides 'NatureWorks™ PLA. In *Polyesters III Applications and Commercial Products*, Vol. 4, Ed. Doi, Y. and Steinbuchel, A., Wiley-VCH, Weinheim, Germany, pp. 235–249.
3. Vink, E.T.H., Davies, S. and Kolstad, J.J. 2010. The eco-profile for current Ingeo polylactide production. *Ind. Biotechnol.*, 6(4): 212–224.
4. Auras, R., Harte, B. and Selke, S. 2004. An overview of polylactides as packaging materials. *Macromol. Biosci.*, 4: 835–864.
5. Franklin Associates. 2011. *Life Cycle Inventory of Foam Polystyrene, Paper-Based, and PLA Foodservice Products*. Study commissioned by the Plastics Foodservice Packaging Group of the American Chemistry Council, February, p. 149.
6. Vannini, C., Fiodelisi, F., Movilli, W. and Lanzani, F. 2005. Polylactic acid-based degradable foams and process for their production. WO 2005/042627, Appl. Coopbox Europe S.p.A.

7. Schut, J.H. 2007. Extrusion close-up: Foamed PLA shows promise in biodegradable meat trays. *Plastics Technology,* December.
8. Thielen, M. 2011. Foam trays in Seattle. *Bioplastics Magazine,* 6(1): 46.
9. Gruber, P.R., Hall, E.S., Kolstad, J.J., Iwen, M.L., Benson, R.D. and Borchardt, R.L. 1992. Continuous process for manufacture of lactide polymers with controlled optical purity. US Patent 5,142,023, ass. Cargill.
10. Jamshidian, M., Tehrany, E.A., Imran, M., Jacquot, M. and Desobry, S. 2010. Poly-lactic acid: Production, applications, nanocomposites, and release studies. *Comprehensive Rev. Food Sci. Food Safety,* 9: 552–571.
11. de Jong, E., Higson, A., Walsh, P. and Wellish, M. 2012. Product developments in the bio-based chemicals arena. *Biofuels, Bioprod. Bioref.* 6: 606–624.
12. Tan, K.L., Lim, S.K., Chan, J.H. and Low, C.Y. 2014. Overview of poly(lactic acid) production with oil palm biomass as potential feedstock. *Int. J. Eng. Appl. Sci.,* 5(4): 1–10.
13. Saeidlou, S., Huneault, M., Li, H. and Park, C.B. 2012. Poly(lactic acid) crystallization. *Prog. Polym. Sci.,* 37: 1657–1677.
14. Bigg, D.M. 2005. Polylactide copolymers: Effect of copolymer ratio and end capping on their properties. *Adv. Polym. Tech.,* 24: 69–82.
15. Kolstad, J.J. 1996. Crystallization kinetics of poly(L-lactide-co-meso-lactide). *J. Appl. Polym. Sci.,* 62: 1079–1091.
16. Witzke, D.R. 1997. *Introduction to Properties Engineering and Prospects of Polylactide Polymers,* PhD dissertation. Michigan State University, Lansing, Michigan.
17. Hartmann, M. 1998. High molecular weight polylactic acid polymers. In *Biopolymers from Renewable Resources,* Ed. D. Kaplan, Springer-Verlag, Berlin/ Heidelberg, pp. 367–411.
18. Ou, X. and Cakmak, M. 2008. Influence of biaxial stretching mode on the crystalline texture in polylactic acid films. *Polymer,* 49(24): 5344–5352.
19. Mihai, M., Huneault, M.A. and Favis, B.D. 2009. Crystallinity development in cellular poly(lactic acid) in the presence of supercritical carbon dioxide. *J. Appl. Polym. Sci.,* 113: 2920–2932.
20. Liao, X., Nawaby, A.V. and Whitfield, P.S. 2010. Carbon dioxide-induced crystallization in poly(L-lactic acid) and its effect on foam morphologies. *Polym. Int.,* 59: 1709–1718.
21. Sato, Y., Yamane, M., Sorakubo, A., Takishima, S., Masuoka, H., Yamamoto, H. and Takasugi, M. 2000. Solubility and diffusion coefficient of carbon dioxide in polylactide. *Proceedings of the 21st Japan Symposium on Thermophysical Properties,* Nagoya, Japan, pp. 196–198.
22. Oliveira, N.S., Dorgan, J., Coutinho, J.A.P., Ferreira, A., Daridon, J.L. and Marrucho, I.M. 2006. Gas solubility of carbon dioxide in poly(lactic acid) at high pressures. *J. Polym. Sci. B: Polym. Phys.,* 44: 1010–1019.
23. Li, G., Li, H., Turng, L.S., Gong, S. and Zhang, C. 2006. Measurement of gas solubility and diffusivity in polylactide. *Fluid Phase Equilib.,* 246: 158–166.
24. Aionicesei, E., Skerget, M. and Knez, Z. 2008. Measurement of CO_2 solubility and diffusivity in poly(l-lactide) and poly(d,l-lactide-co-glycolide) by magnetic suspension balance. *J. Supercrit. Fluids,* 47: 296–301.
25. Mahmood, S.H., Keshtkar, M. and Park, C.B. 2013. Determination of carbon dioxide solubility in polylactide acid with accurate PVT properties. *J. Chem. Thermodyn.,* 70: 13–23.

26. Sato, Y., Sakai, H., Wang, M., Takishima, S. and Masuoka, H. 2004. Solubility of butane and isobutane in molten poly(lactide) and poly(butylene succinate). *PPS 2004*, Akron, OH, June 20–24, Paper 114.

27. Liao, X., Nawaby, V., Whitfield, P., Day, M., Champagne, M. and Denault, J. 2006. Layered open pore poly(l-lactic acid) nanomorphology. *Biomacromolecules*, 7: 2937–2941.

28. Sato, Y., Takishima, S. and Masuoka, H. 2003. Solubility of carbon dioxide in polymers. *Proceedings of Polymer-Supercritical Fluid Systems and Foams* (P-(SF)²), Tokyo, Japan, December 4–5, pp. 139–140.

29. Sato, Y., Takikawa, T., Takishima, S. and Masuoka, H. 2001. Solubilities and diffusion coefficients of carbon dioxide in poly(vinyl acetate) and polystyrene. *J. Supercrit. Fluids*, 19: 187–198.

30. Shen, Z., McHugh, M.A., Xu, J., Belardi, J., Kilic, S., Mesiano, A., Bane, S., Karnikas, C., Beckman, E. and Enick, R. 2003. CO_2-solubility of oligomers and polymers that contain the carbonyl group. *Polymer*, 44: 1491–1498.

31. Kazarian, S.G., Vincent, M.F., Bright, F.V., Liotta, C.L. and Eckert, C.A. 1996. Specific intermolecular interaction of carbon dioxide with polymers. *J. Am. Chem. Soc.*, 118: 1729–1736.

32. Shieh, Y.-T. and Liu, K.-H. 2003. The effect of carbonyl group on sorption of CO_2 in glassy polymers. *J. Supercrit. Fluids*, 25: 261–268.

33. Liao, X. and Nawaby, A.V. 2012. The sorption behaviors in PLLA-CO_2 system and its effect on foam morphology. *J. Polym. Res.*, 19: 9827–9835.

34. Nawaby, V.A. and Zhang, Z. 2005. Ch. 1—Solubility and diffusivity. In *Thermoplastic Foam Processing: Principles and Development*, Ed. R. Gendron, CRC Press, Boca Raton, FL, pp. 1–42.

35. Bao, L., Dorgan, J.R., Knauss, D., Hait, S., Oliveiora, N.S. and Marruccho, I.M. 2006. Gas permeation properties of poly(lactic acid) revisited. *J. Membr. Sci.*, 285: 166–172.

36. Yuryev, Y. and Wood-Adams, P. 2010. Rheological properties of crystallizing polylactide: Detection of induction time and modeling the evolving structure and properties. *J. Polym. Sci. Part B. Polym. Phys.*, 48: 812–822.

37. Randall, J.R., Cink, K. and Smith, J.C. 2009. Branching polylactide by reacting OH or COOH polylactide with epoxide acrylate (co)polymer. US Patent 7,566,753, ass. NatureWorks.

38. Corre, Y.-M., Duchet, J., Reignier, J. and Maazouz, A. 2011. Melt strengthening of poly(lactic acid) through reactive extrusion with epoxy-functionalized chains. *Rheol. Acta*, 50(7): 613–629.

39. Mihai, M., Huneault, M.A. and Favis, B.D. 2010. Rheology and extrusion foaming of chain-branched poly(lactic acid). *Polym. Eng. Sci.*, 50: 629–642.

40. Lehermeier, H.J. and Dorgan, J.R. 2001. Melt rheology of poly(lactic acid): Consequences of blending chain architectures. *Polym. Eng. Sci.*, 41(12): 2172–2184.

41. Utracki, L.A. and Schlund, B. 1987. Linear low density polyethylenes and their blends: Part 2. Shear flow of LLDPE's. *Polym. Eng. Sci.*, 27(5): 367–379.

42. Corre, Y.-M., Maazouz, A., Reignier, J. and Duchet, J. 2014. Influence of the chain extension on the crystallization behavior of polylactide. *Polym. Eng. Sci.*, 54(3): 616–625.

43. Reignier, J., Gendron, R. and Champagne, M.F. 2007. Extrusion foaming of poly(lactic acid) blown with CO_2: Toward 100% green material. *Cell. Polym.*, 26(2): 83–115.

44. Chow, T.S. 1980. Molecular interpretation of the glass transition temperature of polymer-diluent systems. *Macromolecules*, 13(2): 362–364.
45. Nofar, M. and Park, C.B. 2014. Poly(lactic acid) foaming. *Prog. Polym. Sci.*, 39: 1721–1741.
46. Reignier, J. 2015. Personal communication.
47. Gendron, R. and Champagne, M.F. 2004. Effect of physical foaming agents on the viscosity of various polyolefin resins. *J. Cell. Plast.*, 40(2): 131–143.
48. Gendron, R., Champagne, M.F., Delaviz, Y. and Polaski, M.E. 2006. Foaming polystyrene with a mixture of CO_2 and ethanol. *J. Cell. Plast.*, 42(2): 127–138.
49. Beckman, E. and Porter, R.S. 1987. Crystallization of bisphenol A polycarbonate induced by supercritical carbon dioxide. *J. Polym. Sci. Part B Polym. Phys.*, 25(7): 1511–1517.
50. Takada, M., Hasegawa, S. and Ohshima, M. 2004. Crystallization kinetics of poly(L-lactide) in contact with pressurized CO_2. *Polym. Eng. Sci.*, 44(1): 186–196.
51. Reignier, J., Tatibouet, J. and Gendron R. 2009. Effect of dissolved carbon dioxide on the glass transition and crystallization of poly(lactic acid) as probed by ultrasonic measurements. *J. Appl. Polym. Sci.* 112: 1345–1355.
52. Nofar, M., Ameli, A. and Park, C.B. 2014. The thermal behavior of polylactide with different D-lactide content in the presence of dissolved CO_2. *Macromol. Mater. Eng.*, 299(10): 1232–1239.
53. Yu, L., Liu, H. and Dean, K. 2009. Thermal behaviour of poly(lactic acid) in contact with compressed carbon dioxide. *Polym. Int.*, 58: 368–372.
54. Li, D., Liu, T., Zhao, L., Lian, X. and Yuan, W. 2011. Foaming of poly(lactic acid) based on its non isothermal crystallization behavior under compressed carbon dioxide. *Ind. Eng. Chem. Res.*, 50: 1997–2007.
55. Yu, L., Liu, H., Dean, K. and Chen, L. 2008. Cold crystallization and post-melting crystallization of PLA plasticized by compressed carbon dioxide. *J. Polym. Sci. B: Polym. Phys.*, 46: 2630–2636.
56. Nofar, M., Zhu, W. and Park, C.B. 2012. Effect of dissolved CO_2 on the crystallization behavior of linear and branched PLA. *Polymer*, 53: 3341–3353.
57. Nofar, M., Tabatabaei, A., Ameli, A. and Park, C.B. 2013. Comparison of melting and crystallization behaviors of polylactide under high-pressure CO_2, N_2, and He. *Polymer*, 54: 6471–6478.
58. Hirai, T., Nishijima, K. and Ochiai, T. 2013. Polylactic acid-based resin foamed particles for in-mold foam molding and method for producing the same, as well as method for producing polylactic acid-based resin foam-molded article. US Patent 8,372,512, ass. Sekisui Plastics Co. Ltd.
59. Füssi, A., Sampath, B., Hofmann, M., Bellin, I., Nalawade, S., Hahn, K., Künkel, A. and Loos, R. 2013. Method for producing expandable granulates containing polylactic acid. US Patent Appl. 2013/0150468, ass. Badische Anilin- & Sodafabrik BASF SE.
60. Juergen Witt, M.R. and Shah, S. 2012. Methods of manufacture of polylactic acid foams. US Patent 8,283,389, Appl. Biopolymer Network Limited.
61. Pawloski, A., Cernohous, J., Kaske, K. and Van Gorden, G. 2013. Compostable or biobased foams, method of manufacture and use. WO 2013/006781, Appl. Lifoam Industries LLC.
62. Schut, J.H. 2008. PLA biopolymers: New copolymers, expandable beads, engineering alloys and more. *Plastics Technology*, pp. 68–69, November.
63. 2009. Expanded PLA as a particle foam, *Bioplastics Magazine*, 4: 22–23.

64. Noordegraaf, J., de Jong, J., Matthijssen, de Bruijn, P., Baltissen, T. and Molenveld, K. 2013. Bioplastic substrates in horticulture and agriculture. *Bioplastics Magazine*, 8: 34–36.
65. Cink, K. Smith, J.C. Nangeroni, J.F. and Randall, J.R. 2011. Extruded polylactide foams blown with carbon dioxide. US Patent 8,013,031, September 6, ass. NatureWorks LLC.
66. Lee, S.T., Leonard, L. and Jun, J. 2008. Study of thermoplastic PLA foam extrusion. *J Cell. Plast.*, 44(4): 293–305.
67. Mihai, M., Huneault, M.A., Favis, B.D. and Li, H. 2007. Extrusion foaming of semi-crystalline PLA and PLA/thermoplastic starch blends. *Macromol. Biosci.*, 7: 907–920.
68. Pilla, S., Kim, S.G., Auer, G.K., Gong, S. and Park, C.B. 2009. Microcellular extrusion-foaming of polylactide with chain-extender. *Polym. Eng. Sci.*, 49: 1653–1660.
69. Ramesh, N.S. Nawaby, A.V. and Amrutiya, N. 2011. Polylactic acid foam composition. US Patent Appl. 2011/0263732, October 27, ass. Sealed Air Corp.
70. Larsen, A. and Neldin, C. 2013. Physical extruder foaming of poly(lactic acid)— Processing and foam properties. *Polym. Eng. Sci.*, 53(5): 941–949.
71. Wang, J., Zhu, W., Zhang, H. and Park, C.B. 2012. Continuous processing of low-density, microcellular poly(lactic acid) foams with controlled cell morphology and crystallinity. *Chem. Eng. Sci.*, 75: 390–399.
72. Matuana, L. and Diaz, C.A. 2010. Study of cell nucleation in microcellular poly(lactic acid) foamed with supercritical CO2 through a continuous-extrusion process. *Ind. Eng. Chem. Res.*, 49: 2186–2193.
73. Keshtkar, M., Nofar, M., Park, C.B. and Carreau, P.J. 2014. Extruded PLA/clay nanocomposite foams blown with supercritical CO_2. *Polymer*, 55: 4077–4090.
74. Corre, Y.-M. 2010. *Poly(Lactic Acid) Foaming Assisted by Supercritical CO_2*. PhD Thesis, Institut national des sciences appliquées de Lyon, p. 218.
75. Lee, E.K., Guo, Y., Tamber, H., Planeta, M. and Leung, S.N. 2015. Extrusion and thermoforming of polylactic acid foam sheets for hot-fill packaging applications. *SPE Foams 2015 Conference*, Kyoto, Japan, September 10–11.
76. Fogarty, D. and Nangeroni, J. 2007. Processing NatureWorks foams using new screw technology. *Polymer Foam 2007*, Newark, NJ, October 2–3.
77. Nangeroni, J. and Randall, J.R. 2014. Extruded foams made with polylactides that have high molecular weights and high intrinsic viscosities. US Patent 8,722,754, May 13, ass. NatureWorks LLC.
78. Yang, T.C., Lee, K.L. and Lee, S.T. 2002. Dimensional stability of LDPE foams: Modeling and experiments. *J. Cell. Plast.*, 38(2): 113–128.
79. Tabatabaei, A., Barzegari, M.R., Keshtkar, M., Nofar, M. and Park, C.B. 2012. Visualization of PLA crystallization in extrusion process and its effect on the foaming behavior of PLA. *Proceedings PPS*, May 24, pp. 337–338.
80. Tabatabaei, A. 2012. *Visualization of the Crystallization in Foam Extrusion Process*. M. Applied. Sci. Thesis, University of Toronto, Toronto, ON, Canada, p. 118.
81. Coccorullo, I., Di Maio, L., Montesano, S. and Incarnato, L. 2009. Theoretical and experimental study of foaming process with chain extended recycled PET. *eXPRESS Polym. Letters*, 3(2): 84–96.
82. Hassan, M.M. 2013. *A Systematic Study of Solubility of Physical Blowing Agents and Their Blends in Polymers and Their Nanocomposites*. PhD Thesis, University of Toronto, Toronto, ON, Canada.

83. Tatibouët, J., Champagne, M.F., Gendron, R. and Ton-That, T.M. 2008. Effect of dissolved carbon dioxide and organoclay on the crystallization of poly(lactic acid) as probed by ultrasonic measurements. *Proceedings Polym. Proc. Soc., 24th Annual Meeting PPS-24*, Salerno, Italy, June 15–19.

84. Gibson, L.J. and Ashby, M.F. 1997. *Cellular Solids Structure and Properties,* 2nd ed. Cambridge University Press, New York, p. 510.

85. Throne, J.L. 2004. *Thermoplastic Foam Extrusion: An Introduction.* Carl Hanser Verlag, Germany, p. 86.

86. Parker, K., Garancher, J.-P., Shah, S. and Fernyhough, A. 2011. Expanded poly-lactic acid—An eco-friendly alternative to polystyrene foam. *J. Cell. Plas.,* 47(3): 233–243.

87. Bureau, M.N. and Gendron, R. 2003. Mechanical-morphology relationship of PS foams. *J. Cell. Plast.,* 39(5): 353–367.

88. NatureWorks LLC. Ingeo™ Biopolymer 8052D Technical Data Sheet, NW8052D_120213V1.

89. NatureWorks LLC. Ingeo™ Foam Sheet Extrusion Processing Guide, NWPG003_020111.

90. Macro Engineering and Technology Inc. PLA Foam Sheet Extrusion Line Specification Sheet.

91. Fogarty, J. 1998. Thermoplastic foam extrusion screw with circulation channels. US Patent 6,015,227, May 26.

92. Fogarty, J., Fogarty, D., Rauwendaal, C. and Rios, A. 2001. Turbo-screw, new screw design for foam extrusion. *SPE ANTEC Tech. Papers,* pp. 167–172; Fogarty, J., Fogarty, D., Rauwendaal, C. and Rios, A. 2002. Turbo-screw, new screw design for foam extrusion. *Proceedings of Blowing Agents and Foaming Processes 2002*, Paper 19, Heidelberg, Germany, May 27–28, pp. 205–212.

93. Admin. 2006. Screw foams biodegradable polymer. *Plastics Today*, October 31. http://www.plasticstoday.com/screw-foams-biodegradablepolymer/9560978028532.

94. Fogarty, D.J. 2010. Bio-plastics—A sea change for the polystyrene foam pack-aging industry. *Florida Atlantic University (FAU) Executive Forum Lecture Series*, February 4.

95. Admin. 2008. Screw puts new spin on foam PLA. *Plastics Today*, January 31.

96. 2009. Significant extrusion throughput rate increase for PLA foam. *Bioplastics Magazine*, 1, p. 25.

97. Iannace, S., Sorrentino, L. and Di Maio, E. 2015. Ch. 16—Extrusion foaming of biodegradable polymers. In *Foam Extrusion: Principles and Practice*, 2nd ed., Ed. S.T. Lee, CRC Press, Boca Raton, FL, pp. 527–570.

98. Zhang, J.-F. and Sun, X. 2007. Biodegradable foams of poly(lactic acid). I. Extrusion condition and cellular size distribution. *J. Appl. Polym. Sci.,* 106(2): 857–862.

99. Geissler, B., Feuchter, M., Laske, S., Walluch, M., Holzer, C. and Langecker, G.R. 2014. Tailor-made high density PLA foam sheets—Strategies to improve the mechanical properties. *Cell. Polym.,* 33(5): 249–257.

100. Orhan, M. 2010. *Polylactide Foams Reinforced with Wood Fibers or Microfibrillated Cellulose.* Masters Thesis, Uppsala University, Uppsala, Sweden.

101. Diaz-Acosta, C.A. 2011. *Continuous Microcellular Foaming of Polylactic Acid/Natural Fiber Composites.* PhD Thesis, Michigan State University, East Lansing, MI.

102. 2011. The door of the future. *JSP Annual Report*, p. 10.

103. Noordegraaf, J., Matthijssen, P., de Jong, J. and de Loose, P.A. 2011. Comparative LCA of building insulation products. *Bioplastics Magazine*, 6: 30–33.

6

Innovative PLA Bead Foam Technology

Mohammadreza Nofar, Alireza Tabatabaei, and Chul B. Park

CONTENTS

6.1 Introduction and Background

Among plastic foaming technologies, extrusion foaming has generally been known to obtain low-density foam products with simple geometry. On the other hand, foam injection molding has been implemented to manufacture high-density foam products with three-dimensional geometries. Bead foaming is another means of foam manufacturing to produce low-density foam products with complex three-dimensional geometries. In this method, low-density bead foams are molded into the desired final foam product shape [1].

Producing expanded polystyrene (EPS) bead foams has been of one of the earliest bead foams that were developed for bead foam molding to produce cost-effective, three-dimensional low-density foams. This conventional method includes the saturation of PS particles at temperatures below the T_g of PS with a blowing agent, followed by further expansion in a pre-expander machine, and then the molding of the foamed beads (i.e., particles) at high temperature into the desired shape [2]. On the other hand, bead foam technology with a double crystal-melting peak structure has been well established for polyolefins. The double crystal melting peak structure, which is required in the molding stage of the bead foams, generates a strong sintering among the foamed beads and maintains the overall foam structure. In this context, the sintering technique used in expanded polypropylene (EPP) bead foaming [3] can potentially be a promising solution to the sintering problem in several innovative plastics such as biopolymers, specifically polylactide (PLA) bead foaming technology [4–6]. In this technique, the high-temperature melting-peak crystals formed during the isothermal saturation step in a batch-based bead foaming process are utilized to maintain the bead geometry even at a high temperature that is required for good sintering. The formation of this crystal-melting peak at a higher temperature than the original crystals is due to the crystal perfection that is induced during the gas-saturation stage that is conducted nearly around the original melting temperature. After gas saturation, the low-temperature melting peak forms during cooling as foaming occurs. Figure 6.1 shows a schematic of the batch-based EPP bead foaming saturation process together with the generated EPP bead foams with double crystal melting peak structure [7].

In the steam chest molding process (i.e., in the final processing step to induce the sintering of the beads and thereby to produce three-dimensional foam products), the EPP beads are supplied into the cavity of the mold, and then a steam at a high temperature between the low-temperature melting peak and the newly created high-temperature melting peak is supplied to the mold cavity to heat the beads. Figures 6.2 and 6.3 show a steam chest molding machine and a schematic of the EPP manufacturing steps from pellet status to bead-foam manufacturing stage and molding stage to produce three-dimensional EPP bead foam products [7]. When exposed to this high-temperature steam, the low-temperature melting-peak crystals of the

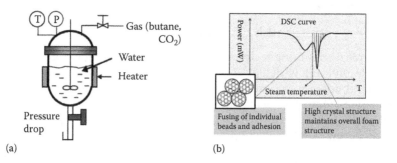

(a)   (b)

FIGURE 6.1

(a) Batch-based EPP bead foaming saturation process and (b) the generated EPP bead foams with double crystal melting peak structure. (From Lee, E.K., *Novel Manufacturing Processes for Polymer Bead Foams*, PhD thesis, University of Toronto, 2010.)

FIGURE 6.2
A steam chest molding machine to produce EPP bead foam products with three-dimensional shape. (From Lee, E.K., *Novel Manufacturing Processes for Polymer Bead Foams*, PhD thesis, University of Toronto, 2010.)

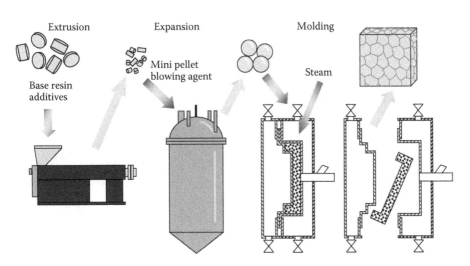

FIGURE 6.3
A schematic of three stages used in manufacturing EPP products: (1) PP micropellet preparation, (2) EPP bead foaming, and (3) steam chest molding. (From Lee, E.K., *Novel Manufacturing Processes for Polymer Bead Foams*, PhD thesis, University of Toronto, 2010.)

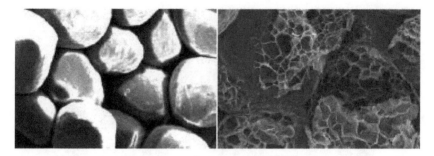

FIGURE 6.4
Scanning electron microscope micrographs of the fracture surfaces of molded EPP samples produced at different steam pressures (i.e., temperatures). (From Zhai, W.T. et al., *Ind. Eng. Chem. Res.* 50, 5523–5531, 2011.)

beads will melt and will thereby cause a sintering of the beads to each other, whereas the unmolten high-temperature melting-peak crystals in the beads will maintain the overall bead geometry.

Since the low-temperature melting-peak crystals will melt, and molecular diffusion will occur between the beads while maintaining the overall geometry of beads, the foam products will have a good geometry after the mold cavity shape with an outstanding sintering of the beads that are formed by molecular diffusion between the beads. This high-quality sintering of the beads will provide the outstanding mechanical properties of the products. Figure 6.4 shows the fracture surfaces of the EPP bead foam products with weak and strong sintering qualities. As seen in the figure, the EPP bead foam products with strong bead-to-bead sintering showed more intrabead fracture than interbead fracture. In other words, the bead foams tended to fail within the bead foam structure rather that the beads' interface [8].

6.2 PLA Molded Beads

Poly (lactic acid) or polylactide is a thermoplastic aliphatic polyester polymer that is derived from renewable resources such as cornstarch and sugarcane. Unlike petroleum-based polymers, it is environmentally friendly and biodegradable [9–12]. Recently, it has received increased attention as a potential substitute for polystyrene (PS) products [13–16]. For example, EPS bead foams are now being widely used in applications such as packaging, cushioning, construction, and thermal and sound insulation. But their weak bead-to-bead sintering property and their nonbiodegradable nature mean that broken EPS beads threaten the environment. In this context, PLA bead foams can provide a suitable and green substitute for these EPS products [1]. As well, the use of green physical blowing agents, such as gaseous or supercritical CO_2/N_2, has been well established in the manufacture of foam products [1]. Therefore, the concerns about ozone depletion, global warming potential, blowing agent toxicity, and flammability, which accompany traditional hydrocarbon and hydrochlorofluorocarbon blowing agents, are eliminated [1,17]. For example, PLA can be saturated with supercritical CO_2 or N_2 under a high pressure at a given temperature. Then, a rapid pressure drop can be induced to generate a thermodynamic instability, which creates a foam structure within the PLA matrix in various foaming processes [1,18,19].

Despite the noted advantages that PLA possesses, it also has drawbacks such as low melt strength and slow crystallization kinetics that suppress the foaming ability of PLA [1,20,21]. Several attempts have been made to improve PLA's poor foaming behavior through continuous processes [1,22–32]. A few studies have shown that the use of a chain extender [23,24,29] or inducing nano-/microsized additives [22,26–28,30–32] can enhance PLA's melt strength and its crystallization kinetics and thereby its foaming behavior. Some of these studies have also demonstrated that the crystallization that can occur during the foam processing can significantly promote the expanding ability and cell morphology of the PLA foams [24–28]. These attempts have been made mainly through extrusion foaming [22–27] or foam injection molding [28–32]. However, these methods are somehow limited to either achieve low-density PLA foam products with simple geometries or to manufacture high-density foam products with three-dimensional complex geometries, respectively.

Bead foaming has been introduced as a continuous foam manufacturing technology to produce low-density foam products with three-dimensional complex geometries [33–45]. In this technology, numerous low-density foam beads are molded into the desired final product shape. Although producing expanded polylactide (EPLA) bead foams has not been well established yet, several attempts have been made to manufacture EPLA bead foams similar to the manufacturing method that is used for EPS [40–45]. This method includes the gas impregnation of PLA particles at temperatures below PLA's T_g, followed by expansion in a pre-expander machine, and then molding of the foamed beads at a high temperature into the desired shape.

Although few companies currently produce EPLA bead foams, the bead sintering needed to manufacture three-dimensional final foam products still remains a serious challenge [40–45]. Currently, EPLA bead foams are costly to manufacture, and the bead-to-bead sintering is weak, as in the case of the EPS. Therefore, expensive and nonbiodegradable adhesives or coatings are being used for bead-to-bead bonding. Despite the promise of EPLA bead foams, their weak bead-to-bead sintering, the high cost of the adhesive/coating and their environmental drawbacks have become a serious impediment to their promotion within the industry and commercial usage [43–45]. In this work, we have developed a new technology that removes the environmental challenges and the expensive practices that are associated with bead sintering. We have created a double crystal melting peak in the EPLA bead foams by using low-melting-temperature peak crystals that melt during the sintering stage (i.e., molding). This causes sintering of the beads, whereas the crystals with a high melting temperature maintain the beads' integrity. This concept has been used for the existing EPP bead foaming technology [33–36] but has not been used for other materials, especially for EPLA bead foams. Specifically, crystals with a high melting temperature form during the isothermal saturation step in the batch-based bead foaming process [33–35], through crystal perfection [3,38]. In the steam-chest molding process (that is, the process used to produce three-dimensional foam products), sintering among the beads occurs when the steam temperature is between the high and low melting peaks. Consequently, the low-melting-temperature crystals will melt and contribute to good bead sintering, whereas the high-melting-temperature crystals will stay unmelted and will maintain the overall cell structure and the foam bead shape [33,35]. Although this double crystal melting peak technology cannot be applied in amorphous foams such as EPS, PLA has the potential of having this structure, despite its very slow crystallization kinetics. To create two-peak crystal characteristics in PLA foams, the fabrication steps will be different from those that are associated with existing PLA foaming technologies [40–45].

Improving PLA crystallization kinetics will not only provide the potential of generating a double crystal melting peak structure but will also enhance the poor foaming feature of PLA products significantly. Specifically, crystals formed during the saturation step (that is, crystals with a high melting temperature) can significantly affect the cell nucleation and expansion of the PLA bead foams. According to heterogeneous cell nucleation theory, cell nucleation can be promoted through the local stress variations [46,47] around the formed crystals [1,47–50]. On the other hand, the molecules connected through the crystals improve the PLA's low melt strength and consequently increase the PLA's ability to expand by minimizing the gas loss and cell coalescence [26,27]. It must also be noted that too high a crystallinity will also depress the foam's expansion ratio (ER) due to the increased stiffness [1].

6.3 Bead Foaming Mechanism of PLA

In this section, the development of EPLA bead foams with a double crystal melting peak structure will be discussed. A branched PLA with reasonably high crystallization kinetics [51,52] was selected. First, the generated double peaks in PLA using a regular and a high-pressure differential scanning calorimeter (DSC and HP-DSC) were characterized. Subsequently, EPLA bead foams were produced with double crystal melting peaks. Then, the EPLA bead foams were characterized, and the dependency of bead foam properties on the high-melting-temperature crystals that were generated during CO_2 saturation and the low-melting-temperature crystals that were formed during cooling while foaming occurred were investigated. Moreover, the DSC was used to analyze the bead foams' crystallization behavior. The effects of various CO_2 pressures on the crystallization kinetics of the crystals with a high melting temperature and thereby on the resultant PLA beads' foam properties were also investigated. In addition, this manufacturing method was introduced as a novel way to produce nanocellular foams.

6.3.1 Procedure to Form Double-Peak Crystals in PLA

Figure 6.5a shows a schematic of the double crystal melting peak generation steps during a bead foaming process, that is, the evolution of the crystals with a high melting temperature during the isothermal saturation step and, subsequently, the evolution of the crystals with a low melting temperature during cooling/foaming. Figure 6.5b shows the procedure of the simulation experiment within the DSC and the actual bead foaming process: (1) first, the generation of crystals with a high melting temperature via isothermal annealing/saturation, (2) second, the generation of crystals with a low melting temperature via cooling/foaming, and (3) third, the characterization of the peaks that are generated in the sample. It should be noted that the gas pressure in the first step was high in order to simulate the gas-saturation process, and the pressure was decreased in the second step to simulate the foaming process. The third step was to investigate the double-peak crystals that developed in the PLA bead foams. This action is described in detail in the following section.

6.3.2 Characterization of Generated Double-Peak Crystals in PLA through DSC and HP-DSC

The evolution of the double crystal melting peak in the PLA samples was investigated by varying the annealing temperature (T_a) and the time (t_a) using a DSC (DSC2000, TA Instruments) in a nitrogen atmosphere. The PLA samples were heated at a rate of 30°C/min to various T_a (around the PLA's T_m) [3] and were isothermally annealed at various t_a. The samples were then cooled at a rate of 20°C/min to room temperature, during which the low-melting-temperature crystals formed. The double crystal melting peak behavior was subsequently analyzed by reheating the samples to 200°C at a rate of 5°C/min. In addition, the plasticizing effect of CO_2 (at 6 MPa) [3,53] on the generated double crystal melting peak was analyzed using an HP-DSC (NETZSCH DSC 204 HP). The annealing temperature (T_a) and the annealing time (t_a) terminologies were used exclusively for dealing with a non-high-pressure gas using a regular DSC. Otherwise, the saturation temperature (T_s) and the saturation time (t_s) were used to deal with a high-pressure gas, when the HP-DSC or foaming apparatus were used.

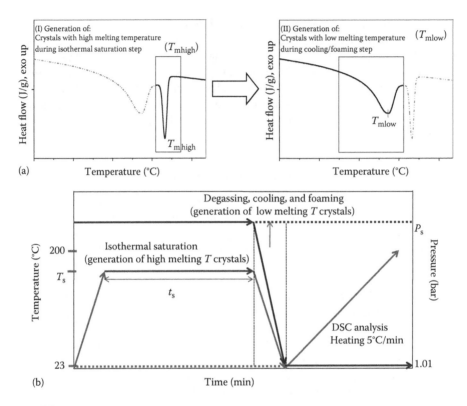

FIGURE 6.5
Schematics of (a) forming the double-peak crystals (i.e., the crystals with a high melting temperature and crystals with a low melting temperature) and (b) the actual experimental procedure used to form these double-peak crystals within the DSC.

Figure 6.6 shows the heating thermograms of the PLA samples that are achieved after isothermal annealing for 60 min at various temperatures. Table 6.1 also reports the generated high and low melting peaks (T_{mhigh} and T_{mlow}), as well as the total crystallinity, the crystallinity of the low melting temperature peak, and the crystallinity of the high melting temperature peak of the samples after annealing.

At a T_a of 140°C and 145°C, most of the original PLA samples' crystals remained unmelted during annealing, as the T_a was still below the PLA's melting temperature (T_m). However, after annealing, the melting peak of these crystals started to shift to higher temperatures (above T_m) due to the crystal perfection that occurred during annealing [3,38].

In the PLA samples that were annealed above 145°C, a portion of the original crystals started to melt, and some remaining crystals became subject to crystal perfection during annealing. Consequently, a smaller area (as the crystals with a high melting temperature), compared with the earlier lower T_a case, was generated. However, the generated high melting temperature

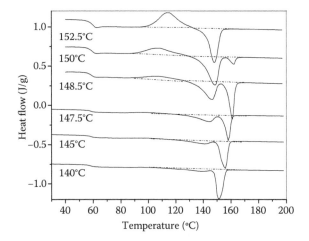

FIGURE 6.6

The evolution of double crystal melting peaks with respect to varying annealing temperatures.

TABLE 6.1

Melting Temperatures and Crystallinities of PLA Samples Annealed at Various Temperatures

T_a	T_{mlow}	T_{mhigh}	Total Crystallinity (%)	Crystallinity of the Low Melting Temperature Peak (%) (Low T_m)	Crystallinity of the High Melting Temperature Peak (%) (High T_m)
140.0°C	138.8 ± 0.5	151.2 ± 0.4	28.8 ± 0.4	2.3 ± 0.6	26.5 ± 0.4
145.0°C	141.6 ± 0.1	156.2 ± 0.5	29.2 ± 0.5	4.4 ± 0.3	24.8 ± 0.6
147.5°C	144.6 ± 0.1	158.2 ± 0.1	29.4 ± 0.3	8.2 ± 0.3	21.2 ± 0.3
148.5°C	146.5 ± 0.4	160.9 ± 0.4	30.1 ± 0.4	12.0 ± 0.4	18.1 ± 0.4
150.0°C	148.5 ± 0.2	162.2 ± 0.3	15.3 ± 0.5	10.7 ± 0.5	4.6 ± 0.4

peak appeared at higher temperatures. This was due to the increased molecular mobility and the easier molecular retraction at a higher T_a that formed more closely packed crystal structures (i.e., better perfection) [3]. By contrast, the samples annealed at higher temperatures revealed a larger peak at lower temperatures as the low-melting-temperature crystals. This was due to the increased amount of melt that is available for crystallization during both cooling and the cold crystallization.

Figure 6.7 shows the crystallization behavior of the annealed PLA samples at 148.5°C while the t_a was varied. When the samples were cooled, without permitting any annealing (t_a = 0 min), the unmelted crystals with a more

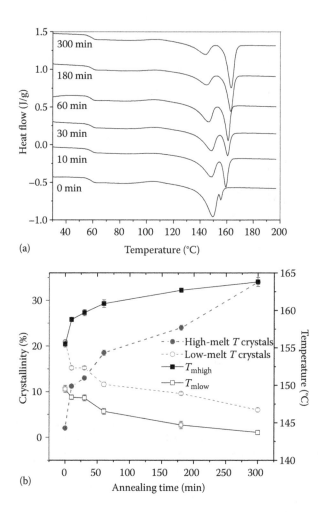

FIGURE 6.7
(a) Double crystal melting peak evolution with respect to various annealing times and (b) the generated lower and higher crystal melting peaks and the crystallinity of the low and high melting temperature peaks of the annealed PLA samples at a fixed annealing temperature of 148.5°C.

closely packed structure appeared as a small peak at a slightly higher temperature (155.5°C). The major content of the crystals formed during cooling and appeared as the original melting peak.

The samples that were annealed longer (10–300 min) had a larger amount of perfect crystals (i.e., crystals with a high melting temperature) due to the longer diffusion time that is allowed for molecular rearrangement. Consequently, the total crystallinity of the samples was also enhanced. Moreover, the crystals' degree of perfection was further increased, and the high melting temperature peak (T_{mhigh}) was formed at higher temperatures when annealed for a longer time. This can provide a wider processing window during the steam chest molding of bead foams. We should also note that after a longer annealing time, the PLA samples revealed a smaller low melting temperature peak area due to the less amount of available melt that could crystallize during the cooling process.

Figure 6.8 shows the generated double crystal melting peak in the PLA samples under 6 MPa CO_2 pressure while the saturation temperature (T_s) was varied with a fixed annealing time of 60 min. Compared to the atmospheric pressure (Figure 6.6), the double-peak structure was generated at a lower range of T_s when a CO_2 pressure was applied. This was due to the plasticization effect of the dissolved CO_2, which suppressed the required T_s range by almost 8°C–10°C, which is consistent with our earlier study [3]. We also demonstrated that the PLA's melting temperature was depressed almost below its original T_m under the same CO_2 pressure [53].

6.3.3 Preparation and Characterization of Foam Samples

The PLA bead foams were prepared by saturating the pellets with gaseous and supercritical CO_2 in a small autoclave foaming chamber. After the samples were loaded and the chamber was vacuumed to remove moisture, it was pressurized with CO_2 using a Teledyne Isco high-pressure syringe pump. The samples were then heated to various annealing temperatures and saturated for 60 min. After the saturation with CO_2, the pressure was rapidly released by opening a ball valve. The chamber was then cooled in a water

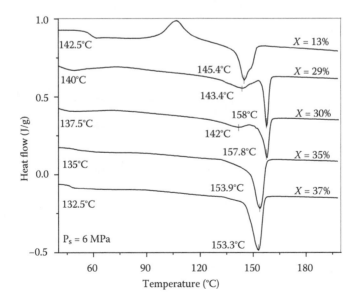

FIGURE 6.8
Double crystal melting peak investigated under 6 MPa CO_2 pressure.

bath, and the PLA bead foams were collected. Figure 6.9 shows a schematic of the autoclave bead foaming chamber that we used.

To achieve various bead foam cellular morphologies and ERs, three different CO_2 saturation pressures were applied: (1) 3 MPa, (2) 6 MPa (gaseous CO_2), and (3) 17.2 MPa (supercritical CO_2). The required range of annealing temperatures needed to generate a double crystal melting peak at each saturation pressure was found by extrapolating the DSC and the HP-DSC results [3] and through the foaming experiments.

The cell morphology and the ER of the bead foams were both analyzed using scanning electron microscopy (SEM) micrographs and by measuring the foam density with the Archimedes method, respectively. The ER of the samples was evaluated using the following equation:

$$ER = \frac{V_{foam}}{V_{polymer}} \sim \frac{\rho_{polymer}}{\rho_{foam}} \tag{6.1}$$

where V_{foam} and $V_{polymer}$ are the measured volumes, and ρ_{foam} and $\rho_{polymer}$ are the calculated density of the foam and unfoamed PLA, respectively.

The crystallization behavior of the bead foams was also characterized using the DSC by heating the foamed samples to 200°C at a rate of 5°C/min. The PLA foam samples' crystallinity was calculated using the following equation:

$$\chi = \frac{\Delta H_m - \Delta H_{cc}}{93.6} \times 100\% \tag{6.2}$$

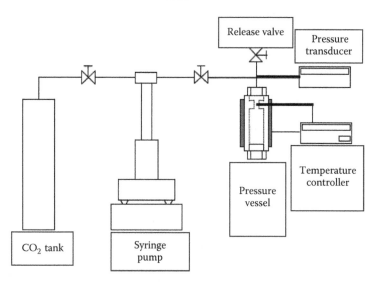

FIGURE 6.9
A schematic of the autoclave bead foaming set up.

where ΔH_m is the melting enthalpy, ΔH_{cc} is the cold crystallization enthalpy, and 93.6 is the melting enthalpy in J/g of 100% crystalline PLA [54].

In this section, the PLA bead foams, which were obtained at CO_2 saturation pressures of 3, 6, and 17.2 MPa, were characterized. At these saturation pressures, the saturation temperature ranges where the double crystal melting peak was generated were found to be between 145°C and 155°C, 135°C and 145°C, and 110°C and 120°C, respectively. The depression rate of the required T_s range versus the increased pressure was almost consistent with the HP-DSC results. The bead foam results showed that the PLA samples saturated at 3 MPa CO_2 pressure were not able to foam. In Sections 6.3.3.1 and 6.3.3.2, we investigated the crystallization behavior of the PLA bead foams, as well as the unfoamed samples that are saturated at 3 MPa. The properties of the bead foams were subsequently analyzed.

6.3.3.1 Crystallization Behavior of the PLA Beads

Figure 6.10 shows the DSC heating thermograms of the PLA beads that are recorded at a heating rate of 5°C/min. Figure 6.11 also shows the generated high melting temperature peak (T_{mhigh}), the total crystallinity, and the high melting temperature peak crystallinity.

As Figure 6.10 shows, the double crystal melting peak was developed in the PLA beads within the selected saturation temperature ranges at the given saturation pressures. In all three of the applied pressures, after saturation at low saturation temperatures, the majority of the foams' total crystallinity was formed as crystals with a high melting temperature. However, in the samples prepared at higher saturation temperatures, the amount of crystals with a high melting temperature was reduced. Thus, the total crystallinity decreased and was mainly governed by the crystallinity of the low melting temperature peak (formed during cooling and foaming). Despite the reduced amount of crystals with a high melting temperature, this high melting temperature peak (i.e., T_{mhigh}) appeared at higher temperatures. As discussed in Section 6.3.2, this was due to the increased mobility of the PLA molecules and better crystal perfection when higher saturation temperatures were applied.

Further, the crystals with a high melting temperature generated in the PLA beads were influenced by the saturation pressure. When a low saturation pressure was applied, the generated high melting temperature peak appeared at higher temperatures with narrower peaks. However, as the saturation pressure increased, the generated high melting temperature peaks became wider and appeared at relatively lower temperatures. This could indicate that at a low saturation pressure, larger crystals and/or crystals with much more perfection were formed in the PLA beads. Moreover, the obtained beads had a higher melting temperature crystallinity portion and a higher total crystallinity. In contrast, with increased saturation pressure (that is, increased dissolved CO_2 content), the generated crystals with a high

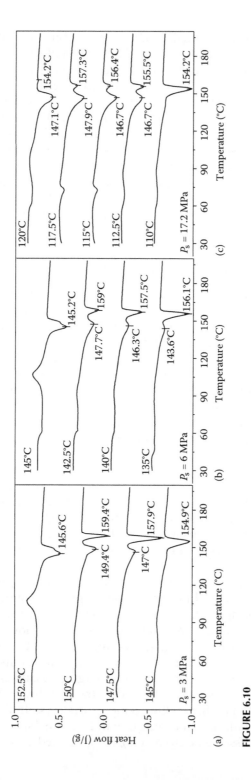

FIGURE 6.10

DSC heating thermograms of the PLA bead foam samples achieved under saturation pressures of (a) 3 MPa, (b) 6 MPa, and (c) 17.2 MPa.

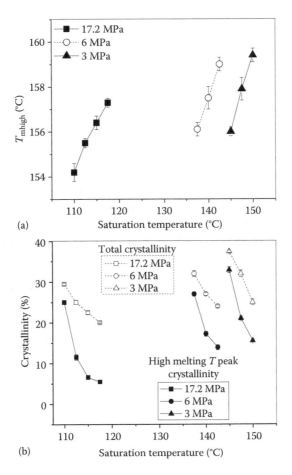

FIGURE 6.11
(a) The high melting temperature peak (T_{mhigh}) and (b) total crystallinity and high melting temperature peak crystallinity of the PLA bead samples after different saturation pressures and temperatures.

melting temperature might have been formed in smaller sizes, and possibly with less perfection, than those that are created during saturation at lower pressures. In our earlier studies [52,53,55,56], we demonstrated that at low CO_2 pressures (that is, low dissolved CO_2 content), larger perfect crystals were more likely to form, which also increases the final crystallinity. On the other hand, in PLA, the increased CO_2 pressure accelerated the crystal nucleation rate. Therefore, the large number of nucleated crystals increased the PLA molecular entanglement and thereby reduced the crystals' perfection and hindered their growth. It seems that the effects of the dissolved CO_2 on the crystallization kinetics of the generated crystals with a high melting temperature during saturation show a trend similar to what we found in our earlier investigations [52,53,55,56].

6.3.3.2 Characterization of PLA Bead Foams

Figures 6.12 and 6.13, respectively, show the ER and the cell morphology of the PLA bead foams that are obtained after applying various saturation pressures and temperatures. With an increased saturation temperature, bead foams with ERs from 3- to 25-fold and from 3- to 30-fold were obtained after CO_2 saturation of 6 MPa and 17.2 MPa, respectively. The corresponding average cell sizes were also observed between 500 nm and 500 µm and 700 nm and 15 µm, respectively. As seen in the figures, nanocellular bead foams with ERs around threefold could also be reached at the low applied saturation temperatures. This was most likely due to the enhanced heterogeneous cell nucleation rate that occurred around the greater amount of the perfected crystals (that is, the crystals with a high melting temperature) that are formed during the CO_2 saturation process. As Figure 6.11b shows, after the saturation of 6 and 17.2 MPa, and at the lowest corresponding saturation temperatures, 25%–27% of the high-melting-temperature crystallinity was induced during CO_2 saturation. These large amounts of induced crystals promoted cell nucleation toward a nanocell structure formation, although the ER was limited to threefold. Therefore, regardless of the bead foaming purposes with a double crystal melting peak, this manufacturing route is also offered as a novel way to produce nanocellular foams for applications such as superinsulation materials [57].

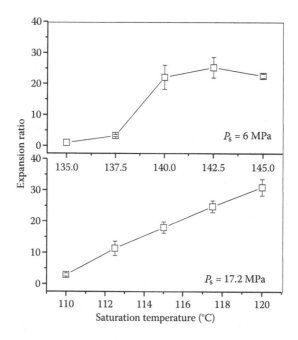

FIGURE 6.12
ER of the PLA bead foams after saturation at various pressures and temperatures.

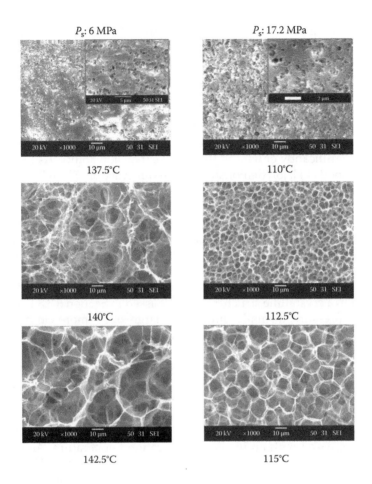

FIGURE 6.13
SEM images of PLA bead foams saturated for 60 min at CO_2 pressures of 6 and 17.2 MPa and at various saturation temperatures.

After saturation at CO_2 pressures of 6 and 17.2 MPa, as the saturation temperature increased, the ERs of the PLA bead foams were enhanced to 25- and 30-fold, respectively. This was due to the reduced stiffness of the PLA matrix, which facilitated cell growth. In other words, with an increased saturation temperature, the amount of crystals with a high melting temperature started to decrease from around 25%–27% to less than 15%. Due to the reduced high-melting-temperature crystallinity in the PLA matrix, cell growth was promoted during foaming, although the heterogeneous cell nucleation rate and, therefore, the cell density were reduced.

It should also be noted that due to the absolute absence of crystals with a high melting temperature at very high saturation temperatures, the PLA's melt strength will be significantly reduced, and this will expedite gas loss during foaming and will also suppress the ER [58,59]. This expansion–collapse trend at high saturation temperatures can be seen in the PLA foamed samples after saturation at 145°C under 6 MPa CO_2 pressure. As Figure 6.10 shows, during saturation at this given condition, the crystals with a high melting temperature did not form at all.

Moreover, Figure 6.12 shows that the ERs of the bead foams improved as the applied saturation pressure increased from 6 MPa to 17.2 MPa. The increased amount of dissolved CO_2 in the PLA at a saturation pressure of 17.2 MPa significantly enhanced the PLA's expansion ability. As Li et al. [60] and Mahmood et al. [61] demonstrated, the solubility of CO_2 in PLA at 180°C, when it is exposed to 6 MPa and 17.2 MPa CO_2 pressures is around 4.5% and 13%, respectively. It should also be mentioned that during the saturation process in bead foaming, the solubility of CO_2 in PLA would become even more complex when the saturation temperature changes [62] and while crystallization occurs [63]. As the temperature decreases, the CO_2 solubility in polymers typically increases. On the other hand, as crystallization occurs, the solubility of CO_2 in a polymer should decrease because it can hardly be dissolved in a crystalline structure. As discussed in Section 6.3.3.1, at a lower saturation pressure, a smaller number of large-sized, more closely packed crystals would be most likely to form during the saturation step, and the high melting temperature crystallinity would further increase. Therefore, due to the presence of larger crystalline domains and a higher melting temperature peak crystallinity, the CO_2 solubility in PLA during saturation at a lower saturation pressure would decrease much more than during saturation at a higher saturation pressure.

As Figure 6.13 shows, after saturation at 6 and 17.2 MPa, the obtained cell density decreased with an increased saturation temperature. According to DSC analyses, after saturation at 137.5°C (6 MPa) and 110°C (17.2 MPa), 27% and 25% of high melting temperature crystallinity were induced during saturation, respectively. The large amount of induced crystallinity significantly promoted the heterogeneous cell nucleation rate through the local stress variations [46,47] around the crystals [1,47–50]. However, cell growth was noticeably hindered due to the increased stiffness of the PLA matrix with developed crystals with a high melting temperature. Alternatively, when saturated at higher temperatures, the amount of high-melting-temperature crystals was reduced, and this facilitated cell growth. But the reduced high-melting-temperature crystals decreased the cell nucleation density because the heterogeneous cell nucleation sites were reduced.

Overall, the SEM images illustrate that at a saturation pressure of 17.2 MPa, very uniform cell morphologies with more closed-cell content were achieved with average cell sizes of 700 nm to 15 μm as the saturation temperature increased from 110°C to 117.5°C. However, at a saturation pressure of 6 MPa, nonuniform, more open-celled structures were obtained as the saturation temperature was varied from 137.5°C to 142.5°C. At 145°C, a uniform closed-cell structure with an average cell size of 500 μm was obtained. Figure 6.14

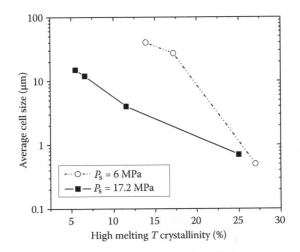

FIGURE 6.14
Cell size variations of PLA bead foams saturated at various pressures as a function of high-melting-temperature crystallinity.

summarizes how the cell size of the PLA bead foams was affected by the amount of crystals with a high melting temperature.

After saturation at a high CO_2 pressure of 17.2 MPa (T_s between 112.5°C and 117.5°C), the increased cell density and the more uniform closed-cell structure can be explained by two mechanisms. First, the increased content of dissolved CO_2 in PLA at a saturation pressure of 17.2 MPa must have promoted the cell nucleation rate through the increased degree of thermodynamic instability [64]. Second, during saturation at higher pressures, the crystals with a high melting temperature induced during saturation were most likely characterized by a larger number of crystal nuclei [52,53,55,56]. The large number of small-sized nucleated crystals could have significantly promoted the heterogeneous cell nucleation around the crystals. Moreover, the PLA molecules connected through these nucleated crystals must have improved the PLA's melt strength, as well as the cell morphology, with a more closed-cell content.

In contrast, at a lower saturation pressure of 6 MPa (T_s between 140°C and 145°C), the crystals with a high melting temperature were most likely induced with a limited number of crystals that were relatively larger in size [52,53,55,56]. Therefore, the number of heterogeneous cell nuclei around the crystals was reduced, and the large-sized closely packed crystals could have suppressed the uniformity of the cell morphology. Consequently, more open-cell content was observed. This can be explained through the hard–soft segment cell opening mechanism that is demonstrated by Lee et al. [65]. In this mechanism, the large soft matrix (i.e., amorphous structure) that exists between the hard domains (i.e., the large-sized closely packed crystals) can initiate cell opening and nonuniformity during cell growth. Figure 6.15 proposes a schematic of this hypothetical mechanism. At various saturation

(1) *Low saturation pressure (140°C)/the resulting foam morphology*

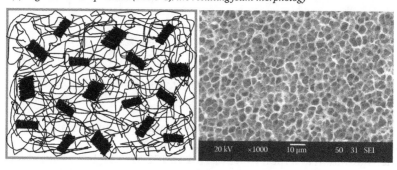

(2) *High saturation pressure (112.5°C)/the resulting foam morphology*

FIGURE 6.15
Schematic of the hypothetical mechanism showing the effects of the crystals with a high melting temperature with different crystal sizes on the cell morphology of the bead foams. The crystal sizes are affected at various CO_2 pressures during saturation. The crystals with a high melting temperature form at high and low CO_2 saturation pressures, respectively, with a large number of small-sized crystals and fewer large-sized crystals.

pressures, it explains how the different crystallization kinetics of crystals with a high melting temperature can affect cell morphology.

6.3.4 Effects of Low-Melting-Temperature Crystals

Figure 6.16 shows the first heating thermograms of the as-received PLA pellets before the experiments. As shown, the added talc and lubricant increased the PLA samples' crystallization rate by expediting the cold crystallization at lower temperatures [66]. All of the as-received samples were amorphous with no crystallinity.

The low-melting-temperature crystals that form during cooling also affect the cellular morphology. As Figure 6.16 shows, the addition of a small amount of talc (0.5 wt%) and lubricant expedited the PLA's crystallization rate. And this change affected the PLA's crystallization rate of the low melting temperature peak during cooling. This, in turn, affected the foam's properties and morphology. In this section, PLA, PLA-T, and PLA-TL bead foams

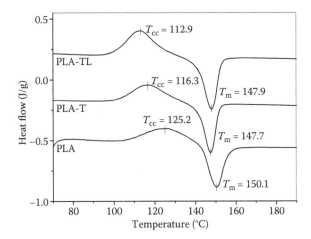

FIGURE 6.16
DSC thermograms of the PLA samples heated at a rate of 5°C/min.

are characterized. The saturation pressure was fixed at 17.2 MPa, and the saturation temperature ranged from 110°C to 120°C. The saturation time was 60 min.

Figure 6.17 shows the DSC heating thermograms of the PLA, PLA-T, and PLA-TL foamed beads. The samples' two-peak formations showed very similar trends with only slight variances. It seems that adding talc and lubricant to the PLA had no significant effect on the formation of the high-melting-temperature-peak crystals. However, due to the faster crystallization rate of PLA samples with talc and lubricant, the formation of the low melting temperature peak during cooling (while foaming occurred) should have been different.

Figure 6.18 shows the total crystallinity and the high-melting-temperature-peak crystallinity of the PLA bead foams, as well as their ER after saturation at various temperatures. As outlined in Section 6.1, the low melting temperature peak forms after the saturation step during the cooling process while foaming occurs. In the PLA samples, crystallization of the low melting temperature peak must have occurred faster during cooling in the following order: PLA-TL > PLA-T > PLA [66,67]. Faster crystallization during cooling can also inhibit foam expansion due to the increased stiffness of the PLA matrix while the samples expand [49]. As Figure 6.18b shows, all of the samples' ERs were enhanced as the saturation temperature increased. This was due to the reduced amount of crystals with a high melting temperature. However, the maximum ERs in the PLA, PLA-T, and PLA-TL were 30-, 20-, and 18-fold, respectively. Yet, despite the faster crystallization rate in the PLA-TL and PLA-T during cooling, the reported total crystallinity in all the samples seemed to be in the same order. This can be explained by the corresponding samples' different ERs. Although the

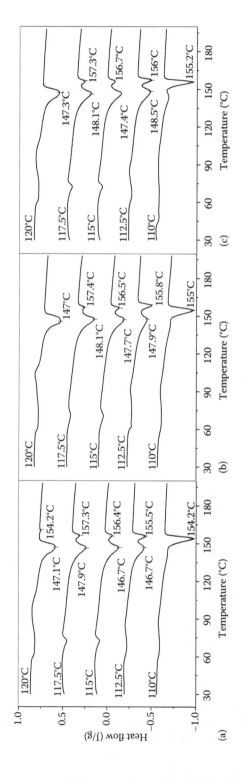

FIGURE 6.17
DSC heating thermograms of PLA bead foams with different crystallization kinetics: (a) PLA, (b) PLA-T, and (c) PLA-TL.

neat PLA's crystallization rate should be slower during cooling, its higher ER could also have contributed to the foam's total crystallinity through the strain-induced crystallization [25–27].

Figure 6.19 shows the cell morphology of the PLA, PLA-T, and PLA-TL bead foams that were saturated at temperatures of 110°C, 112.5°C, and 115°C. Figure 6.20 shows the cell size variations of the PLA bead foams as a function of the high-melting-temperature-peak crystallinity. In all the PLA samples at 110°C, a too-high crystallinity induced during saturation significantly increased the heterogeneous cell nucleation rate and suppressed cell growth. The amount of perfect crystals induced during saturation in the PLA, PLA-T, and PLA-TL bead foams ranged between 20% and 25%, and the average cell sizes were 700 nm, 400 nm, and 350 nm, respectively. The ERs of these samples were all around three-fold. Once again, this indicates that nanocellular

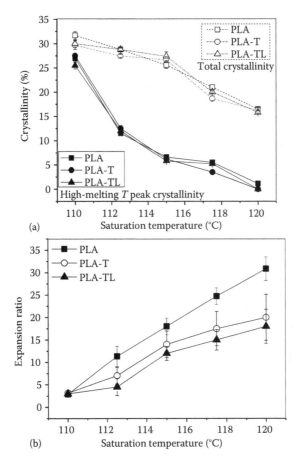

(a)

(b)

FIGURE 6.18
(a) Total crystallinity and high-melting-temperature-peak crystallinity and (b) ER of the foamed beads saturated at 17.2 MPa for 60 min.

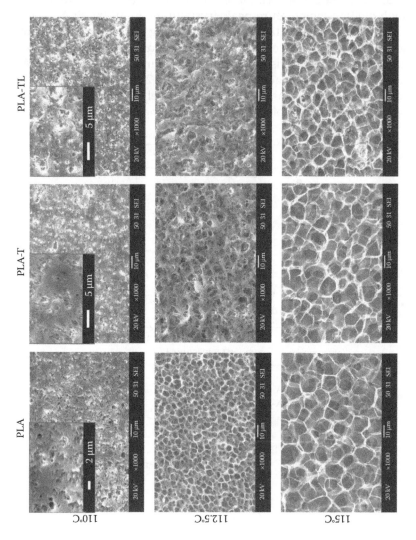

FIGURE 6.19
SEM images of the PLA bead foams saturated at 17.2 MPa CO_2 pressure and various saturation temperatures.

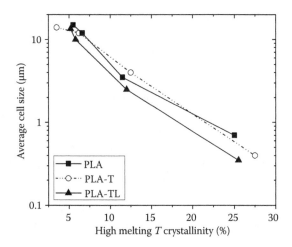

FIGURE 6.20
Cell size variation of the PLA bead foams saturated at 17.2 MPa as a function of a high-melting-temperature crystallinity.

PLA foams can be developed through this bead foaming method, while the cell size can be controlled by optimizing the material composition and the processing conditions.

At 112.5°C, it seems that cell growth in the PLA-T and PLA-TL did not occur as uniformly as it did in the neat PLA, and a network of solidified region can be seen. As discussed in this section, this was most likely due to the faster crystallization of these samples during cooling, which further increased the PLA bead foams' stiffness during cell growth and expansion.

Eventually, at 115°C, the cell's uniformity was improved in all of the PLA samples. This was because at higher saturation temperatures, the reduced amount of crystals with a high melting temperature decreased the stiffness of the PLA matrix, and even faster crystallization in the PLA-T and PLA-TL during cooling could not restrict cell growth and expansion. It should also be noted that the presence of talc does not only help the enhancement of crystallization kinetics, but it also helps the improvement of the cell nucleation during foaming [1].

6.4 Producing Lab-Scale PLA Bead Foams

Section 6.3 demonstrated the mechanism of PLA bead foaming with the generation of a double crystal melting peak structure using a small batch-foaming chamber with the capability of producing a single bead foam. It was demonstrated how the saturation temperature and saturation pressure influence the double crystal melting peak generation and thereby the PLA foaming behavior. It was shown that various CO_2 pressures significantly influence the crystallization kinetics of the crystals with a high melting temperature, which are generated during the isothermal saturation. At high saturation pressures, a larger number of small-sized perfect crystals are generated [52,53,55,56] and significantly promote the heterogeneous cell nucleation around themselves [48–50]. Moreover, the network of these generated perfect crystals improves the cell growth with more uniform morphology. In contrast, at lower saturation pressures, large-sized perfect crystals with fewer numbers are formed [52,53,55,56], and the heterogeneous cell nucleation around them reduces respectively, and the uniformity of PLA foam morphology suppresses.

In this section, using a laboratory-scale autoclave system, a practical way to manufacture a large quantity of microcellular PLA foamed beads with double crystal melting peak structure, followed by using a steam chest molding to further demonstrate the sintering behavior of the manufactured EPLA bead foams, will be demonstrated. The effect of saturation time and temperature on the double crystal melting peak generation and on the subsequent foaming behavior is investigated. The dependency of the closed-cell content of the EPLA bead foams on the processing parameters (i.e., the double crystal melting peak structure with different peak ratios) is also examined under various conditions. Due to the manufacture of a large quantity of EPLA bead foams, water is used (1) as a mixing media during the isothermal saturation and (2) in order to uniformly distribute the heat in the autoclave chamber, and thereby the approaches on how the PLA hydrolysis can be minimized during the bead foaming process are also evaluated. The mechanical properties (i.e., tensile test) of the steam chest-molded EPLA bead foam samples are further studied and are compared with those of EPP products. The sintering quality among the beads is also investigated by studying the fracture surface of the tensile test samples via evaluating the quality of intra- and interbead failure properties of the bead foams.

6.4.1 Experimental Setup, Procedure, and Characterization

A laboratory-scale autoclave system with the chamber volume of 1.2 L was developed in our laboratory for bead foam manufacturing. Figure 6.21 shows a schematic of this system, and the detailed information is provided by Guo et al. [39]. First, the chamber was filled with 850 mL of water, and then 15 g of PLA pellets with an average particle diameter of 1 mm was added into the water. The water was used as a mixing media and to uniformly distribute the heat and the dissolved gas to the pellets while avoiding the agglomeration of the pellets. There are many additives that can be added to the water to prevent the agglomeration of the beads [39], but since PLA is sensitive to hydrolysis, 20 mL of silicon oil was also added to the suspension media as a

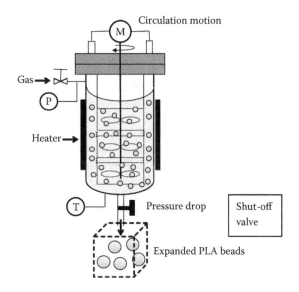

FIGURE 6.21
A schematic of the laboratory-scale autoclave bead foaming chamber.

surfactant to prevent the agglomeration and to postpone the PLA's hydroly-sis. This is because the used nonpolar hydrophobic surfactant can cover the pellets' surface, and the interaction between water and pellets was seen to be significantly minimized.

Subsequently, after sealing the chamber, 5.5 MPa (i.e., 800 psi) CO_2 pressure was supplied into the chamber for gas impregnation into the PLA pellets. Then, the chamber was heated to the set saturation temperature (T_s) and equilibrated for a certain time during which the impregnation occurred and high-melting-temperature crystals formed. The T_s range required to generate a double crystal melting peak was found to be between 120°C and 126°C. The saturation times (t_s) were selected as 15, 30, and 60 min. This processing step will be referred to as the annealing/gas-saturation stage in the subsequent sections.

In the next step, the depressurization occurred by opening the shut-off valve, and the saturated pellets, water, silicon oil, and the blowing agent were discharged from the chamber into a cold water-filled container. Once the saturated pellets evacuated the chamber, foaming occurred due to the thermodynamic instability. Because of the adiabatic expansion effect [7], cooling occurred as foam structures are developed. This processing step will be referred to as the foaming/cooling stage in the subsequent sections. The bead foam samples were then washed by hot water to prepare them for the subsequent steam chest molding stage.

The crystallization behaviors of the EPLA bead foams were studied in a differential scanning calorimeter (DSC) (2000, TA Instruments) by heating the foamed samples to 200°C at a rate of 5°C/min. The final crystallinity and the generated low and high melting temperature peaks (i.e., T_{mlow} and T_{mhigh}) of the foamed samples were analyzed. The final crystallinity of the foamed samples was measured using Equation 6.2.

Due to the chance of hydrolysis that is caused by water during the saturation process, the weight-average molecular weight (M_w) of the bead foam samples could also vary. Therefore, the M_w of the foamed beads was measured using a gel permeation chromatography (GPC). The M_w was analyzed relative to linear PS standards with refractive index (RI) detection in tetrahydrofuran (THF) mobile phase. These experiments were kindly conducted at NatureWorks LLC.

The density of the EPLA bead foams was evaluated using the Archimedes method (i.e., the water-displacement technique) based on ASTM D792-00. The ER of the samples was then evaluated using Equation 6.1.

The cell morphology of the foamed bead samples was also analyzed by using the SEM micrographs that are taken by JEOL JSM-6060. The cell density of each sample was then measured by using the following equation [26,27]:

$$Cell\ density = \left(\frac{\#\ of\ cells}{area} \right)^{3/2} \cdot ER \tag{6.3}$$

The open-cell content of the foamed beads was also measured by using a gas pycnometer (Quantachrome Instrument UltraFoam 1000) in accordance with the ASTM D6226 standard. In this measurement, the pycnometer pressurized the bead foam samples with nitrogen at 41.4 KPa and measured the closed volume content of the beads (V_{closed}). The open-cell content (V_{open}) was then measured using the following equation [68] where the geometric volume of the sample is entitled as $V_{geometric}$:

$$V_{open} = \left(1 - \frac{V_{closed}}{V_{geometric}} \right) \cdot 100\% \tag{6.4}$$

Figure 6.22 compares the images of the EPLA bead foams that are obtained at various T_s and t_s and at CO_2 pressure of 5.5 MPa (P_s). Below and above the selected saturation temperatures, the resulted PLA samples were either unfoamed or molten, respectively. The resulted samples achieved at 126°C after 30 and 60 min of saturation were also molten.

6.4.2 Crystallization Behavior of the EPLA Bead Foams

Figure 6.23 shows the DSC heating thermograms of the EPLA bead foams. Figure 6.24 also presents the total crystallinity and the crystallinity of the generated high-melting-temperature peak, as well as the T_{mlow} and T_{mhigh} at each condition.

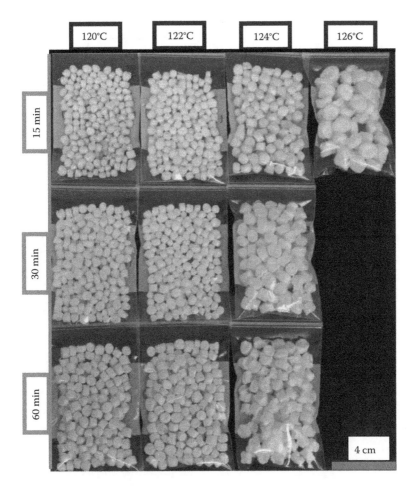

FIGURE 6.22
EPLA bead foams obtained at various saturation conditions. (Each bag contains just a small quantity of foamed beads from each batch.)

At low saturation temperatures, a larger amount of high-melting-temperature crystals were generated during the annealing and gas-saturation stage. This was because the existing unmelted crystals were subjected to crystal perfection and evolution (i.e., the rearrangement of the crystals with a more closed-packed structure) during the saturation, and thereby a new crystal melting peak was created at a higher temperature than the PLA's original melting temperature, as discussed in Section 6.3. By increasing the T_s, the content of high-melting-temperature crystals was decreased, which also caused a reduction of the total crystallinity. This was because during the annealing and gas-saturation stage, more of the existing crystals were melted, and fewer amounts of crystals were subjected to perfection to form high-melting-temperature crystals (demonstrated in Section 6.3). However, the T_{mhigh}

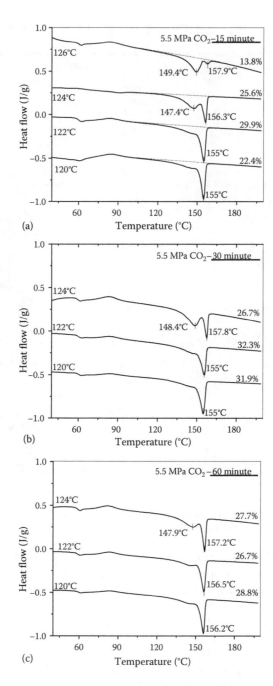

FIGURE 6.23
DSC heating thermograms of the EPLA bead foams obtained at various saturation conditions:
(a) t_s = 15 min, (b) t_s = 30 min, and (c) t_s = 60 min.

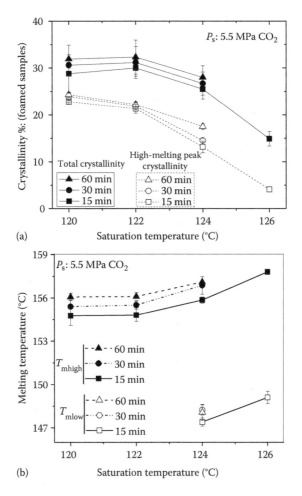

FIGURE 6.24
(a) The total crystallinity and high-melting-peak crystallinity of the EPLA bead foams and (b) the melting temperature of the generated peaks.

appeared at a higher temperature, which was due to the increased molecular mobility of PLA that facilitated the molecular rearrangement and retraction to form crystals with a higher degree of perfection.

On the other hand, after saturation at higher T_s (around 124°C), the T_{mlow}, which forms during the foaming and cooling stage, started to appear more clearly. This was because of the existence of more available melt that could be crystallized during cooling while foaming occurred. In this context, the two distinct generated peaks can be observed at T_s of 124°C, which resembles more to the double crystal melting peak that is used in EPP bead foaming [3–39].

On the other hand, increasing the t_s enhanced the amount of induced perfected crystals as the high-melting-temperature peak. The increased high-melting peak crystallinity promoted the total crystallinity of the foamed beads as well. The increased amount of perfected crystals was followed by an increase in T_{mhigh} because the longer saturation provided further molecular diffusion to form crystals with a higher degree of perfection. On the other hand, as more clearly seen in foam beads saturated at 124°C, the amount of crystals with low-melting temperature was reduced with an increased saturation time. In other words, due to the increased amount of perfected crystals (i.e., crystals with high melting temperature), there was less available melt for crystallization during the foaming and cooling stage, and thereby a smaller amount of crystals could form as the low-melting temperature peak.

6.4.3 Molecular Weight Variations of the EPLA Bead Foams

As noted in Section 6.4.1, during the annealing and gas-saturation stage, water was used as a mixing media and to uniformly distribute the heat and gas to the pellets while avoiding agglomeration. However, the hydrolysis can break down the PLA's molecular chains. In the experiments conducted without the presence of silicon oil, all the saturated PLA pellets were agglomerated with a serious change in their color toward dark brown. This was due to the severe hydrolysis/degradation followed by agglomeration of the PLA pellets. However, we observed that the addition of 2 vol% of silicon oil as a surfactant significantly postponed the hydrolysis, and indeed foamed beads were obtained with no agglomeration and not a significant color change. But yet, there must have been a chance of hydrolysis during the annealing and gas-saturation stage, and therefore, the M_w of the bead foam samples must have been decreased.

Figure 6.25 compares the molecular weight of the EPLA bead foams that are achieved at various saturation temperatures and times using a GPC. As

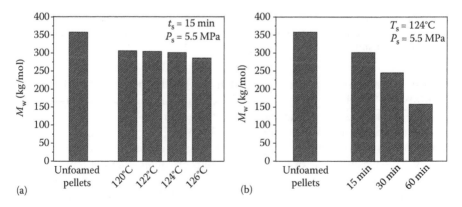

FIGURE 6.25
The average molecular weight of the EPLA bead foams saturated at (a) various temperatures and (b) times.

seen, the used branched PLA relatively possessed a high molecular weight of around 360 kg/mol. At a less-sensitive saturation condition (i.e., low T_s of 120°C and short t_s of 15 min), the M_w was reduced to around 305 kg/mol. However, at the given time, as the T_s increased to 126°C, the hydrolysis was more activated, and the M_w decreased to around 285 kg/mol. On the other hand, at the constant T_s of 124°C, the increased saturation time revealed more influence on reducing the M_w. When the saturation time was increased from 15 min to 60 min, the M_w was reduced from 300 to around 160 kg/mol due to the longer time period during which PLA molecules can be encountered to more hydrolysis/degradation. This indicates that the PLA molecular chains were broken to less than half of their length.

Although this amount of hydrolysis is not even favorable, it could somehow benefit the foaming behavior. As the M_w decreases, the generation of perfect crystals during the annealing and gas-saturation stage can occur with a higher crystal nucleation rate as the molecular chain length reduces. This is because the crystal nucleation mechanism dominates the crystallization kinetics for semicrystalline polymers with lower molecular weight. Therefore, despite the degradation, a larger number of perfected crystals (i.e., high-melting-temperature crystals) that could have been formed during this stage could more profoundly promote the heterogeneous cell nucleation around the crystals. On the other hand, although the reduced M_w could have expedited the gas loss and suppressed foam expansion, the enhanced crystal nucleation rate during the formation of perfect crystals (during annealing and gas saturation) could compensate for the PLA's overall reduced melt strength that is caused by the decreased M_w. But in all, the reduced M_w degrades the mechanical properties, and this undesirable hydrolysis should be minimized. Therefore, since the M_w decrease was more sensitive to the saturation time, it is recommended to minimize the saturation time in the actual manufacturing systems, which also shortens the processing time/cost while providing bead foams with higher quality.

6.4.4 Foaming Behavior of the EPLA Bead Foams

Figure 6.26 shows the ER of the PLA foam beads that are saturated at various T_s and t_s. As the T_s increased from 120°C to 124°C (and to 126°C), the ER of the EPLA bead foams increased from 5–10-fold to around 12–20-fold (and to 30-fold). This must have been due to the reduced amount of perfected crystals (i.e., high-melting-peak crystals) generated during the annealing and gas-saturation stage, which reduced the stiffness of the gas-saturated PLA matrix (demonstrated in Section 6.3.3). According to Figure 6.24, at a T_s of 120°C, around 22%–24% of high-melting-temperature crystals were induced during annealing and gas saturation, whereas at the increased T_s of 124°C (and to 126°C), the second peak crystallinity was reduced to around 15% (and to 4%), respectively. Therefore, cell growth was promoted, and the expanding ability of the beads was increased.

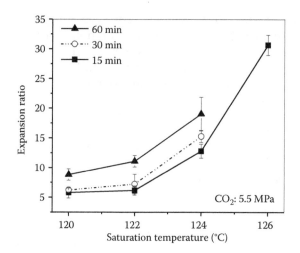

FIGURE 6.26
The ER of the EPLA bead foams obtained after saturation at various conditions.

On the other hand, as the t_s increased from 15 to 60 min, the ER of the EPLA bead foams was also increased despite the increased amount of crystallinity of the perfected crystals. This could indicate that during a longer saturation, the reduced M_w of the amorphous region could have facilitated the cell growth rate, although the high-melting-peak crystallinity (i.e., stiffness) was increased. Therefore, the overall expanding ability of the gas-saturated PLA pellets during the foaming and cooling stage was enhanced despite the increased amount of perfected crystals with a larger number (resulted from the lowered M_w).

Figure 6.27 shows the cell morphology of the EPLA foamed beads, while Figures 6.28 and 6.29 report their corresponding cell density, cell size, and open-cell content, respectively. The obtained cell morphologies in all of the samples, especially for those saturated for 15 min, were uniform, and the achieved average cell sizes were in the range of 6–23 μm (except for the sample saturated at 126°C for 15 min), which are categorized as microcellular foams. As noted, during the low saturation temperatures, a large amount of perfected crystals (around 22%–24%) were induced as the high melting peak. These crystals could have significantly improved the heterogeneous cell nucleation rate during foaming [48–50] and, thereby, increased the cell density of the EPLA foamed bead samples. The reduced effect of high-melting-temperature crystals on the cell nucleation can be observed in the sample saturated at 126°C for 15 min. As Figure 6.24 shows, at this saturation condition, only 4% of high-melting-peak crystals were generated. Therefore, fewer amounts of crystals were involved in heterogeneous cell nucleation, and the reduced stiffness of the PLA matrix facilitated cell growth, and a high ER up to 30-fold was achieved.

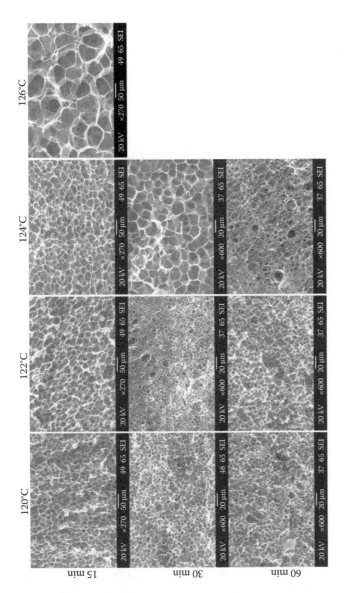

FIGURE 6.27
SEM images of the cell morphology of the EPLA bead foams obtained after saturation at various conditions.

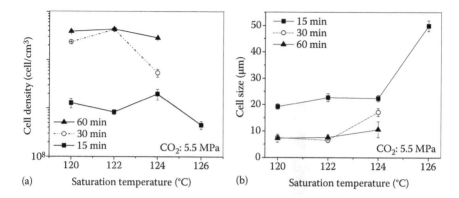

FIGURE 6.28

(a) Cell density and (b) average cell size of the EPLA bead foam samples.

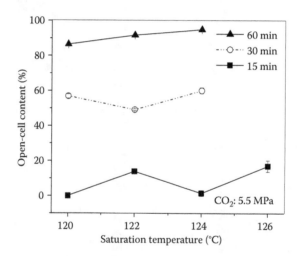

FIGURE 6.29

Open-cell content of the EPLA bead foam samples.

As discussed in Section 6.4.2, the increased t_s further increased the high-melting-temperature crystals, which could additionally promote the heterogeneous cell nucleation rate during foaming through local stress/pressure variations around the crystals [50]. For instance, at T_s of 124°C, the average cell size of the EPLA foam bead samples reduced from 19 to 7 µm at the increased t_s from 15 to 60 min. As we said in Section 6.4.3, the reduced M_w may also have contributed to cell-size reduction through the enhanced crystal nucleation rate during a longer saturation where the high-melting-temperature crystals form.

In order to investigate how the high-melting-temperature peak crystallinity and the PLA molecular chain breakage during the annealing and gas-saturation stage could affect the stability of the nucleated cells, the open-cell content of the EPLA foam beads was measured. Figure 6.29 shows that the increased T_s did not significantly change the open-cell content of the foamed beads. However, the increased t_s had a noticeable influence on the formation of open-cell morphology. The enlarged t_s from 15 to 60 min increased the open-cell content from almost 10% to around 90%. First, this shows that by tailoring the t_s, the stability of the nucleated cells could be controlled. Second, the significance of the t_s effect on the stability of the nucleated cell can be attributed to the impact of the t_s on the decrease in the M_w of PLA chains. As discussed, the reduced M_w could have improved the crystal nucleation rate during a longer saturation period. The larger number of nucleated (perfect) crystals promoted the heterogeneous cell nucleation rate around themselves. However, the reduced melt strength of the amorphous phase with the reduced M_w and the existence of thinner cell walls [65] must also have reduced the cell-wall strength during cell growth. Therefore, the closed-cell content of the bead foams decreased as t_s increased.

6.5 Steam Chest Molding

Laboratory-scale steam chest molding equipment commercially manufactured by Dabo Precision (DPM-0404VS) was used for molding the EPLA bead foams. A mold with cavity dimensions of 15 cm × 6 cm × 5 cm was used to manufacture the molded parts. The mold consists of a fixed side and a moving side. Both of the mold sides have ports for injection of steam into the mold cavity. The basic process of steam-chest molding process consists of three main steps. In the first step, the beads are charged into the mold cavity. In the second step, the steam is injected from the fixed mold at the desired processing steam pressure and temperature (P1). Then, the steam is injected from the moving mold (P2). Finally, the steam is injected from both molds followed by depressurization (P3). The third step consists of the cooling of the mold by water followed by vacuuming to remove the remnant water and then sample ejection. The unit of the steam pressure used in this study is the gauge pressure in bar, which is 1 bar lower than the absolute pressure.

Rectangular specimens were prepared from the molded EPLA part for tensile test experiments. The dimensions of the specimen (i.e., thickness, width, and length) were 14 mm, 19 mm, and 155 mm, respectively. The tensile properties of the specimens were measured using a Micro-tester (Instron 5848) at a crosshead speed of 5 mm/min.

The morphology of the molded EPLA samples was observed by the SEM. For the molded EPLA samples, the product surface, the cut surface, where the product was cut directly by a sharp knife, and the fractured surface after the tensile test were analyzed.

The EPLA bead foams produced at the saturation condition of 124°C and 15 min (ER of 13-fold) were used for steam chest molding and interbead sintering verification among the EPLA bead foams. According to the studies conducted by Zhai et al. [8], we attempted to find the right range of steam chest molding parameters to achieve EPLA bead foam products with reasonable surface quality and proper interbead sintering. Figure 6.30 shows the molded EPLA bead foams at the steam chest–molding condition of: P1: *2 bar, 1 s*, P2: *2 bar, 1 s*, and P3: *2.5 bar, 5 s*.

The tensile properties of the molded EPLA bead foam sample were then analyzed using an Instron machine. Figure 6.31 shows the engineering stress–strain curve of the molded EPLA bead foams and their tensile strengths and Young's moduli, which are compared with the corresponding EPP molded samples' results taken from Zhai et al. [8]. These results show that the tensile properties of the molded EPLA bead foams were even comparable with that of the EPP foam products. In other words, the EPLA bead foam products with double crystal melting peak can not only be considered as a suitable substitute for EPS products but also can potentially be a promising replacement for EPP products in different applications, in case the energy-absorbing ability of the EPLA bead foams can be improved. The Young's modulus is not calculated based on the early loading curve [69] but rather by considering the slope of the tensile stress–strain graph until the yield strength. This is calculated by the *Blue Hill* software, which is designed for mechanical testing.

The bead-to-bead (i.e., interbead) sintering behavior of the EPLA foamed beads was also investigated through SEM images. The surface, cut surface, and fracture surface of the molded EPLA samples were observed, and the results are shown in Figure 6.32. As can be seen, the surface of the molded EPLA samples showed a high surface quality. The cut surface of the molded beads also showed a strong local sintering and high apparent bead-to-bead sintering quality among the bead foams. The fracture surface of the bead foams also showed that the sintered EPLA bead foams elucidated partial intrabead failure, which indicates a strong interbead sintering [7,8].

FIGURE 6.30
Steam chest–molded EPLA bead foam sample.

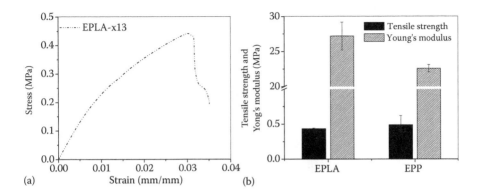

FIGURE 6.31

(a) Engineering stress–strain behavior of the molded EPLA bead foam samples and (b) the corresponding tensile strengths and Young's moduli compared with the EPP molded samples. (From Zhai, W.T. et al. 2011. *Ind. Eng. Chem. Res.* 50, 5523–5531.)

Although further optimization of the steam chest–molding parameters could enhance the interbead sintering and mechanical properties of the EPLA bead foams, this study verified the strong interbead sintering characteristics of the EPLA beads with double melting peak crystals. To further optimize the steam chest molding parameters, a much larger quantity of bead foams is needed, which requires the use of a pilot-scale autoclave system.

6.6 Conclusion and Perspective

Open- and closed-cell microcellular PLA bead foams with a double crystal melting peak structure were developed. The mechanism of double-peak generation in PLA was first investigated through DSC and HP-DSC. Then, PLA bead foams with a double crystal melting peak structure were developed with ERs of 3–30-fold and average cell sizes ranging from 350 nm to 50 μm.

Nanocellular PLA bead foams with a threefold ER were also developed using low saturation temperatures. This development was initiated by the large amount of perfect crystals generated during CO_2 saturation, which significantly improved heterogeneous cell nucleation while hindering cell growth. On the other hand, the bead foams had higher ERs as the saturation temperature increased. In fact, the reduced crystals with a high melting temperature facilitated the expansion of the bead foams. In other words, the cell density was depressed due to the reduced high-melting-temperature peak crystallinity (i.e., fewer cell nucleation sites).

Moreover, the cell density and ER of the foamed beads were promoted as the saturation pressure increased. This was effected through the increased

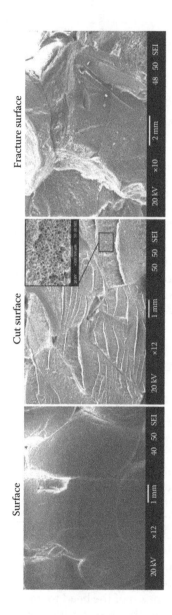

FIGURE 6.32
EPLA bead foams.

degree of thermodynamic instability. However, various CO_2 pressures could also influence the crystallization kinetics of the crystals with a high melting temperature formed during gas saturation. At a high gas-saturation pressure, a larger number of small-sized perfect crystal nuclei were generated, and these could significantly promote the heterogeneous cell nucleation scheme. Moreover, the PLA molecules connected through these crystal nuclei came to have a high melt strength. Consequently, the cell morphology became more uniform with a higher closed-cell content. In contrast, at a lower saturation pressure, the heterogeneous cell nucleation sites around the large-sized crystals were reduced, and the PLA foam structure's uniformity was decreased. Thus, a cell morphology with a higher open-cell content was observed.

The fast crystallization kinetics during cooling also hindered expansion of the PLA-T and PLA-TL foamed beads due to the increased stiffness of PLA matrix. On the other hand, for the neat PLA samples with a slow crystallization rate during cooling/foaming, the final degree of crystallinity could be promoted when the material was subjected to a higher expansion (i.e., an extensional strain). Foaming induces biaxial stretching in cell walls and uniaxial stretching in struts. Both of these strains further promote the final degree of crystallinity. Therefore, a PLA resin with an inherently low crystallization rate could also develop a higher crystallinity with foaming (i.e., a higher degree of expansion).

The steam chest molding of the EPLA foamed beads was also conducted to verify the sintering behavior of the EPLA beads. EPLA molded samples was achieved with high surface quality and excellent interbead sintering. The tensile properties of the molded EPLA bead foams showed that EPLA bead foams with a double crystal melting peak structure can be a promising substitute not only for EPS but also for EPP products.

References

1. Nofar, M.; Park, C.B. **2014**. Poly (lactic acid) foaming. *Prog. Polym. Sci.* 39 (10), 1721–1741.
2. Mills, N. **2007**. Bead foam microstructure and processing. In *Polymer Foams Handbook*, Ed. N. Mills, Butterworth-Heinemann: Oxford, 69–83.
3. Nofar, M.; Guo, Y.; Park, C.B. **2013**. Double crystal melting peak generation for expanded polypropylene bead foam manufacturing. *Ind. Eng. Chem. Res.* 52, 2297–2303.
4. Park, C.B.; Nofar, M. **2014**. A method for the preparation of PLA bead foams. WO 2014158014 A1.
5. Nofar, M.; Ameli, A.; Park, C.B. **2015**. Development of polylactide bead foam with double crystal melting peak structure. *Polymer* 69, 83–94.

6. Nofar, M.; Ameli, A.; Park, C.B. **2015**. A novel technology to manufacture biodegradable polylactide bead foam products. *Materials and Design* 83, 413–421.

7. Lee, E.K. **2010**. *Novel Manufacturing Processes for Polymer Bead Foams*. PhD thesis, University of Toronto.

8. Zhai, W.T.; Kim, Y.W.; Jung, D.W.; Park, C.B. **2011**. Steam-chest molding of expanded polypropylene foams. 2. Mechanism of inter-bead bonding. *Ind. Eng. Chem. Res.* 50, 5523–5531.

9. Grijpma, D.W.; Pennings, A.J. **1994**. (Co)polymers of L-lactide, 2. Mechanical properties. *Macromol. Chem. Phys.* 195, 1649–1663.

10. Perego, G.; Gella, G.D.; Bastioli, C. **1996**. Effect of molecular weight and crystallinity on poly(lactic acid) mechanical properties. *J. Appl. Polym. Sci.* 59, 37–40.

11. Sinclair, R.G. **1996**. The case for polylactic acid as a commodity packaging plastic. *J. Macromol. Sci. Pure Appl. Chem.* 33, 585–590.

12. Tsuji, H.; Ikada, Y. **1998**. Blends of aliphatic polyesters. II. Hydrolysis of solution-cast blends from poly(L-lactide) and poly(E-caprolactone) in phosphate-buffered solution. *J. Appl. Polym. Sci.* 67, 405–410.

13. Fang, Q.; Hanna, M.A. **2001**. Characteristics of biodegradable Mater-Bi®-starch based foams as affected by ingredient formulations. *Ind. Crops. Pdts.* 13 (3), 219–227.

14. Auras, R.; Harte, B.; Selke, S. **2004**. An overview of polylactides as packaging materials. *Macromol. Biosci.* 4, 835–864.

15. Gupta, B.; Revagade, N.; Hilborn, J. **2007**. Poly(lactic acid) fiber: An overview. *Prog. Polym. Sci.* 32, 455–482.

16. Garlotta, D. **2001**. A literature review of poly(lactic acid). *J. Polym. Environ.* 9 (2), 63–84.

17. United Nations Environment Programme (UNEP) Ozone Secretariat. **2006**. *Handbook for the Montreal Protocol Substances that Deplete the Ozone Layer*. UNEP, Kenya.

18. Geissler, B.; Feuchter, M.; Laske, S.; Fasching, M.; Holzer, C.; Langecker, G.R. 2014. Strategies to improve the mechanical properties of high-density polylactic acid foams. *J. Cell. Plast.* 1–21.

19. Peng, J.; Srithep, Y.; Wang, J.; Yu, E.; Turng, L.S.; Peng, X.F. **2012**. Comparisons of microcellular polylactic acid parts injection molded with supercritical nitrogen and expandable thermoplastic microspheres: Surface roughness, tensile properties, and morphology. *J. Cell. Plast.* 48 (5), 433–444.

20. Lim, L.T.; Auras, R.; Rubino, M. **2008**. Processing technologies for poly (lactic acid). *Prog. Polym. Sci.* 33, 820–852.

21. Rasal, R.M.; Janorkar, A.V.; Hirt, D.E. **2010**. Poly (lactic acid) modifications. *Prog. Polym. Sci.* 35, 338–356.

22. Matuana, L.M.; Diaz, C.A. **2010**. Study of cell nucleation in microcellular poly(lactic acid) foamed with supercritical CO_2 through a continuous-extrusion process. *Ind. Eng. Chem. Res.* 49, 2186–2193.

23. Pilla, S.; Kim, S.G.; Auer, G.K.; Gong, S.; Park, C.B. **2009**. Microcellular extrusion-foaming of polylactide with chain-extender. *Polym. Eng. Sci.* 49, 1653–1660.

24. Mihai, M.; Huneault, M.A.; Favis, B.D. **2010**. Rheology and extrusion foaming of chain-branched poly(lactic acid). *Polym. Eng. Sci.* 50, 629–642.

25. Mihai, M.; Huneault, M.A.; Favis, B.D. **2009**. Crystallinity development in cellular poly(lactic acid) in the presence of supercritical carbon dioxide. *J Appl. Polym. Sci.* 113, 2920–2932.

26. Keshtkar, M.; Nofar, M.; Park, C.B.; Carreau, P. **2014**. Extruded PLA/clay nano-composite foams blown with supercritical CO_2. *Polymer* 55 (16), 4077–4090.
27. Nofar, M. **2016**. Effects of nano-/micro-sized additives and the corresponding induced crystallinity on the extrusion foaming behavior of PLA using super-critical CO_2. *Materials and Design* 101, 24–34.
28. Ameli, A.; Nofar, M.; Jahani, D.; Park, C.B. **2015**. Development of high void frac-tion polylactide composite foams using injection molding: Crystallization and foaming behaviors. *Chem. Eng. J.* 262, 78–87.
29. Pilla, S.; Kramschuster, A.; Yang, L.; Lee, J.; Gong, S.; Turng, L.S. **2009**. Microcellular injection-molding of polylactide with chain-extender. *Mater. Sci. Eng. C.* 29, 1258–1265.
30. Pilla, S.; Kramschuster, A.; Lee, J.; Auer, G.K.; Gong, S.; Turng, L.S. **2009**. Microcellular and solid polylactide-flax fiber composites. *Compos. Interfaces.* 16, 869–890.
31. Kramschuster, A.; Gong, S.; Turng, L.S.; Li, T.; Li, T. **2007**. Injection molded solid and microcellular polylactide and polylactide nanocomposites. *J. Biobased. Mater. Bio.* 1, 37–45.
32. Li, K.; Cui, Z.; Sun, X.; Turng, L.S.; Huang, H.X. **2011**. Effects of nanoclay on the morphology and physical properties of solid and microcellular injection molded polyactide/poly(butylenes adipate-co-terephthalate) (PLA/PBAT) nano-composites and blends. *J. Biobased. Mater. Bio.* 5, 442–451.
33. Sasaki, H.; Ogiyama, K.; Hira, A.; Hashimoto, K.; Tokoro, H. **2005**. Production method of foamed polypropylene resin beads. US Patent, US 6838488 B2.
34. Braun, F. **2004**. Impregnating with blowing agent under pressure, heating. US Patent, US 6723760 B2.
35. Sasaki, H.; Sakaguchi, M.; Akiyama, M.; Tokoro, H. **2001**. Expanded polypro-pylene resin beads and a molded article. US Patent, US 6313184 B1.
36. Hira, A.; Hashimoto, K.; Sasaki, H. **2007**. Composite foamed polypropylene resin molding and method of producing same. US Patent, US 7182896 B2.
37. Schweinzer, J.; Fischer, J.; Grave, I.D.; Kogel, W. **1997**. Production of expanded polyolefin beads US Patent, US 5703135 A.
38. Choi, J.B.; Chung, M.J.; Yoon, J.S. **2005**. Formation of double melting peak of poly(propylene-co-ethylene-co-1-butene) during the preexpansion process for production of expanded polypropylene. *Ind. Eng. Chem. Res.* 44, 2776–2780.
39. Guo, Y.; Hossieny, N.; Chu, R.K.M.; Park, C.B.; Zhou, N. **2013**. Critical processing parameters for foamed bead manufacturing in a lab-scale autoclave system. *Chem. Eng. J.* 214, 180–188.
40. Witt, M.; Shah, S. **2012**. Methods of manufacture of polylactic acid foams. US Patent, US 8283389 B2.
41. Shinohara, M.; Tokiwa, T.; Sasaki, H. **2004**. Expanded polylactic acid resin beads and foamed molding obtained therefrom. European Patent, EP1378538 A1.
42. Haraguchi, K.; Ohta, H. **2005**. Expandable polylactic acid resin particles. European Patent, EP1683828 A2.
43. Noordegraaf, J.; Kuijstermans, F.P.A.; Maria De, J.J.P. **2011**. Particulate, expand-able polymer, method for producing particulate expandable polymer, as well as a special use of the obtained foam material. US Patent, US 0218257 A1.
44. Britton, R.N.; Cornelis Van Doormalen, F.A.H.; Noordegraaf, J.; Molenveld, K.; Schennink, G.G.J. **2012**. Coated particulate expandable polylactic acid. US Patent, US 8268901 B2.

45. Britton, R.N.; Maria De, J.J.P.; Kuijstermans, F.P.A.; Molenveld, K.; Noordegraaf, J.; Schennink, G.G.J.; Cornelis Van Doormalen, F.A.H. **2009**. Polymer blend containing polylactic acid and a polymer having a Tg higher than 60°C. European Patent, EP2137249 A2.

46. Wang, C.; Leung, S.N.; Bussmann, M.; Zhai, W.T.; Park, C.B. **2010**. Numerical investigation of nucleating agent-enhanced heterogeneous nucleation. *Ind. Eng. Chem. Res.* 49, 12783–12792.

47. Wong, A.; Guo, Y.; Park, C.B. **2013**. Fundamental mechanisms of cell nucleation in polypropylene foaming with supercritical carbon dioxide—Effects of extensional stresses and crystals. *J. Supercrit. Fluids* 79, 142–151.

48. Lips, P.A.M.; Velthoen, I.W.; Dijkstra, P.J.; Wessling, M.; Feijen, J. **2005**. Gas foaming of segmented poly(ester amide) films. *Polymer* 46, 9396–9403.

49. Marrazzo, C.; Di Maio, E.; Iannace, S. **2008**. Conventional and nanometric nucleating agents in poly(e-caprolactone) foaming: Crystals vs. bubbles nucleation. *Polym. Eng. Sci.* 48 (2), 336–344.

50. Taki, K.; Kitano, D.; Ohshima, M. **2011**. Effect of growing crystalline phase on bubble nucleation in poly(L-Lactide)/CO_2 batch foaming. *Ind. Eng. Chem. Res.* 50, 3247–3252.

51. Nofar, M.; Zhu, W.; Park, C.B.; Randall, J. **2011**. Crystallization kinetics of linear and long-chain-branched polylactide. *Ind. Eng. Chem. Res.* 50, 13789–13798.

52. Nofar, M.; Zhu, W.; Park, C.B. **2012**. Effect of dissolved CO_2 on the crystallization behavior of linear and branched PLA. *Polymer* 53, 3341–3353.

53. Nofar, M.; Tabatabaei, A.; Ameli, A.; Park, C.B. **2013**. Comparison of melting and crystallization behaviors of polylactide under high-pressure CO_2, N_2, and He. *Polymer* 54, 6471–6478.

54. Fischer, E.W.; Sterzel, H.J.; Wegner, G. **1973**. Investigation of the structure of solution grown crystals of lactide copolymers by means of chemical reaction. *Kolloid-Zu Z-Polymer* 251, 980–990.

55. Nofar, M.; Tabatabaei, A.; Park, C.B. **2013**. Effects of nano-/micro-sized additives on the crystallization behaviors of PLA and PLA/CO_2 mixtures. *Polymer* 54, 2382–2391.

56. Nofar, M.; Ameli, A.; Park, C.B. **2014**. The thermal behavior of polylactide with different D-lactide content in the presence of dissolved CO_2. *Macromol. Mater. Eng.* 299 (10), 1232–1239.

57. Thiagarajan, C.; Sriraman, R.; Chaudhari, D.; Kumar, M.; Pattanayak, A. **2010**. Nano-cellular polymer foam and methods for making them. US Patent, US 7838108.

58. Lee, S.T.; Leonard, L.; Jun, J. **2008**. Study of thermoplastic PLA foam extrusion. *J. Cell. Plast.* 44, 293–305.

59. Naguib, H.E.; Park, C.B.; Reichelt, N. **2004**. Fundamental foaming mechanisms governing volume expansion of extruded PP foams. *J. Appl. Polym. Sci.* 91, 2661–2668.

60. Li, G.; Li, H.; Turng, L.S.; Gong, S.; Zhang, C. **2006**. Measurement of gas solubility and diffusivity in polylactide. *Fluid Phase Equilib.* 246, 158–166.

61. Mahmood, S.H.; Keshtkar, M.; Park, C.B. **2014**. Determination of carbon dioxide solubility in polylactide acid with accurate PVT properties. *J. Chem. Thermodyn.* 70, 13–23.

62. Li, G.; Wang, J.; Park, C.B.; Simha, R. **2007**. Measurement of gas solubility in linear/branched PP melts. *J. Polym. Sci., Part B*. 45, 2497–2508.

63. Li, G.; Park, C.B. **2010**. A new crystallization kinetics study of polycarbonate under high-pressure carbon dioxide and various crystallinization temperatures by using magnetic suspension balance. *J. Appl. Polym. Sci*. 118, 2898–2903.

64. Park, C.B.; Baldwin, D.F.; Suh, N.P. **1995**. Effect of the pressure drop rate on cell nucleation in continuous processing of microcellular polymers. *Polym. Eng. Sci*. 35, 432–440.

65. Lee, P.C.; Wang, J.; Park, C.B. **2006**. Extruded open-cell foams using two semi-crystalline polymers with different crystallization temperatures. *Ind. Eng. Chem. Res*. 45 (1), 175–181.

66. Monticelli, O.; Bocchini, S.; Gardella, L.; Cavallo, D.; Cebe, P.; Germelli, G. **2013**. Impact of synthetic talc on PLLA electrospun fibers. *Eur. Polym. J*. 49 (9), 2572–2583.

67. Li, H.; Huneault, A. **2007**. Effect of nucleation and plasticization on the crystallization of poly(lactic acid). *Polymer* 48, 6855–6866.

68. McRae, J.; Naguib, H.E.; Attala, N. **2010**. Mechanical and acoustic performance of compression molded open-cell polypropylene foams. *J. Appl. Polym. Sci*. 116 (2), 1106–1115.

69. Wu, X.L.; Huang, W.M.; Tan, H.X. **2013**. Characterization of shape recovery via creeping and shape memory effect in ether-vinyl acetate copolymer (EVA). *J. Polym. Res*. 20, 150.

7

Nanocellular Foams

Stéphane Costeux

CONTENTS

7.1 Introduction

In recent years, nanostructured materials have been the new frontier for materials and polymer science. Nanoscale fillers contribute to the enhancement of mechanical properties when adequately dispersed [1]. Their presence affects the stability of interfaces, resulting in remarkable properties such as superhydrophobicity [2] that would not be achieved by the use of microscale objects.

This distinction extends to porous or cellular materials. Introducing nanosized voids into solid materials provides benefits that go beyond lightweighting. Unusual properties exist due to nanometer-scale confinement of the solid, or of the medium (air, gas, or liquid) that is contained within the pores or cells, or owing to surface effects at nanostructured interfaces between the polymer and medium [3]. For instance, the presence of nanosized voids and the nanoscale distribution of the solid material in silica aerogels have long been known to be responsible for their unique properties, in particular, their low thermal conductivity (TC) [4].

Quite naturally, various techniques have been considered to achieve polymeric structures with nanopores. Sol–gel techniques have been applied toward the production of organic aerogels with TC or mechanical properties that approach or exceed those of inorganic aerogels [4]. Other methods to produce nanoporous structures involve the use of a porogen component or block copolymers with sacrificial blocks [5–7], colloidal assembly [8,9], microemulsion templating [10], or phase separation [11]. These techniques require solvents that have to be subsequently removed or freeze-dried [12]. Yet an important challenge is to produce such structures by *sustainable* processes that minimize the use of solvents and sacrificial blocks, and minimize the energy or time that is needed to produce the nanoporous structure, such as eliminating solvent exchange and supercritical drying, which is often used for aerogels. Efforts to use CO_2 as a green solvent for the sol–gel process have had limited success [13,14].

Recently, the focus has been to build on the advances in the development of microcellular foams that occurred in the 1980s [15] and became commercial in the 2000s. CO_2-blown microcellular foams, defined as having an average cell size between 1 and 10 μ, were initially produced by a batch foaming process. They showed beneficial properties over regular foams. For instance, microcellular foaming allowed for weight reduction (10%–30%) with a minimal decrease in mechanical strength [16]. Due to the cells' ability to interfere with crack propagation, impact properties and toughness were improved compared to unfoamed polymers. The conversion of this technology to a continuous process by injection molding and its application to the production of lightweight parts with a variety of materials have been one of the most significant advances in foams in the past 30 years.

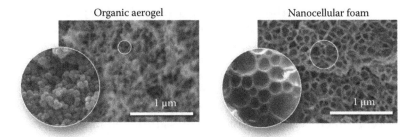

Organic aerogel Nanocellular foam

1 µm 1 µm

FIGURE 7.1
Comparison of nanoscale morphology of aerogels and nanocellular foam, resulting in superior mechanical strength for the latter due to continuous solid structure.

The next frontier is to exploit physical foaming to produce nanocellular foams with cells as small as 100 nm or smaller (Figure 7.1) that will rival the properties of solvent-based nanoporous materials. In particular, research is being devoted to the production of low-density nanofoams, which, in addition to lightweighting, are expected to provide thermal insulation properties that are comparable to aerogels without drawbacks of the latter such as low mechanical strength and a cost-prohibitive process. Although several other physical blowing agents such as nitrogen and hydrocarbons could be envisaged to achieve this goal, CO_2 remains a candidate of choice, being both benign and effective with a wide variety of polymers, as was shown during microcellular foam development. Yet, to successfully produce medium- to low-density nanocellular CO_2-blown foams, it is necessary to generate, grow, and stabilize a prodigious number of cells, typically 10^{15}–10^{16} cells per cubic centimeter of the unfoamed material, compared to 10^9–10^{12}/cm^3 for microcellular foams. Achieving such a breakthrough requires new strategies, processes, and polymer systems.

7.2 Strategies to Generate Nanoscale Cells with CO_2

Conventional CO_2 physical foaming is the pathway by which an equilibrated polymer/CO_2 mixture at high-enough pressure to maintain CO_2 in solution transitions to a stable two-phase (solid/gas) system at ambient temperature and pressure. The pathway differs depending on the time evolution of both pressure and temperature, but the mechanism generally occurs in three stages.

The initial stage is a phase-separation event, which can be triggered by a sudden pressure decrease or by an increase in temperature causing supersaturation. Two major mechanisms are known to lead to new phase formation, such as the formation of a bubble in a polymeric material [17]. One is spinodal decomposition in which a wavelike concentration fluctuation with a periodic wavelength increases over time, producing a co-continuous structure. This mode preferentially occurs when a liquid–liquid phase separation is induced by a temperature change before the pressure is dropped. Since CO_2 solubility in most polymers is higher at lower temperatures, a temperature increase is generally used to trigger spinodal decomposition. The other is nucleation, which occurs when the temperature and pressure in the supersaturated state fall in the metastable region between the binodal and the spinodal. Due to the position of the spinodal in polymer–CO_2 systems [18], a drop in pressure almost always leads to nucleation. The second stage is the growth of newly formed stable nuclei into bubbles or cells. Cell nucleation and growth can occur simultaneously or successively. The third stage is stabilization, where conditions are controlled to freeze the foam structure.

7.2.1 Controlling or Avoiding Nucleation

7.2.1.1 Guidance from the Classical Nucleation Theory

Nucleation is a kinetic phenomenon. The rate of nucleation in foams is often described using the classical nucleation theory (CNT). The theory, initially developed to describe the boiling of liquids and generalized to other liquid/vapor systems, expresses the energy of the system as the sum of a bubble surface contribution, depending on the interfacial tension γ_{bp} between the bubble and the polymer-rich phase, and a volume contribution proportional to the pressure difference ΔP between the bubble (the vapor phase) and the bulk (the polymer-rich phase). Stability analysis yields a Gibbs free energy barrier:

$$\Delta G^*_{hom} = \frac{16\pi}{3\Delta P^2} \gamma_{bp}^3 \tag{7.1}$$

which needs to be overcome for nuclei to grow into stable bubbles of size R exceeding the critical nucleus radius $R^* = 2\gamma_{bp}/\Delta P$. It should be noted that in a pressure quench situation, where the external pressure drops from an initial value P_0 to a final value P_1, ΔP is not constant during this process but varies continuously as the pressure in the bulk evolves to eventually reach the final external pressure P_1. As a result, both R^* and ΔG^*_{hom} also vary during the nucleation process [19,20]. The homogeneous nucleation rate (bubble formation rate) is then [21]

$$J = f_0 C_0 \exp\left[-\frac{\Delta G^*_{hom}}{kT}\right] \tag{7.2}$$

where C_0 is the concentration of CO_2, and f_0 is the rate of production of nuclei [22]. This simplified theory predicts that higher CO_2 concentrations promote nucleation, as they decrease the interfacial tension [18] γ_{bp} and increase J. A higher pressure difference is also beneficial to reduce ΔG^*_{hom}. The rate J is a function of time, which will vanish once nucleation stops. Because of nuclei collapse and the coalescence of growing bubbles, the total number of nuclei produced, defined as $N = \int J(t)\,dt$, will be larger than the number of effective nuclei, N_0, that will survive through the foaming process to form a cell in the final foam.

Qualitative trends for the effect of pressure, temperature, and CO_2 concentration on the nucleation density of binary polymer/CO_2 predicted by the CNT have generally been observed experimentally for microcellular foams [22,23]. These trends are expected to hold to some extent at the nanoscale, and mathematical foaming models using a form of the CNT in which f_0 is used as an adjustable parameter have shown promising results in predicting foaming at the nanoscale [24]. However, the CNT performs poorly in quantitatively predicting the absolute nuclei production rate, the absolute free energy barrier [25,26], or the maximum cell density [27,28] due in part to failure to capture the polymer/CO_2 interactions at the interface of nanoscale bubbles. Additional limitations of the theory have been reviewed by Lubetkin [29] and Tomasko et al. [30].

7.2.1.2 Heterogeneous Nucleation

The addition of nucleating particles is commonly practiced in conventional or microcellular foaming to enhance the nucleation density by providing heterogeneous surfaces on which the nucleation energy barrier is lowered [31]. The smaller particle size promotes higher nucleation density by providing more nucleation sites at the same particle concentration. Interestingly, Ramesh and Lee [32] showed that nanoparticles are rather ineffective in the nucleation of conventional polypropylene foams. A hypothesis is that defects on the surface of larger particles or microvoids inside elastomeric additives that are not fully wetted by the polymer during compounding provide sites where bubbles can nucleate easily (pseudo-classical nucleation [33]). However, the number of defects large enough to exceed the critical radius for microcellular foams is reduced when the solid particles become significantly smaller than 1 μ. Spitael et al. [34] made a similar observation when adding diblock copolymer micelles in an attempt to reduce the cell size of microcellular polystyrene foams. No significant decrease of cell size was noted. Yet, there is evidence that even at the nanoscale, the principle of heterogeneous nucleation may apply [35–37], provided that the right of type of nanoscale nucleating agent and the optimal foaming conditions are chosen for each polymer to maximize nucleation [38]. The mechanism at the nanoscale may be different from that at the microscale, due to CO_2 solvation effects on nanoparticles [39].

7.2.1.3 Effect of Pressure Drop Rate

The efficiency of cell nucleation is affected by nuclei coalescence. Zhu et al. [40] showed theoretically that for nanocellular foams in which nuclei are only tens of nanometers apart, the pressure difference between small and larger nascent cells causes a ripening of the nuclei population by gas diffusion. This suggests that increasing the rate of nucleation, e.g., by increasing the pressure drop rate, may allow for the survival of a larger number of stable nuclei before diffusion effects come into play, due to narrower nuclei size distribution or additional internal cooling. As shown in Figure 7.2, the benefit of higher depressurization and gas concentration in shortening the nucleation process has been shown at the microcellular scale [41] and should still be significant at the nanoscale.

7.2.1.4 Bypassing Nucleation

The difficulty in controlling the generation of nuclei in a homogeneous polymer system has led researchers to seek methods to generate bubbles from templated multiphase polymer systems, in which one phase forms discrete domains in a continuous matrix. Nanoscale templates can readily be produced by block copolymer self-assembly. The choice of properties of the discrete polymer phase (T_g, CO_2 solubility, etc.) can favor the formation of gas bubbles at a predetermined location, in effect eliminating the random character of nuclei birth and survival. The approach was pioneered by Yokoyama et al. [42,43], who used diblock copolymers with a CO_2-philic fluorinated block to produce thin films with organized micelle structure. The high CO_2 solubility in the fluorinated micelles favored the initiation of one bubble per micelle, thus providing an effective control of the cell density. The concept was also adapted to blends of polymers with block copolymers [44,45] or immiscible blends [46,47].

7.2.2 Maximizing Foam Expansion

At the end of the nucleation, the nascent foam consists of a number of stable nuclei that are dispersed into the polymer/CO_2 mixture. To have a chance to produce nanocellular foams with reduced density after the expansion of these nuclei, a successful nucleation stage should yield stable nuclei in excess of $N_0 = 10^{15}$/cm^3 (per cubic centimeter of unfoamed polymer). N_0 will be called the cell nucleation density and refers to the number of effective nuclei that will result in an actual cell in the final foam. (Nuclei that disappear by coalescence or ripening before the end are thus ignored.) It is a better measure of the cell density than N_c (number of cells per cubic centimeter of the final foam) because it is corrected by the expansion ratio r ($N_0 = r\, N_c$) [48]. If cells are approximated by spheres with an average diameter ϕ_{nm}

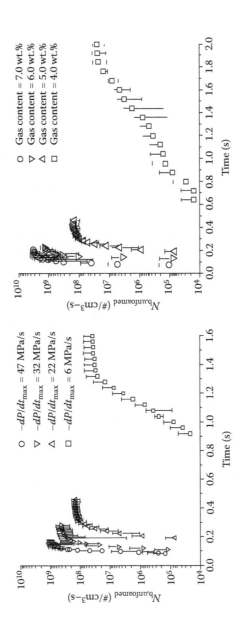

FIGURE 7.2
Nucleation rate determined by visualization experiments at various depressurization rates and CO_2 concentrations. (Reprinted from S. N. Leung, PhD Thesis, University of Toronto, Toronto, Canada, 2009 [© 2009 S. N. Leung]. With permission.)

(in nanometers), the expansion ratio and the foam porosity (void volume fraction), p, are related to cell size and effective nucleation density by

$$r = 1 + \frac{\pi \phi_{nm}^3 N_0}{6 \times 10^{21}}$$

and

$$p = \frac{r-1}{r} \tag{7.3}$$

We note that, here, the term *porosity* is used interchangeably with void volume fraction both for open- and closed-cell foams. The expressions assume distinct cells, as found in most foams, but would not apply to bicontinuous structures, such as aerogels, or certain permeable structures that are produced by spinodal decomposition. To estimate cell wall thickness, it is convenient to model a nascent foam as a regular array of cubic cells. Prior to cell growth, cells in this model foam are nuclei with critical size $2R^*$, typically a few nanometers. The average distance between these nuclei is of the order of 100 nm, and the volume fraction they occupied (porosity p during nucleation) is less than 0.1%. As the model foam expands, each nucleus becomes a growing bubble. Figure 7.3 shows the evolution of the porosity and the distance between the surfaces of adjacent cells (the average wall thickness) as all cells expand simultaneously. It should be noted that N_c decreases during cell growth, whereas, by design, N_0 remains constant.

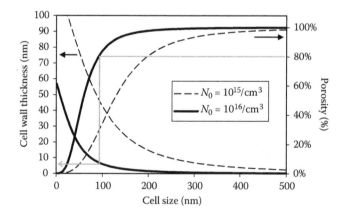

FIGURE 7.3
Relation between average cell size, porosity, and average cell wall thickness for a model foam with cubic cells and a fixed cell nucleation density. Broken line indicates cell wall thickness for 10^{16} cell/cm³ when porosity is 80%. (Reprinted from S. Costeux, *J. Appl. Polym. Sci.*, 131, 23, 41293, 2014 [© 2014 Wiley Periodicals, Inc.]. With permission.)

According to Figure 7.3, for $N_0 = 10^{15}/cm^3$, cells need to expand to a size of 200 nm to achieve 80% porosity, corresponding to an expansion ratio of 5. This requires stretching the cell walls between the cells down to a thickness of approximately 15 nm. For $N_0 = 10^{16}/cm^3$, 80% porosity is reached for 100-nm cells with 5-nm cell walls. These numbers emphasize the magnitude of the challenge in producing polymeric nanofoams. Polymer chains have to deform and become confined at dimensions similar to molecular size, such as their radius of gyration.

Therefore, preventing cell coalescence during expansion is paramount. Cell ripening by diffusion needs to be prevented by reducing the expansion time (fast depressurization), by controlling the temperature to maintain very high viscosities [38,49], or by manipulating the relative properties of the matrix and domains in multiphase polymer systems, for instance, by lowering the temperature below the glass transition temperature (T_g) of the matrix [42]. Lack of integrity of the cell walls can, in some cases, yield open-cell foams, which is desirable in certain applications such as membranes.

In all cases, a stabilizing mechanism is needed to freeze the foam structure. Even if the temperature is not intentionally varied right before or during foaming, the Joule–Thomson effect resulting from the rapid expansion of CO_2 in the cells induces a rapid decrease of the temperature in or around the foam [36]. Another stabilizing effect relates to a decrease of the plasticization of the polymer by CO_2 as the gas diffuses into the growing cells. If the initial foaming temperature is chosen below the T_g of the pure polymer but above the T_g of the polymer that is plasticized by CO_2, the polymer/CO_2 mixture will eventually become glassy as the CO_2 concentration decreases during foaming [50]. The same strategy can be used with the crystallization temperature of semicrystalline polymers.

7.3 Nanofoam Batch Foaming Process

The process to make nanofoam is inspired from the batch foaming process that was initially used to make microcellular foams [15] in the 1980s and 1990s. In its most general form, the equipment consists of an autoclave that can be maintained at a set temperature, connected to a gas source with pressure control (typically a syringe pump), and a valve to release the gas from the vessel. While in theory, a number of gases could be used to produce nanofoams, CO_2 offers the benefit of having high solubility in polymers as the temperature is decreased. Nitrogen, for instance, has a much lower solubility in most polymers, and solubility increases with temperature [51]. As a result, high solubility of nitrogen requires temperatures at which viscosity is low, which limits options to stabilize the foam and restrict cell growth or coalescence. Use of nitrogen as a co-blowing agent with CO_2 has not proven to be particularly effective in improving nanocellular foams [52]. Therefore, the following will focus on CO_2 as the main physical blowing agent. The autoclave containing the sample is conditioned at the *soak temperature* (or sorption temperature), CO_2 is fed into the autoclave, and the pump is activated to progressively increase the CO_2 pressure in the autoclave. The sorption conditions are maintained until the CO_2 concentration in the sample reaches the solubility limit at the soaking conditions. The minimum soaking time has to be determined either empirically or by calculation [53] using the CO_2 diffusion coefficient in the particular polymer at the soaking conditions and the sample thickness. It varies from less than an hour for thin samples (100 μm or less) to several days for thick samples at low temperature.

Upon equilibration, the release valve is opened to remove CO_2. The engineering design of the pressure release system, involving a valve and tubing of various lengths and internal diameters, determines the peak depressurization rate dP/dt typically between 0.1 MPa/s (slow depressurization) and 1 GPa/s (fast depressurization). Systems with back pressure regulators involving slower dP/dt (down to 0.01 MPa/s) have been used for experiments that involve CO_2 swelling rather than typical foaming [43], whereas other systems involve the rapid expulsion of a small sample out of the autoclave to maximize dP/dt [54]. Pressure decrease generally triggers nucleation but not necessarily foam expansion. In the case of microcellular foams, soaking could be done at a temperature higher than the T_g or right above the melting temperature (T_m) of the polymer, and depressurization would induce both nucleation and foam expansion. However, these conditions generally do not produce high-enough cell nucleation densities to generate nanocellular foams (see Figure 7.4a). Therefore, CO_2-blown nanofoams are more likely produced by soaking at temperatures lower than the T_g of the polymer when it is amorphous and slightly below T_m when it is semicrystalline. From these conditions, nucleation can be maximized by controlling the pressure drop rate (Figure 7.4c) or controlled by the presence of soft, CO_2-philic nanodomains (Figure 7.4b). These strategies will be explained in more detail in Section 7.4.

CO_2 solubility in amorphous polymers generally increases as the temperature is decreased so that, at the soaking temperature, the CO_2-laden polymer can be either glassy or rubbery. In the former case, upon depressurization, nucleation will take place, but little expansion will be observed. The specimen retrieved from the autoclave may remain transparent. Foam expansion then requires a second step, where the temperature of the sample is increased (e.g., by immersion in a heated bath) to allow for the expansion of nuclei into cells. Procedures involving this second thermal conditioning step will be called a *two-step process* and denoted 2S in the following. The process is depicted in Figure 7.5b.

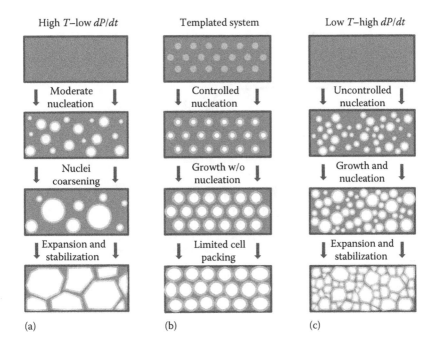

High *T*–low *dP/dt*   Templated system   Low *T*–high *dP/dt*

Moderate nucleation   Controlled nucleation   Uncontrolled nucleation

Nuclei coarsening   Growth w/o nucleation   Growth and nucleation

Expansion and stabilization   Limited cell packing   Expansion and stabilization

(a)   (b)   (c)

FIGURE 7.4
Strategies for microcellular and nanocellular foams: (a) Homogeneous systems foamed under conditions of insufficient nucleation density or favoring nuclei ripening become microcellular foams. (b) Templated systems under similar conditions remain nanocellular but with low expansion ratio (<2). (c) Homogeneous systems foamed under conditions favoring high nucleation and growth simultaneously can yield low-density nanofoams. (Reprinted from S. Costeux, *J. Appl. Polym. Sci.*, 131, 23, 41293, 2014 [© 2014 Wiley Periodicals, Inc.]. With permission.)

If the polymer specimen becomes rubbery upon saturation with CO_2, the internal pressure in the nuclei formed during pressure release will deform the polymer matrix and allow foam expansion to occur. The diffusion of CO_2 out of the polymer matrix will eventually return the matrix to a glassy solid, at which point foam expansion will stop [55]. This procedure is called a *one-step process*, denoted 1S (Figure 7.5a). As will be illustrated by the foaming results of various polymer systems under different conditions, one of the key parameters in determining whether foaming can occur is the saturation pressure. Saturation below the critical pressure of CO_2 (7.4 MPa) generally produces little or no foaming until a second heating step is applied. For pressures above 10 MPa, especially at temperatures in the vicinity of its critical temperature (31°C), CO_2 is a dense supercritical fluid or liquid, the expansion of which will provide a large driving force for the growth of the nuclei. This is particularly effective in the supercritical state, where interfacial tension effects will be reduced.

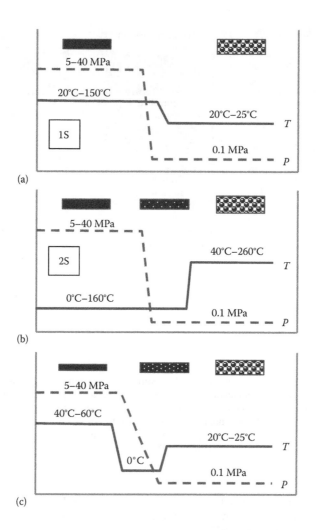

FIGURE 7.5
Temperature/pressure profiles commonly used for the production of nanofoam: (a) 1S process, (b) 2S process, and (c) modified process with temperature quench before depressurization. (Reprinted from S. Costeux, *J. Appl. Polym. Sci.*, 131, 23, 41293, 2014, [© 2014 Wiley Periodicals, Inc.]. With permission.)

Some variations of these two processes exist. For instance, the 2S process can involve expansion in both the depressurization and the thermal conditioning steps if soaking is done when the CO_2-laden polymer is in the rubbery regime [56]. Conversely, soaking can be done in the rubbery regime, but the temperature is decreased to turn the mixture glassy before pressure drop [42] with optionally a second thermal conditioning step to enhance expansion [36] (see Figure 7.5c).

7.4 Polymer Systems for Nanofoams

For the successful production of nanofoams, the choice of the polymer system will often limit options for the process that is used, and vice versa. In this section, three general systems are considered:

1. Single-phase polymer systems, which can be a single polymer or a blend of miscible polymers. Both amorphous polymers and semicrystalline polymers will fall in this category.

2. Nucleated systems and nanocomposites, which include a nonpolymeric second phase, usually an inorganic nanoparticle such as clay or silica, into a single-phase polymer system.

3. Multiphase polymer system, designed to manipulate the partition of CO_2 between several phases to enhance or bypass homogeneous nucleation.

Successful attempts to produce nanocellular foams with CO_2 with the three systems are reported in Tables 7.1 through 7.3. Each table lists the following:

- *The type of polymer system used.* A code indicates the type of sample that is used: [CF] for solvent cast film (usually a thin film, often a few microns thick, on a substrate such as a silicon wafer), [SF] for self-standing film (thickness < 1 mm), and [TS] for thick sheet (thickness > 1 mm, usually produced by compression or injection molding).

- *The process conditions leading to nanocellular foams, starting with the soaking pressure and temperature range.* A code is used to classify the depressurization time: [SD] for slow pressure drop (>1 min), [FD] for fast pressure drop (<1 sec), and [ID] for times intermediate between [FD] and [SD]. A second code refers to the type of process: [1S] for one-step foaming and [2S] for two-step foaming (with the temperature of thermal conditioning). A temperature quench before depressurization is indicated in the soaking temperature condition (e.g., 40C\0C indicates that after CO_2 saturation at 40°C, the temperature was dropped to 0°C before the pressure drop).

- *A symbol is used to represent the foam density via the void volume fraction* (porosity, *p*): ● for dense nanofoams ($p < 40\%$), ◐ for intermediate-density nanofoams ($40 \leq p \leq 70\%$), and ○ for high-expansion foams ($p > 70\%$).

- *The key characteristics of the nanofoams:* their average cell size ϕ_{nm} (or a range if multiple examples of nanofoams were produced), porosity *p* (calculated from the reported foam density), and cell nucleation density, N_0, in cells per cubic centimeter of unfoamed polymer. Values of

TABLE 7.1

CO_2-Blown Nanocellular Foams from Single-Phase Polymer Systems Reported in the Literature

Polymer Type	Process Conditions		Nanofoam Characteristics	Reference
Amorphous Polymers				
PS [SF]	25 MPa/180°C, [FD], [2S at −10°C]	◑	<1000 nm/~50% (10^{12})	Janani [57]
PMMA [TS]	27.6–34.5 MPa/40°C, [SD], [1S]	●	550 nm/<40% (~10^{12})	Goel [50]
	3.4 MPa/0°C, [ID], [2S at 80°C]	○	350 nm/~90% (~3.8×10^{13})	Handa [58]
	5 MPa/−30°C, [ID], [2S at 50°C]	◑	50 nm/68% (5×10^{14})	Guo [59]
	5 MPa/−20°C, [ID], [2S at 70°C−90°C]	○	120–240 nm/77% (1.5×10^{14})	Guo [59]
	13.8 MPa/0°C, [ID], [2S at 80°C]	○	40 nm/77% (1.2×10^{16})	Liao [60]
PEMA [TS]	33 MPa/30°C, [FD], [2S at 60°C]	○	345 nm/79% (1.7×10^{14})	Costeux [61]
P(MMA-co-EMA) [TS]	30 MPa/35°C, [FD], [2S at 50°C−70°C]	○	80 nm/80%−82% (1.6×10^{16})	Costeux [24,62]
P(MMA-co-tBMA) [TS]	30 MPa/35°C, [FD], [2S at 70°C]	○	145 nm/78% (2×10^{15})	Costeux [24,62]
Other MMA copolymers [TS] (PFMA, BMA; MSMA)	30 MPa/35°C, [FD], [1S]	◑	70–100 nm/50%−60% (~5×10^{15})	Costeux [24,62]
PES [SF]	5 MPa/25°C, [FD], [2S at 150°C−200°C]	●	40–120 nm/~30% (10^{13}−2×10^{14})	Krause [23,63]
PEI [SF]	4.6 MPa/25°C, [FD], [2S at 160°C−240°C]	●	50–80 nm/~30% (10^{13}−3×10^{14})	Krause [63]
PEI [TS]	5 MPa/21°C, [ID], [2S at 145°C−180°C]	◑	30–120 nm/30%−53% (2×10^{14})	Miller [64], Chatchaisucha [65]

(Continued)

TABLE 7.1 (CONTINUED)

CO$_2$-Blown Nanocellular Foams from Single-Phase Polymer Systems Reported in the Literature

Polymer Type	Process Conditions		Nanofoam Characteristics	Reference
PEI [TS]	20 MPa/45°C, [2S at 165°C–205°C, 0.1–10 MPa]	◐	40–100 nm/30%–60% (>10^{15})	Aher [66]
PI [CF]	5 MPa/25°C, [FD], [2S at 260°C–270°C]	●	20–50 nm/<35% (~10^{15})	Krause [67,68]
PC [SF]	5 MPa/–30°C, [ID], [2S at 60°C–110°C]	◐	20–30 nm/38%–62% (2 × 10^{15})	Guo [69]
PSU [SF]	5 MPa/–10°C–0°C, [ID], [2S at 150°C]	◐	20–30 nm/47% (1.7 × 10^{15})	Guo [70]
Miscible Blends				
PEMA/SAN [TS]	33 MPa/30°C, [FD], [2S at 60°C]	◐	95–110 nm/59%–68% (3 × 10^{15})	Costeux [49]
P(MMA-co-EA)/SAN [TS]	33 MPa/30°C, [FD], [2S at 60°C]	◐◐	86–110 nm/60%–69% (2–5 × 10^{15})	Costeux [49]
PSU/PI [CF]	5 MPa/25°C, [FD], [2S at 210°C–240°C]	◐	200–500 nm/–50% (~10^{13})	Krause [68]
Semicrystalline Polymers				
PP [SF]	15 MPa/135°C–143°C, [FD], [1S]	●	150–300 nm/<25% (~10^{12})	Bao [71]
PET	6 MPa/25°C, [ID or FD], [2S at 235°C]	●	193 nm/<20% (3.4 × 10^{13})	Li [53]

Note: Explanations for process conditions and foam characteristics (cell size, porosity, and cell nucleation densities in cells/cm^3) are given in the text.
Sample type: [CF]: solvent cast film; [SF]: self-standing film (<1 mm); [TS]: thick sheet (>1 mm).
Depressurization: [SD]: slow (>1 min); [FD]: fast (<1 sec); [ID]: intermediate between [FD] and [SD]; Foaming process: [1S]: one-step foaming; [2S]: two-step foaming (with temperature condition).
Foam void volume (porosity): ●: low (<40%) ◐: intermediate (40%–70%) ○: high (>70%).

TABLE 7.2

CO$_2$-Blown Nanocellular Foams from Nucleated Polymer/Nanocomposite Systems Reported in Literature

Polymer Type	Process Conditions		Nanofoams Characteristics	Reference
Semicrystalline Polymers				
PLA+clay [TS]	10 MPa/165°C, [FD], [2S at 165°C]	●	360 nm/~42% (~1.2 × 10^{14})	Fujimoto [72]
	28 MPa/110°C, [FD], [1S]	●	200 nm/~<20% (2 × 10^{13})	Ema [35]
HDPE/clay [TS]	8.3 MPa/50°C, [1D or FD], [2S at 125°C]	●	250 nm/17% (2.9 × 10^{13})	Lee [73]
PP+MWCNT [TS]	30 MPa/142°C–144°C, [1D or FD], [1S]	●	70–150 nm/10%–30% (<2 × 10^{14})	Ameli [74]
Amorphous Polymers				
PC+clay	18 MPa/80°C, [FD], [2S at 160°C]	●	600 nm/~20% (~10^{13})	Ito [75]
PC+SiO$_2$ [TS]	20 MPa/50°C, [FD or 1D], [2S at 120°C]	●	300–500 nm/<20% (6 × 10^{12})	Zhai [76]
PI+SiO$_2$ [CF]	20 MPa/60°C, [1D], [2S at 250°C]	◐	500 nm/45% (~10^{12})	Li [77]
PMMA+clay [TS]	20 MPa/70°C↘20°C, [FD], [2S at 50°C]	◐	330 nm/50% (3.2 × 10^{13})	Realinho [78]
PMMA+SiO$_2$ [TS]	30 MPa/60°C, [FD], [1S]	○	280 nm/77% (3 × 10^{14})	Costeux [38,79]
PMMA-co-EA+SiO$_2$ [TS]	33 MPa/40°C, [FD], [2S at 80°C]	○	180 nm/82% (1.5 × 10^{15})	Costeux [38,79]
	30 MPa/40°C, [FD], [2S at 70°C]	○	95 nm/79% (8 × 10^{15})	Costeux [38,79]
PMMA-co-EMA+POSS [TS]	33 MPa/40°C–50°C, [FD], [2S at 70°C]	○	100–120 nm/85% (~10^{16})	Costeux [38,79]
	33 MPa/35°C, [FD], [2S at 55°C]	○	65 nm/74% (2 × 10^{16})	
SAN+clay [TS]	30 MPa/40°C↘0°C, [1D], [2S at 100°C]	●	300 nm/~40% (~8 × 10^{13})	Urbanczyk [36]
SAN+POSS [TS]	33 MPa/30°C, [FD], [2S at 60°C]	●	200 nm/70% (5.5 × 10^{14})	Costeux [38,79]

Note: Explanations for process conditions and foam characteristics (cell size, porosity, and cell nucleation densities in cells/cm³) are given in the text.

Sample type: [CF]: solvent cast film; [SF]: self-standing film (<1 mm); [TS]: thick sheet (>1 mm).

Depressurization: [SD]: slow (>1 min); [FD]: fast (<1 sec); [1D]: intermediate between [FD] and [SD]; Foaming process: [1S]: one-step foaming; [2S]: two-step foaming (with temperature condition).

Foam void volume (porosity): ●: low (<40%) ◐: intermediate (40%–70%) ○: high (>70%).

TABLE 7.3

CO$_2$-Blown Nanocellular Foams from Multiphase Polymer Systems Reported in Literature

Polymer Type	Process Conditions		Nanofoams Characteristics	Reference
Diblock Copolymer				
PS-*b*-PFMA [CF]	7.5–30 MPa/60°C↘0°C, [SD], [1S]	●	10–30 nm/~30% (~3 × 10^{16})	Yokoyama [42,43]
PS-*b*-PFS [CF]	10–30 MPa/60°C↘0°C–40°C, [SD], [1S]	●	20–40 nm/<20% (5 × 10^{15})	Yokoyama [43]
PS-*b*-PMMA [CF]	4.2–8.6 MPa/40°C↘0°C–40°C, [SD],[1S]	●	15–40 nm/<20% (9 × 10^{15})	Taki [80]
PS-*b*-PFDA [SF]	30 MPa/0°C, [SD], [1S]	●	~100 nm/30% (7 × 10^{14})	Reglero Ruiz [81]
PMMA-*b*-PFMA [CF]	8–30 MPa/45°C↘0°C, [SD], [1S]	●	30–70 nm/<25% (~10^{15})	Dutriez [82]
PMMA + PMMA-*b*-PFMA	34.5 MPa/40°C, [FD or ID], [1S]	●	~300 nm/<25% (~10^{12})	Siripurapu [37]
Triblock Copolymer				
PFMA-*b*-PMMA-*b*-PFMA [CF]	8–30 MPa/45°C↘0°C, [SD], [1S]	●	20–40 nm/<20% (~10^{16})	Dutriez [82]
MAM [TS]	30 MPa/23°C, [ID], [1S]	○	120 nm/55% (~10^{15})	Pinto [83,84]
PMMA+MAM [TS]	30 MPa/23°C, [ID], [2S at 80°C–100°C]	●	200–300 nm/<30% (~10^{14})	Reglero Ruiz [56]
	30 MPa/23°C, [ID], [1S]	○	~300 nm/~30% (<10^{14})	Reglero Ruiz [85]
	30 MPa/23°C, [ID], [1S]	○	~200 nm/60% (4 × 10^{14})	Reglero Ruiz [86]
	10–30 MPa/23°C–50°C, [SD or ID], [1S]	○	100–200 nm/42%–58% (4 × 10^{14})	Pinto [87,88]
	30 MPa/23°C–40°C, [ID], [1S]	○	130–150 nm/35%–45% (~4 × 10^{14})	Pinto [89]

(Continued)

TABLE 7.3 (CONTINUED)

CO$_2$-Blown Nanocellular Foams from Multiphase Polymer Systems Reported in Literature

Polymer Type	Process Conditions		Nanofoams Characteristics	Reference
	3.2 MPa/0°C, [SD], [2S at 90°C]	◑	70–90 nm/45% (1.5 × 10^{15})	Forest [90]
	3.2 MPa/0°C, [SD], [2S at 60°C–80°C]	○	110 nm/70% (5.5 × 10^{14})	Forest [91]
Immiscible blends				
ABS [SF]	3.4 MPa/0°C, [SD], [2S at 60°C]	●	470 nm/~25% (~3 × 10^{12})	Nawaby [92]
PS/PMMA [CF]	8.6 MPa/40°C↘20°C, [SD], [1S]	●	40–50 nm <20% (~8 × 10^{14})	Otsuka [46]
PPE/SAN/SBM	5 MPa/40°C, [ID], [2S at 180°C]	◑	400 nm, ~40% (~10^{13})	Ruckdäschel [44,93]
		●	350 nm, <30% (2 × 10^{13})	Ruckdäschel [93]
PEEK/PEI	20 MPa/40°C, [ID], [2S at 200°C]	●	40 nm/~10% (~10^{14})	Nemoto [94]
PP/hSIS [SF]	20 MPa/25°C, [ID], [2S at 120°C]	●	250 nm/<20% (7 × 10^{13})	Nemoto [45]
PP/EP rubber [SF]	20 MPa/25°C, [ID], [2S at 120°C]	●	500 nm < 10% (~10^{11})	Nemoto [47]
PP/SEBS [SF]	20 MPa/25°C, [ID], [2S at 120°C]	●	250 nm/<20% (7 × 10^{13})	Sharudin [95]
PEMA/SAN (32 wt% AN) [TF]	33 MPa/30°C, [FD], [2S at 60°C]	○	360 nm/74% (1.3 × 10^{14})	Costeux [49]

Note: Explanations for process conditions and foam characteristics (cell size, porosity, and cell nucleation densities in cells/cm^3) are given in the text.

Sample type: [CF]: solvent cast film; [SF]: self-standing film (<1 mm); [TS]: thick sheet (>1 mm).

Depressurization: [SD]: slow (>1 min); [FD]: fast (<1 sec); [ID]: intermediate between [FD] and [SD]; Foaming process: [1S]: one-step foaming; [2S]: two-step foaming (with temperature condition).

Foam void volume (porosity): ●: low (<40%) ◑: intermediate (40%–70%) ○: high (>70%).

porosity or cell nucleation density not reported in the literature reference were estimated using either Equation 7.3 for p or Equation 7.4 for N_0:

$$N_0 = \frac{6.10^{21}}{\pi\phi_{nm}^3} \frac{p}{1-p} \tag{7.4}$$

An overview of the literature results follows, organized per polymer type (amorphous versus semicrystalline) and per the strategy that is used (homogeneous system versus nanocomposite or phase-separated system) rather than chronologically. Polymer acronyms are defined in the "Abbreviations" section at the end of this chapter.

7.4.1 Single-Phase Polymers

7.4.1.1 Polymethylmethacrylate and Retrograde Vitrification

The first polymer considered for the production of foams with very high cell nucleation densities was polymethylmethacrylate (PMMA) [55]. This choice was based on the great affinity of PMMA for CO_2. Under moderate pressure conditions (above 6 MPa), sorption of over 15 wt% CO_2 can be achieved, which provides sufficient plasticization to depress T_g below ambient temperature. Condo et al. [96] predicted that this could trigger a peculiar behavior that they named *retrograde vitrification*, whereby T_g is completely suppressed above a pressure threshold (i.e., PMMA remains in a rubbery regime above this pressure, irrespective of the temperature), whereas under the threshold pressure, PMMA is glassy below a classical upper T_g but returns to a rubbery state once the temperature is decreased below a lower *retrograde T_g*. This behavior is depicted in Figure 7.6 (line labeled PMMA). Using creep compliance measurements, Condo and Johnston [97,98] confirmed their findings experimentally, as did Handa and Zhang [58] using stepwise differential scanning calorimetry (DSC) experiments. This behavior was only observed for a few polymer/CO_2 systems, in particular, for PMMA, polyethylmethacrylate (PEMA), and acrylonitrile–butadiene–styrene copolymer (ABS) [99]. It should be noted that the lower T_g branch for these systems is very close to the liquid–vapor transition of CO_2, across which CO_2 density and heat capacity vary sharply, which could interfere with creep and DSC results in this vicinity. Using a quartz crystal microbalance (QCM), Dutriez et al. [82] observed for PMMA/CO_2 a behavior that resembles that of polystyrene, namely, only an upper T_g branch and a leveling of T_g above a pressure of approximately 7 MPa (see Figure 7.6, lines labeled PMMA [QCM] and PS). According to their results, CO_2 plasticization depresses T_g to approximately 10°C up to 30 MPa, compared to approximately 35°C for PS.

The initial approach used by Goel and Beckman [55] did not rely on retrograde vitrification, since their first successful attempt at producing foams

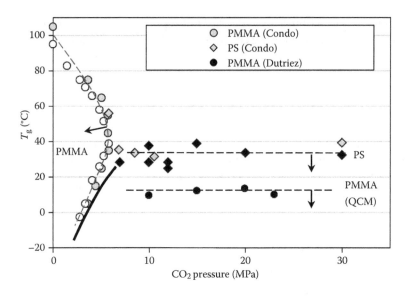

FIGURE 7.6

T_g of PS and PMMA. Solid curve represents the liquid–vapor equilibrium for CO_2. Dotted lines are qualitative trends. Arrows indicate the glassy regions. (Adapted from C. Dutriez et al., *RSC Adv.*, 2, 2821–2827, 2012 [© 2012 The Royal Society of Chemistry]. With permission.)

with cells below 1 μ used a 1S process with CO_2 sorption around 30 MPa and 40°C, i.e., clearly in the rubbery state for PMMA/CO_2 [50]. As shown in Table 7.1, this method produced foams with cell nucleation density close to 10^{12}/cm³ and 550-nm average cell size. Handa and Zhang [58] exploited the retrograde vitrification of PMMA by choosing pressure and temperature conditions in the narrow region below the lower T_g branch but above the liquid–vapor phase boundary of CO_2. At –0.2°C, the corresponding pressure is 3.4 MPa, and, upon depressurization, the PMMA sample came out unfoamed from the autoclave. The authors explained that this is not indicative of the PMMA being glassy under such conditions but of the low vapor pressure of CO_2 close to the transition line. Upon further exposing the CO_2-laden specimen to a second annealing step at 80°C for 2 min, PMMA foamed and produce low-density foams (up to 90% porosity) and cell sizes as low as 350 nm. The improvement in porosity compared to the 1S process of Goel shows the benefit of adding a thermal annealing step. This remarkable achievement (cell density N_0 near 4×10^{13}/cm³) was seen as a promising sign toward the production of low-density nanofoams with cell size approaching 100 nm by homogeneous nucleation. However, the range of conditions and types of polymers for which this strategy can be used are rather limited. Nawaby et al. [99] used the same approach with syndiotactic PMMA and ABS, both found to have the retrograde vitrification behavior, but failed to produce low-density nanofoams under these conditions.

7.4.1.2 High T_g Engineering Plastics

Interest shifted from PMMA to high T_g polymers for the production of separation membranes. Starting in 2001, the pioneering work of Krause [100] showed the benefit of the 2S process where sorption conditions are chosen at a temperature much lower than the T_g of the polymer, in which plasticization is insufficient to render the polymer rubbery. The second thermal conditioning step at a temperature close to the T_g of the polymer is thus essential to get nuclei to expand. As shown in Table 7.1, Krause et al. used CO_2 sorption at 25°C under moderate pressures (around 5 MPa) to foam thin films (approximately 100-μm thick) of PES [23,63], PEI [63], and PI [65,66] and produce foams with cells as small as 20 nm. In all cases, however, porosities were very low as the outer surfaces of the membranes were generally left unfoamed. Interestingly, despite the absence of a retrograde behavior for these polymers, the most successful sorption conditions were just above the liquid–vapor line of CO_2, as in Handa and Zhang's [58] work. The cell nucleation densities were of the order of 10^{14}/cm³ for PES and PEI, and close to 10^{15}/cm³ for PI, but the conditions and polymer properties did not allow expansion beyond approximately 30% porosity (1.4 expansion ratio). Despite the low expansion, Krause et al. [63] identified conditions of high CO_2 solubility in PES and PEI under which the core of the membrane could be made bicontinuous (i.e., have a continuous, open-pore network). Similar work has been carried out since 2006 by Chatchaisucha and Kumar [65] and Miller et al. [64,101] with thicker PEI sheets, at pressures and temperature close to values that were previously used by Krause, but with a slower pressure drop rate. Optimization of the 2S foaming conditions led to a slight improvement in foam porosity. A trade-off between cell size and foam expansion was observed, with porosities of approximately 30% for 30-nm cell size and 53% for foams with 120-nm cells.

Very recently, Aher et al. [66] used higher sorption pressures and temperatures (20 MPa/45°C), introduced a clamping force during the high-temperature conditioning step, and adjusted the desorption time (time lapse before the sample is exposed to the high-temperature foaming bath). On each side, foams showed a dense skin, an intermediate, low-density microcellular layer, and a nanocellular core. The lowest-density PEI foams had 52-nm cells in the core and an average porosity (including the dense and microcellular layers) of 64%. (The porosity of the nanocellular core was not reported.) The idea of constraining the surfaces of the sample during foaming was first applied by Siripurapu et al. [102] to PMMA, although no foam with cell nucleation density higher than 10^{12}/cm³ was reported at that time.

Guo et al. [69] investigated a 2S process to produce PSU foams, with CO_2 saturation temperatures as low as –10°C. Such low temperatures allowed CO_2 concentrations approaching 15 wt% at a pressure of 5 MPa. Saturation temperatures of 0°C and –10°C yielded nanofoams, with the average cell size around 30 nm and 47% porosity (cell density of 1.7×10^{15}/cm³). In contrast, microcellular foams were obtained at higher saturation temperatures for which the CO_2 solubility in PSU was less than 13 wt%. The authors noted that the transition from micro- to nanocellular foam was close to the vapor–liquid transition temperature of CO_2. They also observed nanoporous structures on the cell walls of some of the microcellular foams that they attributed to a secondary-phase separation by spinodal decomposition, which is induced in the cell walls by biaxial stress that is generated during expansion. This hypothesis has not yet been verified.

These attempts at producing high-cell-nucleation-density foams showed the difficulty of producing a sufficient number of nuclei by homogeneous nucleation while expanding these nuclei sufficiently to achieve high porosities ($p > 70\%$). Invariably, when choosing conditions promoting foam expansion such as slow depressurization [102] or higher foaming temperatures, cell size increased dramatically, presumably due to coalescence of the nuclei.

7.4.1.3 High-Porosity Nanofoams

Starting in 2009, Costeux and coworkers focused on approaches to increase the porosity of nanofoams. Their initial strategy relied on nucleating additives [103] (discussed in Section 7.4.2). This initial work helped map out the role not only of processing conditions but also of polymers' structure and properties [38,79]. The key lessons were that high-enough CO_2 levels (in excess of 25 wt% for acrylic polymers) are essential to provide a high level of nucleation while maintaining enough dissolved CO_2 to plasticize the polymer and allow cell growth and that a very fast pressure drop rate was beneficial. (Their foaming system can be depressurized in a fraction of a second.) This strategy is depicted in Figure 7.4c.

They designed a series of MMA-based random copolymers containing comonomers with higher CO_2 affinity than MMA, such as ethyl acrylate (EA), ethyl methacrylate (EMA), terbutylmethacrylate (tBMA), etc., covering a range of T_g from 60°C to 120°C [24]. Under the sorption conditions (30 MPa and 35°C), all polymers were in the amorphous state and readily foamed upon depressurization (1S process). Copolymers with T_g between 85°C and 105°C produced nanofoams with cell size between 100 and 200 nm with porosities between 60% and 80%, i.e., expansion ratio from 2.5 to 5.0 [24,62]. Some higher T_g polymers with high CO_2 solubility, such as MMA-co-PFMA, gave foams with 70-nm cell and 60% porosity and remarkably high cell nucleation densities. ($N_0 = 5 \times 10^{15}/cm^3$). When adding a second thermal conditioning step below the T_g of the polymer, porosities further increased (from 55% to 78% for the MMA-co-tBMA foam conditioned at 70°C). Costeux et al. then applied the same strategy to a high-molecular-weight MMA-co-EMA copolymer to take advantage of the high viscosity to minimize cell coalescence. The optimized 2S process produced a high-porosity foam ($p = 82\%$, i.e., 0.21 g/cm³ or 5.5 expansion ratio) with 80-nm cells. The cell nucleation density of $1.6 \times 10^{16}/cm^3$ is the highest reported for high-porosity foams relying solely on homogeneous nucleation. The MMA-co-EMA copolymers

(T_g = 96°C) were found to be more effective than PEMA (T_g = 65°C), which under similar conditions produced foams with 345-nm cells and 79% porosity [61]. The studies concluded that the homogeneous nucleation theory was consistent with the experimental trends observed, which was captured in a foaming model [62] that was later improved [104].

Guo et al. [59,105] expanded the 2S approach, which is used to produce PSU nanofoams [70] to PMMA. They carried out a thorough study of solubility and diffusivity of CO_2 in PMMA over a range of temperatures at the pressure of 5 MPa. They identified various regimes for PMMA/CO_2 properties that can be related to the multiple transitions between rubbery and glassy states on the T_g plot of Figure 7.6. To maximize CO_2 solubility in their foaming experiments without using pressures above 5 MPa, they reduced the saturation temperature to −30°C where CO_2 solubility exceeds 39 wt%. This resulted in the finest cell morphologies for PMMA foams (50-nm cells and 68% porosity), although the T_g of their commercial PMMA material (103°C) points to a copolymer rather than a homopolymer (T_g ~ 115°C). They related the PMMA/CO_2 regimes that were previously observed to various foam morphologies: foams were found to be microcellular at low CO_2 concentration and nanocellular at high CO_2 but with different morphologies depending on the foaming temperature. As T increased, foams went from closed nanocellular to open nanoporous structures, and finally to unusual foam morphologies (*nanoworms*), as shown in Figure 7.7. The authors proposed that spinodal decomposition or a localized phase separation could explain these observations.

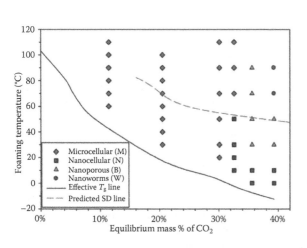
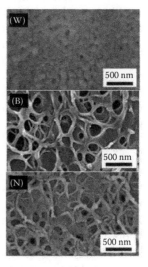

FIGURE 7.7
PMMA foaming map. Saturation performed at 5 MPa and various temperatures that control equilibrium CO_2 concentration. (Adapted from H. Guo et al., *Polymer*, 70, 231–241, 2015 [© 2015 Elsevier]. With permission.)

Guo and Kumar [69] were also able to produce PC nanocellular foams under very similar conditions (2S process with CO_2 saturation at 5 MPa and $-30°C$). The PC/CO_2 system showed lower solubility overall than the $PMMA/CO_2$ system, but similar solubility and diffusivity regimes exist, and similar foam morphological transitions are observed, albeit at different temperatures or CO_2 concentrations. Nanofoams with 20–30-nm cells and porosity up to 62% (2.6 expansion ratio) were obtained for saturation at $-30°C$. Increasing the saturation temperature to $-20°C$ and optimizing the foaming temperature to 70°C–90°C gave low-density nanofoams with 77% porosity and a cell size of 120–240 nm. Most recently, Liao et al. [60] used a similar 2S process (saturation at 13.8 MPa and $-10°C$ to 10°C, and foaming at 80°C) and showed that higher-molecular-weight PMMA produced finer cell structures (down to 37 nm) with cell densities in excess of $10^{16}/cm^3$.

7.4.1.4 Miscible Polymer Blends

Miscible blends are often used as a way to produce intermediate properties between those of the blend components. This may be beneficial toward the production of nanocellular foams, either to optimize the foaming behavior or to provide a compromise between foamability and foam properties. However, this approach has not been very popular to this day. The first study on miscible blends was reported by Krause et al. [68], who used PSU/PI blends as a way to investigate the role of CO_2 solubility on microcellular and nanocellular foam properties. They found a monotonous dependence of foam properties on the blend composition, and determined that nanofoams could be produced for blends of various compositions, as long as the sorption pressure was sufficient to ensure a minimum CO_2 solubility. With a 2S process, at sorption conditions of 5 MPa and 25°C, they obtained nanofoams with cell size between 200 and 500 nm and thus with lower cell nucleation densities than with PI alone.

Costeux et al. [49] used blends of styrene–acrylonitrile copolymer (SAN) with PEMA or MMA-co-EA in which miscibility could be controlled by the acrylonitrile level in SAN. Using a 2S foaming process with sorption at 33 MPa and 30°C, followed by thermal conditioning at 60°C, they observed that foams with cell size around 90–100 nm and porosities around 60%–69% could be produced from miscible blends. Interestingly, in several cases, these foams had more homogeneous cell size distribution and smaller cells than foams that were obtained from either blend component, as shown in Figure 7.8. Addition of as little as 10 wt% MMA-co-EA to SAN increased cell nucleation density by several orders of magnitude. Also of interest was the finding that blends with higher CO_2 solubility often gave lower cell densities, a reverse trend compared to predictions of the homogeneous nucleation theory. They concluded that homogeneous miscible blends may form separate phases under high CO_2 pressures, responsible for enhanced nucleation densities. Recent neutron scattering studies on the phase behavior of similar

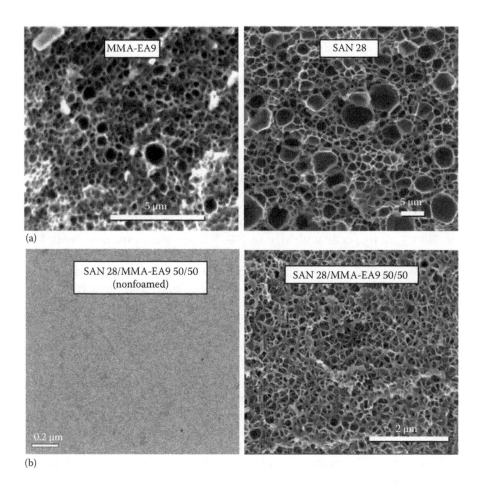

FIGURE 7.8

Unusual synergy in miscible SAN/PMMA blends: (a) scanning electron microscopy (SEM) images of foams produced by 2S process (33 MPa/30°C, post-expansion at 60°C) with MMA-co-EA (9 wt% EA) and SAN (28 wt% AN) and (b) transmission electron microscopy (TEM) of 50–50 blend showing miscibility (left) and SEM of foamed blend with 100-nm cells, 60% porosity (right). (Adapted from S. Costeux et al., *J. Mater. Res.*, 28, 17, 2351–2365, 2013 [© 2013 Cambridge University Press]. With permission.)

blends with CO_2 have confirmed the existence of a phase separation, the exact nature of which remains uncertain [106,107].

7.4.1.5 Semicrystalline Polymers

An alternative approach to avoid typical nucleation involves semicrystalline polymers. Polymer morphology in the confined amorphous phase between crystalline lamellae has been shown to be suitable for the generation of nanoscale cells [71]. In particular, a change in the crystalline morphology during foaming, which can be obtained by quenching the polymer from the melt into a metastable state, may generate local stress in the amorphous phase that favors the formation of voids around growing spherulites [108]. Local stresses are also generated in the cell walls during the cell growth of microcellular foams, leading to the creation of nanoscale voids in the cell walls [64,109–111]. This behavior has been observed both for high-density polyethylene (HDPE) [112] and for polypropylene in which crystal nucleation is promoted by a sorbitol additive [111]. Such foams do not fit the definition of nanocellular foams but rather that of open-cell microcellular foams.

Examples of nanofoams made from pure semicrystalline polymers are few. One recent example is due to Bao et al. [71], who used uniaxially oriented isotactic PP to induce cells in the amorphous space between the crystalline lamellae of shish–kebab structures by a one-step foaming process (15 MPa, ca. 140°C). Sparse cells with size of 150–300 nm were produced, with low porosity ($p < 25\%$). Li et al. [53] used PET, saturated with CO_2 for up to 15 days at 6 MPa and 25°C, followed by a foaming step at 235°C to produce a nanocellular foam with 193-nm cells and 3.4×10^{13} cells/cm^3. Expansion was very limited ($p < 20\%$). Foams produced under different conditions had sandwich structures with microcellular and nanocellular layers.

7.4.2 Nanocomposites

Literature results for nanofoam produced from nanocomposites are compiled in Table 7.2. Addition of nanoparticles to foam compositions may offer the dual benefit of assisting nucleation and improving foam mechanical properties. Clays can be easily modified to facilitate intercalation or exfoliation in various polymer matrices and provide interfaces likely to promote cell nucleation. This motivated a number of foaming studies by the Toyota Technological Institute (Japan) on PLA/clay nanocomposites [35,72] and later on PC/clay nanocomposites [75]. Fujimoto et al. [72] examined the effect of the dispersion of two different organically modified clays on the foaming behavior of PLA by a 2S high-temperature process. Under the optimal foaming temperature (165°C for both sorption and thermal conditioning, which is close to the melting point of PLA) and 10-MPa CO_2 pressure, the nanocomposite material in which modified clay showed the better dispersion produced foams with 360 nm and approximately 42% porosity, compared to 2.6-µm cells for the nanocomposite containing the other clay.

In an attempt to reduce cell size further, Ema et al. [35] foamed the same nanocomposites at higher pressure (28 MPa) and lower temperature (100°C–110°C) with a 1S process. Interestingly, the behavior of the two clays was similar, producing foams with 200-nm cells. However, the low temperature (approximately 60°C below the melting point of PLA) prevented foam expansion significantly. Porosities were very low, and the 200-nm cells were rather sparse ($N_0 = 2 \times 10^{13}$/cm^3 compared to 1.2×10^{14}/cm^3 for Fujimoto et al.). They observed by transmission electron microscopy (TEM) that nanocells were produced at the surface of clay particles, which confirms the heterogeneous nucleation mechanism.

Organically modified clay was also used by Lee et al. [73] to create nanocellular foams in HDPE nanocomposites that are compatibilized by maleic anhydride-grafted HDPE (15 wt%). They observed that both nanocomposite viscosity and cell nucleation increased with increasing clay levels, whereas the low CO_2 solubility was largely unaffected and crystallinity decreased. Smallest cells (ca. 250 nm) were obtained for nanocomposites with 3 wt% clay saturated with CO_2 at 8.3 MPa and 50°C and foamed at 125°C. Porosity remained low (17%). Heterogeneous nucleation was identified as the primary mechanism explaining the high cell nucleation density (3×10^{13}/cm^3), while the high viscosity suppressed cell expansion.

Urbanczyk et al. [36] compared one-step and modified two-step (see Figure 7.5c) foaming of SAN containing modified clays. An exfoliated clay functionalized with poly(ϵ-caprolactone) was shown to produce a significant increase in cell nucleation in both foaming processes when the temperature was reduced to 40°C and when the saturation pressure was higher (30 MPa) and depressurization was faster. The 2S process led to 300-nm cells with foam porosity around 40%, thanks to thermal conditioning close to the T_g of SAN.

Realinho et al. [78] examined the foaming behavior of PMMA that is filled with an organically modified montmorillonite (oMMT). A modified 2S process (Figure 7.5c) was used, with a saturation at 20 MPa/70°C, followed by a quench to ambient temperature before depressurization. This prevented foaming of the sample in this first step. A heating bath was used to induce foam expansion. Addition of 10wt% oMMT led to a foam with 330-nm cells. Porosity was close to 50% after the removal of solid skins.

Ito et al. [75] also used modified clays in PC nanocomposites that are compatibilized with SMA. Submicron cells (600 nm) were produced in a 2S process. Once again, porosities were below 20%. They observed that a higher level of clay restricted expansion due to stiffness of the nanocomposite. Zhai et al. [76] examined the effect of nanosized silica particles in PC nanocomposites. SiO_2 domains of approximately 50 nm promoted heterogeneous cell nucleation, leading to foams with 300–500-nm cells but low porosity. They concluded that higher SiO_2 levels (9 wt%) were beneficial owing to the relative inefficiency of the heterogeneous nucleation process with SiO_2, evidenced by the cell nucleation density in the foams being only 0.01% of the number density of SiO_2 particles in the polymer. Li et al. [77] also used silica nanoparticles that were produced in-situ in a PI matrix. Foams with 500-nm cells and presenting a secondary nanostructure in the cell walls were obtained. They concluded that the secondary pore morphology could be caused by nucleation at the surface of the silica nanoparticles. Note that this bimodal porous structure is similar to those that were observed by Guo et al. [70] on PSU in the absence of particles.

These previous nanocomposite systems make use of polymer matrices that are not ideally suited for the production of nanofoam. Homogeneous nucleation occurs more readily in polymers with a higher affinity for CO_2, which led Costeux and Zhu [79] to assess the benefit of nanoparticles in the foaming of acrylic polymers. Previously, Siripurapu et al. [37] had produced foams with cells around 1 μm by foaming PMMA/SiO_2 nanocomposites that were prepared in a solution at 34.5 MPa and 120°C. Under conditions where CO_2 solubility in PMMA is higher, Costeux and Zhu [79] observed that in a one-step foaming process, pure PMMA produced very inhomogeneous foams with cells around 900 nm but that the addition of 3 wt% silica nanoparticles led to homogeneous nanofoams with 280-nm cells and 77% porosity [38]. Silica particles with a diameter of 5 nm were added to PMMA by melt blending directly from a silica sol or gel to ensure good dispersion. The same benefit was obtained for MMA copolymers; applying a 2S foaming process to MMA-co-EA/SiO_2, with sorption under conditions (30–33 MPa, 40°C) that lead to CO_2 solubility of 30 wt%, Costeux and Zhu [38] produced foams with cell size around 100–180 nm and porosities up to 82%. They observed an optimum SiO_2 concentration at 0.5 wt% and a trend showing smaller cell size (i.e., higher cell nucleation density) for smaller SiO_2 particle size approaching the size of the critical nuclei radius. Similarly to Ito et al. [75], the effectiveness of particles as nucleation sites was also found to be approximately 0.1%, presumably due to nuclei coalescence [40].

The addition of POSS was found to be more effective than SiO_2 with a highly CO_2-philic MMA-co-EMA copolymer. POSS cages are much smaller in size (0.1 nm) and may not act as a true heterogeneous nucleation center. Yet, addition of 0.25 wt% POSS to MMA-co-EMA led to low-density nanofoams with cell size around 100 nm and record porosities up to 85% (0.17 g/cm^3 or expansion of 7). An example is shown in Figure 7.9. Foams with 65-nm

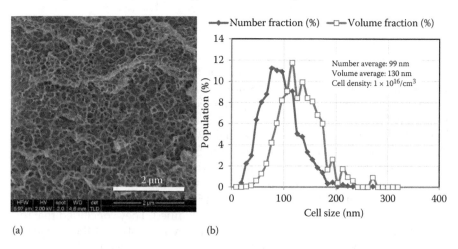

(a)   (b)

FIGURE 7.9
Example of nanocellular foam from MMA-co-EMA (50 wt% EMA) + 0.25 wt% POSS, produced by 2S process with saturation at 33 MPa/40°C and postexpansion in heat bath at 70°C. (a) SEM image of foam with 99-nm average cell size and 84% porosity. (b) Cell size distribution. (Adapted from S. Costeux et al., *Polymer,* 54, 11, 2785–2795, 2013 [© 2013 Elsevier]. With permission.)

cells and 2×10^{16} cells/cm^3 were also made, albeit with lower porosities (74%). Interestingly, the effect of POSS was found to be more dramatic with SAN in which the same POSS concentration (0.25 wt%) improved cell nucleation density by three orders of magnitude, resulting in 200-nm SAN/POSS foams with 70% porosity.

Ameli et al. [74] turned to multiwall carbon nanotubes (MWCNTs) to produce PP nanocomposite foams with microcellular and nanocellular morphologies with improved electrical conductivity. Under optimal conditions (30 MPa, saturation temperature of 142°C–143°C) in a 1S process, foams with cell size around 70–150 nm were produced but with low porosity.

In summary, the addition of nanoparticles such as clay, SiO_2, or POSS enhances nucleation density. The benefit is often more significant at low temperatures. However, to maximize the benefit of nanocomposites, it is preferable to choose conditions where the nucleation rate would be high in the matrix in the absence of nanoparticles. This includes high pressures and/or low temperatures, fast depressurization, and the choice of a CO_2-philic matrix.

7.4.3 Multiphase Polymer Systems

Nanofoams produced from multiphase polymers (immiscible blends or block copolymers) are listed in Table 7.3.

7.4.3.1 Organized Templated Systems

As shown in Figure 7.4b, two-phase polymer systems provide a mechanism to confine nuclei inside nanodomains, especially if a self-assembled structure is used. The concept was first applied with success by Yokoyama et al. [42,43] and Li et al. [113,114] using diblock copolymers of PS with a fluorinated block. PS-*b*-PFMA [42,43,113,114] and PS-*b*-PFS [43] films were cast from the solution to generate spherical morphologies with fluorinated domains inside a PS matrix. Porosity was introduced with a modified 2S process by swelling the diblock films with CO_2 at 60°C and pressures up to 30 MPa to achieve a high concentration of CO_2 in the fluorinated domains. The temperature was quenched to 0°C to freeze the PS matrix, the plasticized T_g of which is approximately 35°C (see Figure 7.6) before a very slow pressure drop. As CO_2 leaves the highly swollen domains, a 10–40-nm void is created in each domain and is maintained stable by the rigidity of the matrix. Very dense foam with 10-nm cells reached a record cell nucleation density of 3×10^{16}/cm^3. The porosity of other foams is also limited (ca. 30% for PS-*b*-PFMA and less than 20% for PS-*b*-PFS). This is ultimately a limit of the approach based on pure diblocks, where swelling cannot expand the domains enough for their volume to occupy a large void fraction during saturation without triggering a phase inversion. Expansion is also hindered by the frozen PS matrix. Other diblock systems, such as PS-*b*-PMMA [80] and PS-*b*-PFDA [81], were used in a similar process and yielded similar foam structures.

Dutriez et al. [82] also used various PMMA-*b*-PFMA diblocks and PFMA-*b*-PMMA-*b*-PFMA triblocks to produce nanocellular foams with cell sizes in the range of 20–70 nm and porosities below 25%. Triblock copolymers gave slightly lower cell size and porosity than diblocks. They concluded that the PMMA block needs to be rubbery during saturation and glassy during depressurization to allow for maximum CO_2 swelling and minimum cell collapse. Siripurapu et al. [37] previously compared the effect of graft (PMMA-*g*-PDMS) and block copolymers (PMMA-*b*-PFMA) as additives in PMMA. They observed that both diblocks form micelles in the polymer matrix but also play a role in stabilizing growing bubbles. The finest cellular morphology (ca. 300-nm cells) was obtained for a low-molecular-weight PMMA-*b*-PFMA added at 2 wt%. The cell nucleation density (and thus the porosity) remained low (ca. $10^{12}/cm^3$).

Reglero-Ruiz et al. [56,85,86] and later, Pinto et al. [83,87–90,115] conducted a thorough investigation of blends of PMMA with a PMMA-*b*-PBA-*b*-PMMA triblock copolymer (MAM). Addition of 10 wt% MAM was shown to significantly increase CO_2 solubility of PMMA [83,84]. The likely explanation is the higher affinity of the PBA block for CO_2, although blends also showed bulk densities lower than both PMMA and MAM [85] which may enhance CO_2 sorption. Reglero-Ruiz et al. foamed blends of PMMA with 10 wt% MAM, which led to dense foams ($p < 30\%$) with 200–300-nm cells using a 2S process [56] with thermal conditioning at 80°C–100°C and 300 nm with a 1S process [101] at 30 MPa and 23°C. Foams made with the same 1S process conditions, after removal of their thick densified skin, revealed higher cell density foams ($4 \times 10^{14}/cm^3$) with 200-nm cells and approximately 60% porosity in the foam core [86].

Pinto et al. [87] adopted the same PMMA/MAM system and systematically studied the effect of saturation pressure from 10 to 30 MPa and MAM content up to 20 wt%, and the relationship between the nanostructure of the blend (density of MAM micelles or vesicles in the PMMA matrix) and the cell nucleation density of the foam. The size of the PBA domains from MAM varied between 20 and 50 nm. They concluded that PBA domains' density is fairly constant at low MAM levels and is very close to the foam cell density (ca. $4 \times 10^{14}/cm^3$), indicating that nuclei start within MAM micelles. While increasing pressure with PMMA decreased cell size dramatically (as expected from the homogeneous nucleation theory), cell size became almost independent of pressure (150–190 nm) once MAM was added at 10 wt%. This important result shows that nanostructured blends with discrete CO_2-philic domains can control nucleation effectively and eliminate homogeneous nucleation in the PMMA matrix and nuclei coalescence (Figure 7.10a). In separate studies, MAM contents from 25 to 75 wt% [89] and 100 wt% [105] were examined. Foams with skinless porosities between 35% and 58% and cell size between 100 and 200 nm were produced with these systems around room temperature. Continuous open-pore structures could be obtained by tuning the MAM levels toward 75 wt% (Figure 7.10b).

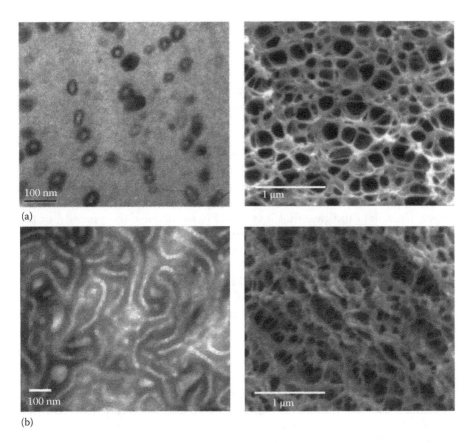

(a)

(b)

FIGURE 7.10
Influence of PMMA/MAM composition ([a] 10 wt% MAM; [b] 75 wt% MAM) on blend morphologies (TEM, left) and foam structures (SEM, right). (Adapted from J. Pinto et al., *J. Phys. Chem. C*, 118, 9, 4656–4663, 2014 [© 2014 American Chemical Society]. With permission.)

Forest et al. [90] used the same approach and compared the effectiveness of the linear MAM triblock copolymer with a core–shell copolymer with a PBA core and a PMMA shell, both blended into PMMA. They observed that linear MAM produced smaller PBA domains than the core–shell copolymer (20 nm versus 60 nm) and higher-domain densities. Foams were produced by a 2S process with CO_2 saturation under low-pressure and low-temperature conditions similar to those that were used by Handa and Zhang [58] and Nawaby [99] to exploit retrograde vitrification (3.2 MPa, 0°C) and foaming at 90°C. It is worth noting that due to low pressure and temperature during saturation and a slower depressurization rate, no foaming occurred before exposure to 90°C. This is in contrast with Pinto et al., who obtained foams after saturation at 10–30 MPa and ambient temperature without further heat treatment. Forest et al. confirmed the correlation between domain density in the blend and cell density in the foam and concluded that there was an inherent limitation in the cell density achievable with PMMA+MAM ($<1.5 \times 10^{15}$/cm³). Foams with 70–90-nm cells and 45% porosity revealed an interesting phenomenon when annealed just above the T_g of PMMA: the cell structure disappeared, and the specimens recovered the shape and translucence that they had before foaming. This led to the conclusion that during foaming, the polymer behaves as a viscoelastic solid that deforms in a reversible manner. This feature may be specific to saturation that is performed just under the retrograde vitrification (low T_g) branch shown in Figure 7.6, as upon temperature increase, the material saturated with CO_2 should remain in the vicinity of its glass transition temperature. Forest et al. [108] also investigated the effect of the molecular weight (MW) of the PMMA matrix. Upon introduction of a lower MW PMMA-B into the PMMA-A used above, their data indicate that it is possible to increase the porosity of the foams but with an increase in cell size. Nevertheless, 65 wt% PMMA-B and 35 wt% PMMA-A yielded foams with 110-nm cells and a porosity around 70%, which is unique for MAM-based systems (0.32 g/cm³). Medium-density foams with high levels of PMMA-B did not recover their shape upon annealing, indicating that the polymer underwent viscous flow during foaming due to the presence of a low MW fraction with higher mobility than highly entangled PMMA-A molecules.

7.4.3.2 Two-Phase Blends

A number of immiscible blend systems were used as an alternative to self-assembled block copolymers. Nawaby et al. [99] observed that ABS displayed the retrograde vitrification behavior that produced high-density foams with 470-nm cells under the conditions used by Handa and Zhang [58] for PMMA. Forest et al. [116] recently revisited this approach with various ABS produced either by the mass polymerization process or the emulsion process, the latter (ABS-C) producing smaller (0.1 μm), monodisperse rubber domains whereas the former gives large rubber particles with SAN inclusions. ABS-C produced rather homogeneous closed-cell foams with 400–600-nm cells similar to Nawaby's, whereas the mass ABS produced open-cell foams with larger cells but secondary structures (100-nm voids) in the cell walls.

Elastomeric domains (hydrogenated SIS, EP rubber, or SEBS) were also used by Nemoto et al. [45,47] and Sharudin and Ohshima [112] to modify polypropylene. Foams remain dense ($p < 25\%$) with cell size between 250 and 500 nm but with improved mechanical properties compared to pure PP.

Otsuka et al. [46,117] produced nanoscale PS/PMMA interpenetrating networks by in-situ polymerization of MMA in PS. PMMA domains were 200 nm in size in blends with 2:1 PS/PMMA ratio. Foaming with the process used by Taki [80] for PS-*b*-PMMA, in which the system is cooled below the T_g of the PS phase before pressure release, yielded irregular 50-nm voids that the authors attributed to cells forming in the PMMA phase.

Ruckdäschel et al. [44,93] engineered blends of SAN and PPE that are compatibilized by SBM triblocks, in which PPE form discrete domains, and the triblock butadiene blocks form interfacial domains between SAN and PPE phases. A 2S process with sorption at 5 MPa and 60°C and foaming at 160°C produced foams with 400-nm cells and 10^{13} cells/cm^3. Slightly smaller cells were obtained with (PPE/PS)/SAN blends that are compatibilized with SBM [93].

7.4.4 Summary

Nanocellular foam characteristics from various approaches are summarized graphically in Figure 7.11, showing a map of expansion ratio (with corresponding porosity) and average cell size. Contour plots for constant cell nucleation density are also represented. The graph illustrates the limitation of pure block copolymer approaches to produce high expansion ratios, even though cell nucleation densities are very high (10^{15}–10^{16}/cm^3) and controlled by templating. High T_g polymers provide foams with similar nucleation density that have recently reached expansion ratios close to 2.5. Acrylic copolymers and their nanocomposites with silica or POSS have resulted in similarly high nucleation densities but with expansion ratios up to 7.0. PMMA results show the impact of process conditions, such as low saturation temperatures, and the benefit of a thermal conditioning step to maximize expansion. Other approaches such as PMMA+MAM offer the potential to control nucleation

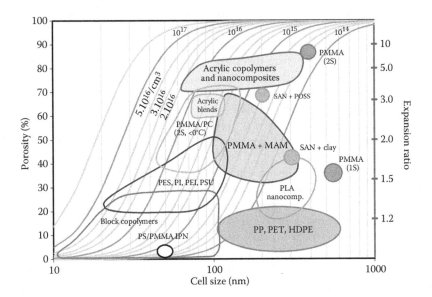

FIGURE 7.11
Overview of characteristics (average cell size, porosity, expansion ratio, and cell nucleation density) of nanocellular foams achieved from various polymer systems at the end of 2015. (From S. Costeux, *J. Appl. Polym. Sci.*, 131, 23, 41293, 2014 [© 2014 Wiley Periodicals, Inc.].)

density while achieving higher porosity than pure block copolymer (2.5 times expansion).

7.5 From Batch to Continuous Production

Solid-state batch foaming provides the flexibility of controlling pressures, temperatures, and the timing of the foaming event. Scale-up is limited to the production of small samples (typically less than 50 mm in size and less than 5 mm in thickness) on which certain properties can be measured. This is a valuable research tool to screen formulations, but it does not offer a viable pathway to enable large-scale continuous production, without which the applications of nanofoams will ultimately be very limited.

Unlike microcellular foams, which can readily be produced by a high temperature, one-step foaming process suitable for extrusion or injection molding, stable nanofoam structures are more likely to be produced in processes that involve a low temperature step. Most of the examples in Tables 7.1 through 7.3 use a 2S process, where the first step is low temperature saturation that ends with a depressurization event, during which no foaming occurs. Heating is needed to expand the polymer into a foam. For such situations, the method initially used at the Massachusets Institute of Technology (MIT) for semicontinuous microcellular foam production could be suitable for nanocellular foamed sheet production. It consists in saturating a roll of film with CO_2 in an autoclave, using a permeable spacer to allow CO_2 to access the surface of the film throughout the roll, followed by unwinding of the film into a heated bath or an oven to induce foaming. This process was brought to commercialization scale for the production of thermoformed cups from microcellular PET sheets. The same could be applied to produce nanoporous sheets, e.g., by exposing a polymer film to the low-temperature, low-pressure saturation method that was proposed by Guo et al. [59] for PMMA, followed by heat conditioning to produce a foamed sheet. It is likely that such a process would be limited to producing sheets at most a few millimeters thick. A related approach could involve saturating polymer beads with CO_2, and expanding these beads into a mold by heating, as is done to produce expandable bead foams with semicrystalline polymers [118]. No examples of production of nanocellular sheet or board by either process have yet been reported in the open literature, although there is one example of dense nanoporous monofilament PEI fibers being produced by a similar semicontinuous 2S process [119].

Continuous extrusion of thicker parts or sheets is difficult for nanocellular foams. Traditional extrusion foaming or foam injection molding processes, which involve mixing the blowing agent in the molten polymer and cooling to temperature slightly above the T_g or the T_m of the polymer before exiting through a die or nozzle where pressure drop induces foaming, are seemingly incompatible with the conditions that are commonly used to produce nanofoams. Indeed, in the vast majority of examples in Tables 7.1 through 7.3, polymer and gas come in contact at saturation conditions below the T_g or the T_m of the polymer, often closed to ambient, whereas extrusion requires that the polymer be molten at least 50°C above these temperatures to be able to flow. Furthermore, most of these examples use a 2S process in which the polymer remains unfoamed after depressurization.

Given these constraints, it is not a surprise that, as explained by Okolieocha et al. [120], as of 2015, there has been only one report in the peer-reviewed literature of a successful attempt to adapt extrusion foaming to produce nanocellular foams and no example of nanofoam injection molding. A recent patent application [121] discloses a modified injection molding process that is designed to provide rapid depressurization, which produced foams with 450–600-nm cells and porosities between 70% and 80%. A continuous extrusion process, developed by Costeux and Foether [122,123], succeeded in foaming a commercial PMMA (actually MMA-co-EA with 9 wt% EA) into nanocellular foams with cell size ca. 300 nm and 70% porosity. They used a custom extrusion foaming system that is equipped with a static cooling device (series of static mixers) and a generic slit die. Having previously shown [62] that nanofoams can be produced with this material and CO_2 in a 1S batch foaming process at 30 MPa and 40°C, they aimed at achieving these conditions just before the die. This is shown by the square symbol in Figure 7.12a, which represent a map of iso-solubility lines as a function of P and T. CO_2 solubility at these conditions is 25 wt%. However, melting of the polymer in the extruder can

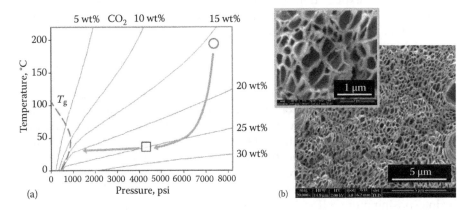

(a)

(b)

FIGURE 7.12
Nanofoam extrusion: (a) Pressure–temperature CO_2 solubility map in PMMA. Thin lines represent iso-solubility lines, and dashed line shows the retrograde vitrification temperature. Thick lines and symbols represent the ideal P–T path along the extrusion foaming line. (b) Nanocellular foam produced with 23 wt% CO_2. (Insert is a magnification.) (Adapted from S. Costeux et al., *Proceedings of the SPE FOAMS 2014 Conference*, Iselin, NJ [Sept 10–11, 2014] [© 2014 Society of Plastics Engineers]. With permission.)

only be done close to 200°C, at a maximum pressure that is indicated by the circle symbol, where the CO_2 solubility is only 16 wt%. Adding an excess of blowing agent is not recommended if homogeneous foams are desired [124], yet by adding 23 wt% CO_2 and achieving a very efficient temperature drop in the cooling element while maintaining sufficient back pressure, Costeux et al. succeeded in bringing the polymer/CO_2 mixture to P–T conditions close to batch foaming conditions, with sufficient residence time in the cooler to allow full dissolution of the excess CO_2 (see pathway on Figure 7.12a). Figure 7.12b shows an SEM of the 300-nm cell morphology with very few micron-sized cells. Adding silica nanoparticles produced similar nanofoams free of these larger cells, also with 70% porosity (expansion ratio of 3.3).

Further improvements are possible by way of secondary expansion with heat, use of additives such as MAM that allow control of nucleation at reduced pressures, or optimization of the die design to tailor the pressure drop rate [41]. This is an encouraging step toward a future commercial-scale process, but several technology breakthroughs are still needed to produce nanofoams in dimension and at output rates that can rival other foam products.

7.6 Properties of Nanocellular Foams and Potential Applications

A significant effort has been dedicated to the production of nanocellular foams with the expectation of superior performance compared to conventional foams. Recent research has turned to the characterization of properties of these materials to aid the discovery of new properties and applications.

7.6.1 Thermal Conductivity

Heat transfer through porous polymeric materials proceeds via conduction, convection, and radiation. Convection is negligible when pores or cells are below 1 mm. Conduction occurs through the solid polymer that constitutes the struts and walls of the cells and through the gas that is contained within the cells. Radiation described the phenomenon by which cell walls emit a radiation spectrum that is centered in the infrared, which is absorbed by other cell walls and reemitted. Reducing cell size not only affects radiation by interposing more cell walls in the path of heat transfer but also causes cell walls to be thinner, and thus more transparent to infrared radiation. It should also affect gas conductivity but only when cells are below 1 μm; smaller cells reduce the number of elastic collisions between gas molecules, in particular, when the cell size reaches the *mean free path* (i.e., the average traveling distance of gas molecules) between successive collisions (Knudsen effect). The mean free path of air is approximately 75 nm at ambient temperature and pressure so that, for porous materials with cells below 100 nm, air conduction should be reduced by approximately 2.5 times compared to still air (25 mW/m-K). Nanoscale dimensions may also contribute to change in the solid conduction and radiation by mechanisms that are negligible in regular foams.

Hrubesh and Pekala [4] applied a TC model to nanoporous organic aerogels and confirmed experimentally the presence of the Knudsen effect and the effect of tortuosity on solid conductivity. They determined that there is an optimal density to minimize the TC of organic aerogels, corresponding to a relative density of approximately 0.15 ($p \sim 85\%$). At lower density, heat transfer by conduction through the polymer is reduced, but the materials become more transparent to infrared radiation, which increases TC. Therefore, assuming that similar mechanisms control heat transfer in nanofoams, a common target for low-TC nanofoams is to minimize cell size below 100 nm (preferably below 75 nm) and porosity of 85% or more [8,125], corresponding to expansion ratios above 7.

Recent progress was made in refining this target. First, Notario et al. [126] published the first experimental validation of the Knudsen effect in nanocellular foams. They measured TC on a series of open-cell nanofoams from PMMA/MAM blends with medium to high density. The choice of an open-cell structure made it possible to impose a vacuum in all the cells and to extract the gas contribution by comparison with measurement without vacuum. As shown in Figure 7.13, their experimental data follow the cell-size dependence of the Knudsen equation, as do aerogels [127]. This confirms the cell size target (100 nm or less) to achieve a reduction of gas conduction by a factor 2 to 3 compared to conventional foams.

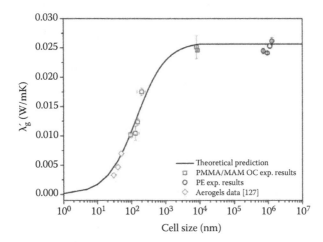

FIGURE 7.13
Comparison of gaseous contribution to TC predicted from the Knudsen equation with experimental data on open-cell nanocellular foams and aerogels. (Reprinted from B. Notario et al., *Polymer*, 56, 57–67, 2015 [© 2015 Elsevier]. With permission; X. Lu et al., *J. Non-Cryst. Solids*, 188, 226–234, 1995.)

No conclusive data are available yet to confirm the density target. Several approaches listed in Tables 7.1 through 7.3, such as foaming of diblock and triblock copolymers, of high T_g polymers or at sub-zero temperatures, have led to cells below 75 nm, but with expansion ratios usually below 2 ($p < 50\%$). The highest porosity demonstrated for nanofoams with cells of approximately 100 nm or less is actually 85% [24,38,62,79], which is expected to meet the minimum requirements. However, no data on the total TC have been published to date on such foams.

Forest et al. [128] applied a simple TC model to nanofoam structures that resulted in the conclusion that production of PS foams with cells smaller than 100 nm and densities between 0.1 and 0.2 g/cm^3 is required to obtain a minimum TC approximately half that of still air. Sundarram and Li [129], who used modeling in combination with experiment to extract the TC contribution of a thin nanoporous PEI layer in microcellular foams, concluded that nanofoams could have TC as low as 15 mW/m-K, although the measuring technique used does not directly measure the TC of the nanofoam layer. Notario et al. [126] also used a simplified model and predicted that a relative density < 0.05 (expansion ratio over 20) would be needed for PMMA foams with 100-nm cells to achieve 17 mW/m-K. More refined models accounting for coupling between different TC contributions may be needed at the nanoscale to successfully predict heat transfer in low-density nanofoams [130].

7.6.2 Mechanical Behavior

Gold nanofoams have been shown to become stronger as the solid is distributed in thinner ligaments, which ultimately produces nanofoams with higher modulus than the nonporous metal [131,132]. The behavior of polymers cannot be transposed from metals, and thus attempts have been made to confirm the mechanical strength benefit of polymer nanofoams. Chen and Pugno [133] explained that in nanostructured foams, surface effects (residual stress and surface elasticity) have increasingly stronger contribution as the cell size is decreased [3]. The large surface area of nanocellular foam may contribute to the modulus or yield stress sufficiently that it could offset the loss of mechanical properties that are usually associated with density reduction. In addition, confinement of the solid in the cell walls can result in a reduction of free volume, as evidenced by an increase in the T_g of the polymer in the cell wall [86,134], which can also contribute to mechanical strength. Miller and Kumar [101] compared the tensile and impact properties of PEI nanofoams with microcellular foams. They determined that at the same foam density, decreasing cell size did not affect the tensile modulus and the yield stress but led to a significant increase of toughness and strain at break. PEI nanofoams were also found more ductile, resulting in better impact resistance than both the unfoamed polymer and microcellular foams over the density range studied, as shown in Figure 7.14.

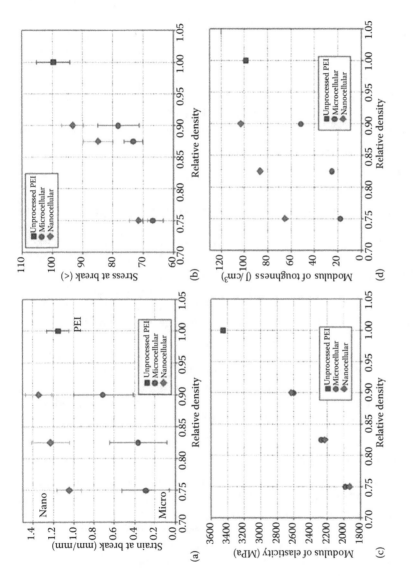

FIGURE 7.14

Comparison of selected mechanical properties of PEI nanocellular foams with microcellular and unfoamed PEI as a function of relative density. (a) Stain at break, (b) stress at break, (c) elastic modulus, and (d) toughness. (Reprinted from D. Miller and V. Kumar, *Polymer*, 52, 13, 2910–2919, 2011 [© 2011 Elsevier]. With permission.)

Conversely, Sharudin and Ohshima [94] observed for PP and PP/SEBS blends that the yield stress and stress at break were increased compared to unfoamed materials, whereas the strain at break was unaffected. Notario et al. [134] showed a 25% increase of the modulus of elasticity of nanocellular foams at high strain rate and an improvement of the Shore hardness by 15%. However, the mechanical response (strain at break, toughness) at low strain rates was worse for nanocellular foams than for the unfoamed polymer, likely due to increased rigidity of the polymer matrix and of a decrease in cell wall thickness.

Thus nanocellular foams have generally superior properties compared to microcellular foams at the same porosity, but the nature of the confined polymer, its crystallinity, the density, and likely the morphology of the foam determine whether the nanofoams' mechanical properties will approach or exceed those of the solid.

7.6.3 Filtration, Separation, and Storage

Open nanoporous membranes made from solvent-based processes have found application as battery separators in which the transport of ions can be controlled by hindered diffusion in the narrow channels. Films and membranes with connected nanopores have also been used as catalyst support due to their high surface area or as filtration or separation media in which the balance of properties, such as particle rejection and flux, can be tuned by adjusting the pore size distribution. High-surface-area nanoporous materials can also be useful for gas capture and storage [135], catalysis, or light harvesting, as reviewed by Dawson et al. [136].

For CO_2-blown nanofoams to be used in such applications, it is necessary to produce bicontinuous foams, as was done by Krause [63,68] and Pinto et al. [88], and to form a foam without integral skin [57]. While such features are easily obtained with solvent-based processes, no successful production of bicontinuous CO_2-blown nanofoams with open surface has been reported.

7.6.4 Other Potential Applications

Low κ dielectric materials have also been produced with PI using the porogen approach [137,138]. Krause et al. [67] demonstrated that a CO_2 foaming process could produce PI nanofoams with a dielectric constant below 1.8.

Yokoyama et al. [42] showed that optically transparent nanocellular foams can be made by producing homogeneous nanofoams with cell size well below the visible wavelength. The refractive index can be tuned between 1.5 and 1.23 by control of the porosity. Such properties have not yet been exploited.

Finally, other applications such as the production of bioengineered scaffolds, sensors, and acoustic absorbers [139] have been proposed for nanocellular foams.

7.6.5 Patent Activity

The first patents that mention production of foams with cells below 1 μ emanated from the MIT in the 1990s as an extension of microcellular foam activities. Figure 7.15 shows patent applications that are related to gas-blown nanocellular foam composition and processes that were filed over the years. The graph clearly indicates a growing interest since 2005, in particular, from industry, with 18 applications filed in that period by three companies (BASF, Dow, and Sabic) out of a total of 33 since 1990.

These applications cover a range of compositions and methods, which are broadly categorized in Table 7.4. Processes include a generic approach based on a two-step foaming (film saturation followed by expansion), using specific conditions such as low temperature saturation to maximize CO_2 uptake (Handa at Canadian National Research Council [CNRC] and University of Washington). Improvements are proposed, such as the use of systems that are capable of an ultrafast depressurization rate of 15 to 200 GPa/s (BASF). Equipment for the continuous production of submicrocellular foams was also proposed by MIT, while Dow used methods to increase CO_2 level to produce nanofoams with higher expansion ratios. Batch systems in which film expansion is restricted as a way to control cell structure were proposed in academia (University of Washington and University of North Carolina), while Dow used a semicontinuous process to produce nanofoams by

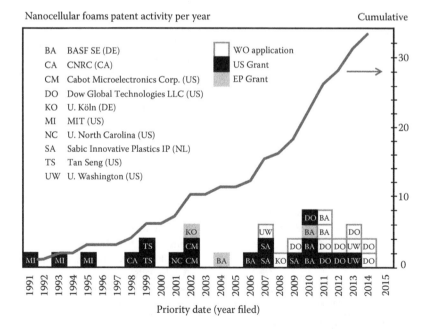

FIGURE 7.15
Patent activity related to gas-blown nanocellular foams. Each box represents one patent application. Shading indicates grant status of these applications.

TABLE 7.4

Main Patents or Patent Applications Relevant to CO_2-Blown Nanofoams

Category	Patent/Application #	General Area Shown in Examples
Generic process and foams	USRE37932E1 (MI-1991)	Processes for (sub)microcellular foams
	US5955511A (CA-1998)	Low temperature saturation
	US7842379B2 (SA-2007)	[No foam examples]
	WO2014210523A1 (UW-2013)	Low temperature saturation
Explosive foaming	US8529808B2 (BA-2010)	Ultrafast depressurization process
	EP2571929B1 (BA-2010)	Ultrafast depressurization process
	US8636929B2 (BA-2010)	Pharmaceutical composition
	WO2013075991A1 (BA-2011)	Open-cell nanoporous foams
Continuous process	US5866053A (MI-1993)	Fast depressurization continuous device
	US6005013A (MI-1995)	Gear throttle as a nucleation device
	WO2013075994A1 (BA-2011)	Ultrafast depressurization
	US9145478B2 (DO-2011)	High CO_2 process
Constrained foaming	US7658989B2 (NC-2001)	Constrained nanofoam film
	US9203070B2 (UW-2013)	Constrained nanofoam film
	WO2015183639 (DO-2014)	Variable volume injection molding
Alternate process	EP1576038B1 (KO-2002)	Reactive microemulsions
	US20110174192A1 (KO-2008)	Sugar microemulsions
	US8206626B2 (BA-2006)	Multiphase nanostructured polymer mixtures
Polymer compositions	EP1732975B1 (BA-2004)	Multiphase nanostructured polymer mixtures
	US7842379B2 (SA-2007)	Nanofoams from specific phthalimidine compound
	US8283390B2 (SA-2009)	Polysiloxane block copolymer nanofoams
	US9187611B2 (DO-2010)	Methacrylate copolymers nanofoams
	WO2013048761A3 (DO-2011)	Blend composition with acrylonitrile for nanofoams
	WO2014078215A1 (DO-2012)	Nanofoams from POSS-grafted polymer
Additives	WO2011066060A1 (DO-2009)	Nanoscale nucleators
	US9145477B2 (DO-2012)	Nanofoams with CO_2-philic additives
	WO2015096127A1 (DO-2014)	Nanofoams with flame-retardant (FR) additives
	WO2015148193A1 (DO-2014)	Nanoscale FR
Partially transparent foams	US6555589B1 (TS-1999)	Transparent, high T_g foams
	US6555590B1 (TS-1999)	Transparent, high T_g foams
	US7267607B2 (CM-2002)	TPU Chemical Mechanical Planarization (CMP) pads
	US7435165B2 (CM-2002)	Transparent CMP pads
	WO2008112821A2 (UW-2007)	Partially transparent nanofoams

Note: The code in parenthesis refers to Figure 7.15.

injecting a polymer/CO_2 mixture in a variable size mold. Foams were also made within a mold by swelling microemulsions that are designed for CO_2 (U. Köln) or two-phase polymer systems with a selective solvent (BASF).

Preferred polymer compositions range from acrylic and acrylonitrile copolymers and blends (Dow) to specific monomers for high-performance foams (Sabic). Additives (nanoscale particles or small molecules) were shown to favor nucleation, reduce density, or provide improved fire performance in nanofoams (Dow). Finally, foams with nanoscale cells and relatively high density were shown to maintain transparency to some wavelength and be useful for polishing pads (Cabot Microelectronics) or high-performance foams (Tan Seng, Wright Materials Research).

7.7 Conclusion

Significant progress has been made in recent years in the development of CO_2-blown nanocellular foams. New strategies have been devised to reduce cell size, increase cell homogeneity, and lower foam density. Research has been focused on optimizing foaming processes and designing the polymer system to maximize the density of nuclei and control their coalescence. Three main directions have been explored for polymer systems: (1) highly CO_2-philic single-phase polymers, (2) thermoplastic composites with nanofillers, and (3) templated, multiphase polymers.

Pure block copolymer-templated systems have successfully produced high-cell-density foams (10^{14}–10^{16}/cm^3) by confining nucleation events within dispersed domains but result in low-porosity foams due to minimal cell expansion (10–50 nm). Single polymers with very high T_g also produce very high density foams ($p < 30\%$) with small cells and cell densities close to 10^{14}/cm^3. Such materials may become useful for their dielectric or optical properties.

Highly CO_2-philic polymers can easily produce medium-density foams ($p < 50\%$) with cell nucleation density around 10^{13}/cm^3 and cell size in the 300–500-nm range. However, reducing both cell size and foam density is essential for applications that target insulation. Using nanofillers or by designing the chemistry of the polymers, in particular, of those based on methylmethacrylate, it has now become possible to produce 100-nm cell foams with porosities as high as 85% (seven-fold expansion) and cell densities up to 10^{16}/cm^3. Such foams have been shown to exploit the Knudsen effect and constitute a major step toward the production of nanocellular superinsulation.

Hybrid systems, where CO_2-philic domains are distributed within a polymer matrix, produce foams with intermediate properties. They offer the potential benefit of controlling nucleation events (as templated systems) while allowing more foam expansion, at lower pressure or pressure drop rates than necessary for single-phase systems. This benefit is yet to be fully realized, as foam expansion remains limited (less than three-fold).

Significant challenges remain for the manufacture of nanocellular foams at a commercial scale, the more substantial of which being the lack of scalability of the widely used solid-state batch foaming process. Continuous nanofoam production is becoming an area of interest for academia, and recent patent literature suggests that the foam industry is also progressing rapidly toward this objective.

Abbreviations

ABS	acrylonitrile–butadiene–styrene copolymer
BMA	butylmethacrylate
EA	ethylacrylate
EP	ethylene–propylene rubber
HDPE	high-density polyethylene
hSIS	hydrogenated polystyrene-*b*-polyisoprene-*b*-polystyrene
MAM	PMMA-*b*-PBA-*b*-PMMA
MSMA	trimethoxysilylpropyl methacrylate
PBA	polybutylacrylate
PC	polycarbonate
PEEK	polyether ether ketone
PEI	polyetherimide
PEMA	polyethylmethacrylate
PEP	poly(ethylene propylene)
PES	polyethersulfone
PET	polyethylene terephthalate
PFDA	poly(1,1,2,2-tetrahydroperfluorodecylacrylate)
PFMA	perfluoro-octylethylmethacrylate
PFS	polystyrene-*b*-poly[4-(perfluorooctylpropyloxy)styrene]
PI	polyimide
PLA	poly lactic acid
PMMA	polymethylmethacrylate
POSS	polyhedral oligomeric silsesquioxane
PP	polypropylene
PPE	polyphenylene ether
PS	polystyrene
PSU	polysulfone
SAN	styrene–acrylonitrile copolymer
SBM	styrene-*b*-butadiene-*b*-methylmethacrylate
SEBS	styrene-*b*-ethylene-butene-*b*-styrene
SMA	styrene–maleic anhydride copolymer

References

1. M. Šupová, G. S. Martynková and K. Barabaszová. 2011. Effect of nanofillers dispersion in polymer matrices: A review. *Sci. Adv. Mater.,* 3, 1–25.

2. R. B. Pernites, R. R. Ponnapati and R. C. Advincula. 2011. Superhydrophobic–superoleophilic polythiophene films with tunable wetting and electrochromism. *Adv. Mater.,* 23, 3207–3213.

3. Z. X. Lu, C. G. Zhang, Q. Liu and Z. Y. Yang. 2011. Surface effects on the mechanical properties of nanoporous materials. Thermal properties of organic and inorganic aerogels. *J. Phys. D: Appl. Phys.,* 44, 395404.

4. L. W. Hrubesh and R. W. Pekala. 1994. Thermal properties of organic and inorganic aerogels. *J. Mater. Res.,* 9, 731–738.

5. A. S. Zalusky, R. Olayo-Valles, J. H. Wolf and M. A. Hillmyer. 2002. Ordered nanoporous polymers from polystyrene–polylactide block copolymers. *J. Am. Chem. Soc.,* 124, 12761–12773.

6. J. L. Hedrick, T. P. Russell, J. Labadie, M. Lucas and S. Swanson. 1995. High temperature nanofoams derived from rigid and semi-rigid polyimides. *Polymer,* 36, 2685–2697.

7. J. L. Hedrick, C. J. Hawker, R. DiPietro, R. Jerome and Y. Charlier. 1995. The use of styrenic copolymers to generate polyimide nanofoams. *Polymer,* 36, 4855–4866.

8. D. Schmidt, V. I. Raman, C. Egger, C. du Fresne and V. Schädler. 2007. Templated cross-linking reactions for designing nanoporous materials. *Mater. Sci. Eng., C,* 27, 1487–1490.

9. D. F. Schmidt, C. du Fresne von Hohenesche, A. Weiss and V. Schädler. 2008. Colloidal gelation as a general approach to the production of porous materials. *Chem. Mater.,* 20, 2851–2853.

10. C. du Fresne von Hohenesche, D. F. Schmidt and V. Schädler. 2008. Nanoporous melamine–formaldehyde gels by microemulsion templating. *Chem. Mater.,* 20, 6124–6129.

11. B. S. Lalia, V. Kochkodan, R. Hashaikeh and N. Hilal. 2013. A review on membrane fabrication: Structure, properties and performance relationship. *Desalination,* 326, 77–95.

12. M. L. Deng, Q. Zhou, A. K. Du, J. van Kasteren and Y. Z. Wang. 2009. Preparation of nanoporous cellulose foams from cellulose-ionic liquid solutions. *Mater. Lett.,* 63, 1851–1854.

13. A. K. Hebb, K. Senoo, R. Bhat and A. I. Cooper. 2003. Structural control in porous cross-linked poly(methacrylate) monoliths using supercritical carbon dioxide as a "pressure-adjustable" porogenic solvent. *Chem. Mater.,* 15, 2061–2069.

14. M. Schwan, L. G. A. Kramer, T. Sottmann and R. Strey. 2010. Phase behaviour of propane- and scCO$_2$-microemulsions and their prominent role for the recently proposed foaming procedure POSME (Principle of Supercritical Microemulsion Expansion). *Phys. Chem. Chem. Phys.,* 12, 6247–6252.

15. J. E. Martini. 1981. *The Production and Analysis of Microcellular Foam.* MS in Mechanical Engineering Thesis, Massachusetts Institute of Technology, Cambridge, MA.

16. J. R. Youn and N. P. Suh. 1985. Processing of microcellular polyester composites. *Polym. Compos.,* 6, 175–180.

17. M. Hatanaka and H. Saito. 2004. In-Situ investigation of liquid–liquid phase separation in polycarbonate/carbon dioxide system. *Macromolecules,* 37, 7358–7363.

18. X. Xu, D. E. Cristancho, S. Costeux and Z.-G. Wang. 2012. Density-functional theory for polymer-carbon dioxide mixtures: A perturbed-chain SAFT approach. *J. Chem. Phys.,* 137, 054902–054908.

19. S. N. Leung, A. Wong, Q. Guo, C. B. Park and J. H. Zong. 2009. Change in the critical nucleation radius and its impact on cell stability during polymeric foaming processes. *Chem. Eng. Sci.,* 64, 4899–4907.

20. S. N. Leung. 2009. *Mechanisms of Cell Nucleation, Growth, and Coarsening in Plastic Foaming: Theory, Simulation, and Experiment.* Ph.D. Thesis, University of Toronto, Toronto, Canada.

21. M. Blander and J. L. Katz. 1975. Bubble nucleation in liquids. *AIChE J.,* 21, 833–848.

22. S. K. Goel and E. J. Beckman. 1994. Generation of microcellular polymeric foams using supercritical carbon dioxide. I: Effect of pressure and temperature on nucleation. *Polym. Eng. Sci.,* 34, 1137–1147.

23. B. Krause, R. Mettinkhof, N. F. A. van der Vegt and M. Wessling. 2001. Microcellular foaming of amorphous high-Tg polymers using carbon dioxide. *Macromolecules,* 34, 874–884.

24. S. Costeux, H. Jeon, S. Bunker and I. Khan. 2012. Nanocellular foams from acrylic polymers: Experiments and modeling. *10th International Conference on Foam Materials & Technology (SPE FOAMS 2012),* Barcelona, Spain.

25. Y. Kim, C. B. Park, P. Chen and R. B. Thompson. 2011. Origins of the failure of classical nucleation theory for nanocellular polymer foams. *Soft Matter,* 7, 7351–7358.

26. X. Xu, D. E. Cristancho, S. Costeux and Z.-G. Wang. 2013. Bubble nucleation in polymer-CO2 mixtures. *Soft Matter,* 9, 9675–9683.

27. Y. Kim, C. B. Park, P. Chen and R. B. Thompson. 2011. Towards maximal cell density predictions for polymeric foams. *Polymer,* 52, 5622–5629.

28. Y. Kim, C. B. Park, P. Chen and R. B. Thompson. 2012. Maximal cell density predictions for compressible polymer foams. *Polymer,* 54, 841–845.

29. S. D. Lubetkin. 2003. Why is it much easier to nucleate gas bubbles than theory predicts? *Langmuir,* 19, 2575–2587.

30. D. L. Tomasko, A. Burley, L. Feng, S.-K. Yeh, K. Miyazono, S. Nirmal-Kumar, I. Kusaka and K. Koelling. 2009. Development of CO_2 for polymer foam applications. *J. Supercrit. Fluids,* 47, 493–499.

31. J. S. Colton and N. P. Suh. 1987. The nucleation of microcellular thermoplastic foam with additives. Part I. Theoretical considerations. *Polym. Eng. Sci.,* 27, 485–492.

32. N. S. Ramesh and S. T. Lee. 2005. Do nanoparticles really assist in nucleation of fine cells in polyolefin foams? *Cell. Polym.,* 24, 269–277.

33. S. F. Jones, G. M. Evans and K. P. Galvin. 1999. Bubble nucleation from gas cavities—A review. *Adv. Colloid Interface Sci.,* 80, 27–50.

34. P. Spitael, C. W. Macosko and R. B. McClurg. 2004. Block copolymer micelles for nucleation of microcellular thermoplastic foams. *Macromolecules,* 37, 6874–6882.

35. Y. Ema, M. Ikeya and M. Okamoto. 2006. Foam processing and cellular structure of polylactide-based nanocomposites. *Polymer,* 47, 5350–5359.

36. L. Urbanczyk, C. Calberg, C. Detrembleur, C. Jérôme and M. Alexandre. 2010. Batch foaming of SAN/clay nanocomposites with scCO$_2$: A very tunable way of controlling the cellular morphology. *Polymer,* 51, 3520–3531.

37. S. Siripurapu, J. M. DeSimone, S. A. Khan and R. J. Spontak. 2005. Controlled foaming of polymer films through restricted surface diffusion and the addition of nanosilica particles or CO2-philic surfactants. *Macromolecules,* 38, 2271–2280.

38. S. Costeux and L. Zhu. 2013. Low density thermoplastic nanofoams nucleated by nanoparticles. *Polymer,* 54, 2785–2795.

39. X. Xu, D. E. Cristancho, S. Costeux and Z.-G. Wang. 2014. Nanoparticle solvation in polymer–CO2 mixtures. *J. Phys. Chem. B,* 118, 8002–8007.

40. Z. Zhu, C. B. Park and J. H. Zong. 2008. Challenges to the formation of nanocells in foaming processes. *Int. Polym. Proc.,* 23, 270–276.

41. C. B. Park, D. F. Baldwin and N. P. Suh. 1995. Effect of the pressure drop rate on cell nucleation in continuous processing of microcellular polymers. *Polym. Eng. Sci.,* 35, 432–440.

42. H. Yokoyama, L. Li, T. Nemoto and K. Sugiyama. 2004. Tunable nanocellular polymeric monoliths using fluorinated block copolymer templates and supercritical carbon dioxide. *Adv. Mater.,* 16, 1542–1546.

43. H. Yokoyama and K. Sugiyama. 2005. Nanocellular structures in block copolymers with CO$_2$-philic blocks using CO$_2$ as a blowing agent: Crossover from micro- to nanocellular structures with depressurization temperature. *Macromolecules,* 38, 10516–10522.

44. H. Ruckdäschel, P. Gutmann, V. Altstädt, H. Schmalz and A. H. E. Müller. 2007. Development of micro- and nano cellular polymers. *MRS Proceedings,* 1056, HH11-75.

45. T. Nemoto, J. Takagi and M. Ohshima. 2008. Nanoscale cellular foams from a poly(propylene)-rubber blend. *Macromol. Mat. Eng.,* 293, 991–998.

46. T. Otsuka, K. Taki and M. Ohshima. 2008. Nanocellular foams of PS/PMMA polymer blends. *Macromol. Mat. Eng.,* 293, 78–82.

47. T. Nemoto, J. Takagi and M. Ohshima. 2008. Control of bubble size and location in nano-/microscale cellular poly(propylene)/rubber blend foams. *Macromol. Mat. Eng.,* 293, 574–580.

48. V. Kumar and N. P. Suh. 1990. A process for making microcellular thermoplastic parts. *Polym. Eng. Sci.,* 30, 1323–1329.

49. S. Costeux, S. P. Bunker and H. K. Jeon. 2013. Homogeneous nanocellular foams from styrenic-acrylic polymer blends. *J. Mater. Res.,* 28, 2351–2365.

50. S. K. Goel and E. J. Beckman. 1994. Generation of microcellular polymeric foams using supercritical carbon dioxide. II: Cell growth and skin formation. *Polym. Eng. Sci.,* 34, 1148–1156.

51. Y. Sato, K. Fujiwara, T. Takikawa, Sumarno, S. Takishima and H. Masuoka. 1999. Solubilities and diffusion coefficients of carbon dioxide and nitrogen in polypropylene, high-density polyethylene, and polystyrene under high pressures and temperatures. *Fluid Phase Equilib.,* 162, 261–276.

52. K. A. Patankar, J. D. Moore, A. Wong and S. Costeux. 2015. Evaluation of nitrogen as a co-blowing agent in nanocellular foam. *Annu. Tech. Conf.—Soc. Plast. Eng.,* Orlando, FL.

53. D. Li, T. Liu, L. Zhao and W. Yuan. 2012. Controlling sandwich-structure of PET microcellular foams using coupling of CO$_2$ diffusion and induced crystallization. *AIChE J.,* 58, 2512–2523.

54. D. Tammaro, V. Contaldi, M. P. Carbone, E. Di Maio and S. Iannace. 2015. A novel lab-scale batch foaming equipment: The mini-batch. *J. Cell. Plast.*

55. S. K. Goel and E. J. Beckman. 1995. Nucleation and growth in microcellular materials: Supercritical CO_2 as foaming agent. *AIChE J.*, 41, 357–367.

56. J. A. Reglero Ruiz, J.-M. Tallon, M. Pedros and M. Dumon. 2011. Two-step micro cellular foaming of amorphous polymers in supercritical CO_2. *J. Supercrit. Fluids*, 57, 87–94.

57. H. Janani and M. H. N. Famili. 2010. Investigation of a strategy for well controlled inducement of microcellular and nanocellular morphologies in polymers. *Polym. Eng. Sci.*, 50, 1558–1570.

58. Y. P. Handa and Z. Zhang. 2000. A new technique for measuring retrograde vitrification in polymer-gas systems and for making ultramicrocellular foams from the retrograde phase. *J. Polym. Sci., Part B: Polym. Phys.*, 38, 716–725.

59. H. Guo, A. Nicolae and V. Kumar. 2015. Solid-state poly(methyl methacrylate) (PMMA) nanofoams. Part II: Low-temperature solid-state process space using CO_2 and the resulting morphologies. *Polymer*, 70, 231–241.

60. Z.-E. Liao, S.-K. Yeh, C.-C. Chu and T.-W. Tseng. 2016. Critical parameters of generating PMMA nanocellular foam. *Annu. Tech. Conf.—Soc. Plast. Eng.*, Indianapolis, IN.

61. S. Costeux, S. Bunker and H. Jeon. 2013. Homogeneous nanofoams from synergistic blends. *11th International Conference on Foam Materials & Technology (SPE FOAMS 2013)*, Seattle, WA.

62. S. Costeux, I. Khan, S. Bunker and H. Jeon. 2015. Experimental study and modeling of nanofoams formation from single phase acrylic copolymers. *J. Cell. Plast.*, 51, 197–221.

63. B. Krause, H. J. P. Sijbesma, P. Münüklü, N. F. A. van der Vegt and M. Wessling. 2001. Bicontinuous nanoporous polymers by carbon dioxide foaming. *Macromolecules*, 34, 8792–8801.

64. D. Miller, P. Chatchaisucha and V. Kumar. 2009. Microcellular and nanocellular solid-state polyetherimide (PEI) foams using sub-critical carbon dioxide I. Processing and structure. *Polymer*, 50, 5576–5584.

65. P. Chatchaisucha and V. Kumar. 2006. Micro and nano scale-state PEI foams. *SPE ANTEC Conference*, 64th, 2790–2794.

66. B. Aher, N. M. Olson and V. Kumar. 2013. Production of bulk solid-state PEI nanofoams using supercritical CO_2. *J. Mater. Res.*, 28, 2366–2373.

67. B. Krause, G.-H. Koops, N. F. A. van der Vegt, M. Wessling, M. Wubbenhorst and J. Van Turnhout. 2002. Ultralow-k dielectrics made by supercritical foaming of thin polymer films. *Adv. Mater.*, 14, 1041–1046.

68. B. Krause, K. Diekmann, N. F. A. van der Vegt and M. Wessling. 2002. Open nanoporous morphologies from polymeric blends by carbon dioxide foaming. *Macromolecules*, 35, 1738–1745.

69. H. Guo and V. Kumar. 2015. Some thermodynamic and kinetic low-temperature properties of the PC-CO_2 system and morphological characteristics of solid-state PC nanofoams produced with liquid CO_2. *Polymer*, 56, 46–56.

70. H. Guo, A. Nicolae and V. Kumar. 2015. Solid-state microcellular and nanocellular polysulfone foams. *J. Polym. Sci., Part B: Polym. Phys.*, 53, 975–985.

71. J.-B. Bao, T. Liu, L. Zhao, D. Barth and G.-H. Hu. 2011. Supercritical carbon dioxide induced foaming of highly oriented isotactic polypropylene. *Ind. Eng. Chem. Res.*, 50, 13387–13395.

72. Y. Fujimoto, S. S. Ray, M. Okamoto, A. Ogami, K. Yamada and K. Ueda. 2003. Well-controlled biodegradable nanocomposite foams: From microcellular to nanocellular. *Macromol. Rapid Commun.*, 24, 457–461.

73. Y. H. Lee, C. B. Park, K. H. Wang and M. H. Lee. 2005. HDPE-clay nanocomposite foams blown with supercritical CO_2. *J. Cell. Plast.*, 41, 487–502.

74. A. Ameli, M. Nofar, C. B. Park, P. Potschke and G. Rizvi. 2014. Polypropylene/ carbon nanotube nano/microcellular structures with high dielectric permittivity, low dielectric loss, and low percolation threshold. *Carbon*, 71, 206–217.

75. Y. Ito, M. Yamashita and M. Okamoto. 2006. Foam processing and cellular structure of polycarbonate-based nanocomposites. *Macromol. Mat. Eng.*, 291, 773–783.

76. W. Zhai, J. Yu, L. Wu, W. Ma and J. He. 2006. Heterogeneous nucleation uniformizing cell size distribution in microcellular nanocomposites foams. *Polymer*, 47, 7580–7589.

77. X. Li, H. Zou and P. Liu. 2015. Structure and dielectric properties of polyimide/ silica nanocomposite nanofoam prepared by solid-state foaming. *J. Appl. Polym. Sci.*, 132, 42355.

78. V. Realinho, M. Antunes, A. B. Martínez and J. I. Velasco. 2011. Influence of nanoclay concentration on the CO2 diffusion and physical properties of PMMA montmorillonite microcellular foams. *Ind. Eng. Chem. Res.*, 50, 13819–13824.

79. S. Costeux and L. Zhu. 2011. Thermoplastic nanocellular foams with low relative density using CO_2 as the blowing agent. *9th International Conference on Foam Materials & Technology (SPE FOAMS 2011)*, Iselin, NJ.

80. K. Taki, T. Otsuka, Y. Waratani and M. Ohshima. 2006. Nano cellular foams of block copolymer thin film. *5th International Conference on Foam Materials & Technology (SPE FOAMS 2006)*, Chicago, IL.

81. J. A. Reglero Ruiz, E. Cloutet and M. Dumon. 2012. Investigation of the nanocellular foaming of polystyrene in supercritical CO_2 by adding a CO_2-philic perfluorinated block copolymer. *J. Appl. Polym. Sci.*, 126, 38–45.

82. C. Dutriez, K. Satoh, M. Kamigaito and H. Yokoyama. 2012. Nanocellular foaming of fluorine containing block copolymers in carbon dioxide: The role of glass transition in carbon dioxide. *RSC Adv.*, 2, 2821–2827.

83. J. Pinto, J. A. Reglero Ruiz, M. Dumon and M. A. Rodríguez-Pérez. 2014. Foaming behavior of poly(methyl methacrylate)-based nanocellular and microcellular foams. *J. Supercrit. Fluids*, 94, 198–205.

84. J. Pinto. 2014. *Fabrication and Characterization of Nanocellular Polymeric Materials from Nanostructured Polymers*, PhD Thesis, University of Valladolid, Spain.

85. J. A. Reglero Ruiz, M. Pedros, J.-M. Tallon and M. Dumon. 2011. Micro and nano cellular amorphous polymers (PMMA, PS) in supercritical CO_2 assisted by nanostructured CO_2-philic block copolymers—One step foaming process. *J. Supercrit. Fluids*, 58, 168–176.

86. J. A. Reglero Ruiz, M. Dumon, J. Pinto and M. A. Rodríguez-Pérez. 2011. Low-density nanocellular foams produced by high-pressure carbon dioxide. *Macromol. Mat. Eng.*, 296, 752–759.

87. J. Pinto, M. Dumon, M. Pedros, J. Reglero and M. A. Rodríguez Pérez. 2014. Nanocellular CO_2 foaming of PMMA assisted by block copolymer nanostructuration. *Chem. Eng. J.*, 243, 428–435.

88. J. Pinto, M. Dumon, M. A. Rodriguez-Perez, R. Garcia and C. Dietz. 2014. Block copolymers self-assembly allows obtaining tunable micro or nanoporous membranes or depth filters based on PMMA; fabrication method and nano-structures. *J. Phys. Chem. C*, 118, 4656–4663.

89. J. Pinto, E. Solórzano, M. A. Rodríguez-Pérez, J. A. de Saja and M. Dumon. 2012. Thermal conductivity transition between microcellular and nanocellular polymeric foams: Experimental validation of the Knudsen effect. *10th International Conference on Foam Materials & Technology (SPE FOAMS 2012)*, Barcelona, Spain.

90. C. Forest, P. Chaumont, P. Cassagnau, B. Swoboda and P. Sonntag. 2015. CO_2 nano-foaming of nanostructured PMMA. *Polymer*, 58, 76–87.

91. C. Forest, P. Chaumont, P. Cassagnau, B. Swoboda and P. Sonntag. 2015. Nanofoaming of PMMA using a batch CO_2 process: Influence of the PMMA viscoelastic behaviour. *Polymer*, 77, 1–9.

92. A. V. Nawaby and Y. P. Handa. 2004. Fundamental understanding of the ABS-CO_2 interactions, its retrograde behavior and development of nanocellular structures. *SPE ANTEC Conference*, 62nd, 2532–2536.

93. H. Ruckdäschel, P. Gutmann, V. Altstädt, H. Schmalz and A. H. E. Müller. 2010. Foaming of microstructured and nanostructured polymer blends. *Adv. Polym. Sci.*, 227, 199–252.

94. T. Nemoto, J. Takagi and M. Ohshima. 2010. Nanocellular foams-cell structure difference between immiscible and miscible PEEK/PEI polymer blends. *Polym. Eng. Sci.*, 50, 2408–2416.

95. R. W. Sharudin and M. Ohshima. 2011. CO_2-induced mechanical reinforcement of polyolefin-based nanocellular foams. *Macromol. Mat. Eng.*, 296, 1046–1054.

96. P. D. Condo, I. C. Sanchez, C. G. Panayiotou and K. P. Johnston. 1992. Glass transition behavior including retrograde vitrification of polymers with compressed fluid diluents. *Macromolecules*, 25, 6119–6127.

97. P. D. Condo and K. P. Johnston. 1992. Retrograde vitrification of polymers with compressed fluid diluents: Experimental confirmation. *Macromolecules*, 25, 6730–6732.

98. P. D. Condo and K. P. Johnston. 1994. In situ measurement of the glass transition temperature of polymers with compressed fluid diluents. *J. Polym. Sci., Part B: Polym. Phys.*, 32, 523–533.

99. A. V. Nawaby, Y. P. Handa, X. Liao, Y. Yoshitaka and M. Tomohiro. 2007. Polymer–CO_2 systems exhibiting retrograde behavior and formation of nanofoams. *Polym. Int.*, 56, 67–73.

100. B. Krause. 2001. *Polymer Nanofoams*. Thesis, University of Twente, the Netherlands.

101. D. Miller and V. Kumar. 2011. Microcellular and nanocellular solid-state polyetherimide (PEI) foams using sub-critical carbon dioxide II. Tensile and impact properties. *Polymer*, 52, 2910–2919.

102. S. Siripurapu, J. A. Coughlan, R. J. Spontak and S. A. Khan. 2004. Surface-constrained foaming of polymer thin films with supercritical carbon dioxide. *Macromolecules*, 37, 9872–9879.

103. S. Costeux. 2009. *Nanoporous Polymeric Foams Having High Porosity*. WO2011066060, assigned to Dow Global Technologies LLC, filed November 25.

104. I. Khan, D. Adrian and S. Costeux. 2015. A model to predict the cell density and cell size distribution in nano-cellular foams. *Chem. Eng. Sci.*, 138, 634–645.

105. H. Guo and V. Kumar. 2015. Solid-state poly(methyl methacrylate) (PMMA) nanofoams. Part I: Low-temperature CO_2 sorption, diffusion, and the depression in PMMA glass transition. *Polymer*, 57, 157–163.

106. S. Inceoglu, N. P. Young, A. J. Jackson, S. R. Kline, S. Costeux and N. P. Balsara. 2013. Effect of supercritical carbon dioxide on the thermodynamics of model blends of styrene-acrylonitrile copolymer and poly(methyl methacrylate) studied by small-angle neutron scattering. *Macromolecules*, 46, 6345–6356.

107. N. P. Young, S. Inceoglu, G. M. Stone, A. J. Jackson, S. R. Kline, S. Costeux and N. P. Balsara. 2014. Thermodynamic interactions and phase behavior of multicomponent blends containing supercritical carbon dioxide, styrene–acrylonitrile random copolymer, and deuterated poly(methyl methacrylate). *Macromolecules*, 47, 8089–8097.

108. K. Taki, D. Kitano and M. Ohshima. 2011. Effect of growing crystalline phase on bubble nucleation in poly(L-lactide)/CO_2 batch foaming. *Ind. Eng. Chem. Res.*, 50, 3247–3252.

109. L. Sorrentino, M. Aurilia and S. Iannace. 2011. Polymeric foams from high-performance thermoplastics. *Adv. Polym. Tech.*, 30, 234–243.

110. P. Gong and M. Ohshima. 2013. Nanoporous structure on the cell wall of polycarbonate foams. *11th International Conference on Foam Materials & Technology (SPE FOAMS 2013)*, Seattle, WA.

111. R. Miyamoto, S. Yasuhara, H. Shikuma and M. Ohshima. 2014. Preparation of micro/nanocellular polypropylene foam with crystal nucleating agents. *Polym. Eng. Sci.*, 54, 2075–2085.

112. D. Miller and V. Kumar. 2009. Fabrication of microcellular HDPE foams in a sub-critical CO_2 process. *Cell. Polym.*, 28, 25–40.

113. L. Li, H. Yokoyama, T. Nemoto and K. Sugiyama. 2004. Facile fabrication of nanocellular block copolymer thin films using supercritical carbon dioxide. *Adv. Mater.*, 16, 1226–1229.

114. L. Li, T. Nemoto, K. Sugiyama and H. Yokoyama. 2006. CO_2 foaming in thin films of block copolymer containing fluorinated blocks. *Macromolecules*, 39, 4746–4755.

115. J. Pinto, M. A. Rodríguez-Pérez, J. A. de Saja, M. Dumon, R. García and C. Dietz. 2011. Relationship between the nano-structured morphology of PMMA/MAM blends and the nanocellular structure of foams produced from these materials. *9th International Conference on Foam Materials & Technology (SPE FOAMS 2011)*, Iselin, NJ.

116. C. Forest, P. Chaumont, P. Cassagnau, B. Swoboda and P. Sonntag. 2015. Generation of nanocellular foams from ABS terpolymers. *Eur. Polym. J.*, 65, 209–220.

117. T. Otsuka, K. Taki and M. Ohshima. 2007. Preparation of nanoblend polymer by polymerization in polymer and its nano cellular foams. *SPE ANTEC Conference*, 65th, 3002–3005.

118. D. Raps, N. Hossieny, C. B. Park and V. Altstädt. 2015. Past and present developments in polymer bead foams and bead foaming technology. *Polymer*, 56, 5–19.

119. B. Krause, M. Kloth, N. F. A. van der Vegt and M. Wessling. 2002. Porous monofilaments by continuous solid-state foaming. *Ind. Eng. Chem. Res.*, 41, 1195–1204.

120. C. Okolieocha, D. Raps, K. Subramaniam and V. Altstädt. 2015. Microcellular to nanocellular polymer foams: Progress (2004–2015) and future directions—A review. *Eur. Polym. J.*, 75, 500–519.

121. S. Costeux, D. A. Beaudoin, H. Kim and D. Foether. 2014. *Mold Process for Making Nanofoams.* PCT application WO2015183639, assigned to Dow Global Technologies LLC, filed May 28.

122. S. Costeux. 2011. *Continuous Process for Extruding Nanoporous Foam.* US patent 9,145,478, assigned to Dow Global Technologies LLC, filed September 29.

123. S. Costeux and D. Foether. 2015. Continuous extrusion of nanocellular foam. *Annu. Tech. Conf.—Soc. Plast. Eng.,* Orlando, FL.

124. S. T. Lee and C. B. Park. 2014. In *Foam Extrusion: Principles and Practice,* Second Edition, CRC Press, p. 441.

125. S. Liu, J. Duvigneau and G. J. Vancso. 2015. Nanocellular polymer foams as promising high performance thermal insulation materials. *Eur. Polym. J.,* 65, 33–45.

126. B. Notario, J. Pinto, E. Solorzano, J. A. de Saja, M. Dumon and M. A. Rodríguez-Pérez. 2015. Experimental validation of the Knudsen effect in nanocellular polymeric foams. *Polymer,* 56, 57–67.

127. X. Lu, R. Caps, J. Fricke, C. T. Alviso and R. W. Pekala. 1995. Correlation between structure and thermal conductivity of organic aerogels. *J. Non-Cryst. Solids,* 188, 226–234.

128. C. Forest, P. Chaumont, P. Cassagnau, B. Swoboda and P. Sonntag. 2015. Polymer nano-foams for insulating applications prepared from CO_2 foaming. *Prog. Polym. Sci.,* 41, 122–145.

129. S. S. Sundarram and W. Li. 2013. On thermal conductivity of micro- and nano-cellular polymer foams. *Polym. Eng. Sci.,* 53, 1901–1909.

130. P. Ferkl, R. Pokorny, M. Bobak and J. Kosek. 2013. Heat transfer in one-dimensional micro- and nano-cellular foams. *Chem. Eng. Sci.,* 97, 50–58.

131. A. M. Hodge, J. Biener, J. R. Hayes, P. M. Bythrow, C. A. Volkert and A. V. Hamza. 2007. Scaling equation for yield strength of nanoporous open-cell foams. *Acta Mater.,* 55, 1343–1349.

132. H. L. Fan and D. N. Fang. 2009. Modeling and limits of strength of nanoporous foams. *Materials & Design,* 30, 1441–1444.

133. Q. Chen and N. M. Pugno. 2012. Mechanics of hierarchical 3-D nanofoams. *EPL,* 97, 26002.

134. B. Notario, J. Pinto and M. A. Rodríguez-Pérez. 2015. Towards a new generation of polymeric foams: PMMA nanocellular foams with enhanced physical properties. *Polymer,* 63, 116–126.

135. J. Germain, J. M. J. Fréchet and F. Svec. 2009. Nanoporous polymers for hydrogen storage. *Small,* 5, 1098–1111.

136. R. Dawson, A. I. Cooper and D. J. Adams. 2012. Nanoporous organic polymer networks. *Prog. Polym. Sci.,* 37, 530–563.

137. Y. Charlier, J. L. Hedrick, T. P. Russell, A. Jonas and W. Volksen. 1995. High temperature polymer nanofoams based on amorphous, high Tg polyimides. *Polymer,* 36, 987–1002.

138. J. L. Hedrick, K. R. Carter, H. J. Cha, C. J. Hawker, R. A. DiPietro, J. W. Labadie, R. D. Miller et al. 1996. High-temperature polyimide nanofoams for microelectronic applications. *React. Funct. Polym.,* 30, 43–53.

139. B. Notario, A. Ballesteros, J. Pinto and M. A. Rodríguez-Pérez. 2016. Nanoporous PMMA: A novel system with different acoustic properties. *Mater. Lett.,* 168, 76–79.

8

Rigid Structural Foam
and Foam-Cored Sandwich Composites

Wenguang Ma and Kurt Feichtinger

CONTENTS

8.1 Introduction

8.1.1 Development of Foam-Cored Sandwich Composites

Sandwich composites have been used in modern industry since World War II (WWII). Balsa wood was the first core material that was used for building the Mosquito air bomber's fuselage and wings (Figure 8.1). The first foam core material was polyvinyl chloride (PVC) rigid foam that was made by mixing isocyanate blend with PVC, first commercialized in Germany by Dr. Lindemann from 1930s to 1940s. It is rumored that this early version of PVC foam was used in the German E-boats and even in the famous *Bismarck* battleship. After WWII, PVC foam production continued, and its development continued for new applications. The primary application for PVC foam core is in the construction of recreational boats and yachts. The foam-cored sandwich structure can be used for building the hull, bulkhead, stringer, deck, superstructure, and even furniture (Figure 8.2). The advantage of foam compared with balsa wood is its good water resistance behavior.[1,2]

Polyurethane (PUR) rigid foam was used as the core of sandwich composites starting in the 1950s. More expensive than balsa but cheaper than PVC

FIGURE 8.1
Balsa core was used for making the fuselage of the Mosquito air bomber during WWII in an Australian plant.

FIGURE 8.2
Boat hull built by foam core sandwich laminates. (From http://www.bavaria-spain.com.)

foam, PUR foam has been used historically in the construction of the more expensive boats.[3]

In the early 1990s, styrene–acrylonitrile copolymer (SAN) foam was introduced into the market to compete with PVC foam core. It is claimed to have good thermal stability and no outgassing problem during the resin curing of sandwich facings at elevated temperatures.

The first thermoplastic rigid foam core, polyethylene terephthalate (PET) foam, emerged on the composite market in 1998. A special format of PET *strand* foam was introduced to the market in 2000. This was the first type of structural core product that was made by a conventional foaming extrusion process, but a special die design, and is recyclable by thermal melt processing. The PET foam core can be used with all types of liquid thermosetting resins and is compatible with vacuum infusion, adhesive bonding, prepreg, resin transfer molding (RTM), compression molding, and thermoforming.[4]

Other high-performance foam cores are polymethacrylimide (PMI) and polyetherimide (PEI) foams. PMI foam is a closed-cell rigid foam based on PMI. The field of application of PMI foam is aircraft, rail vehicles, and racing car construction, which need processing temperatures up to 130°C (266°F).[5] PEI foam is a closed-cell, thermoplastic polymer foam that combines fire resistance with low smoke and toxicity, along with good dielectric properties. It has a good strength/weight ratio, has very low moisture absorption, and is a core material for the structural lightweight applications that demand high fire resistance, radar transparency, or operation in extremely hot or cold environments.[6]

Polystyrene (PS) foam is used as insulation in most applications. However, in recent years, PS foam with greater density has been promoted as a structural core material in the composite industry for wind turbine blade construction, using epoxy resin systems. Its low chemical resistance limits it from laminating with unsaturated polyester and vinyl ester resins.

After applications in the marine industry, the structural foam cores have been used in wind turbine blade construction on a large scale since early 2000. Wind blades are typically manufactured as two half-shells, which are secondarily assembled along with a shear web into a whole blade. Each half of the blade is a sandwich structure, often featuring a combination of different core materials, each used in a particular location. The other components, such as shear webs as shown in Figure 8.3, are also foam-cored sandwich constructions. The wind energy industry is growing fast, and blades are getting longer to increase turbine efficiency. Subject to greater loads and increasingly located offshore where they are less easily maintained and serviced, these massive structures are now built in greater volume.[7]

8.1.2 Significances of Sandwich Composites in Modern Industries

Composite designers determined early on that sandwiching a low-density, lightweight core material between thin face sheets can dramatically increase a laminate's stiffness with little added weight. A sandwich structure is cost-effective because the relatively low-cost core replaces more expensive composite reinforcement material. And the stiffer but lighter sandwich panel requires less supporting structure than a solid laminate.[1]

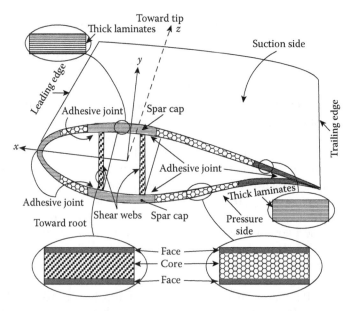

FIGURE 8.3
Wind blade construction. (From http://www.sciencedirect.com.)

In a sandwich panel, the core functions like the connecting web of an I-beam, separating the face skins at a constant distance, while the skins themselves function as the I-beam flanges, as seen in Figure 8.4. The sandwich panel's bending stiffness is roughly proportional to the core thickness cubed, in the same way that an I-beam is stiffer as the distance between the flanges increases. Doubling the panel thickness by using light-density core between the solid lamination yields a panel 3.5 times stronger and 7 times stiffer; in the same way, doubling the panel thickness again yields the panel 9 times stronger and 37 times stiffer, and with very little weight increase, as described in Figure 8.5.

When a bending force works on a sandwich panel, as shown in Figure 8.6,[8] the core transfers the force from one skin to the other and bears the shear and compression stress. A core also helps distribute loads and stresses on the skins, which makes a cored sandwich an excellent design for absorbing impact stresses. A core's compressive strength prevents the thin skins from wrinkling (buckling) failure, while its shear modulus keeps the skins from sliding independent of each other when subjected to bending loads.

FIGURE 8.4
I-beam and sandwich structure comparison.

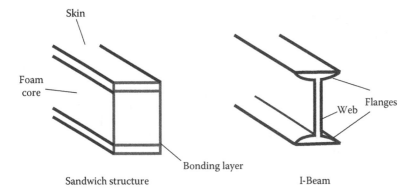

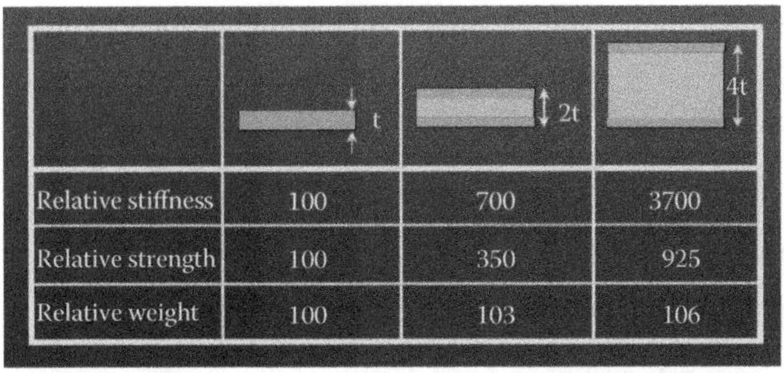

			t			2t			4t
Relative stiffness		100			700			3700	
Relative strength		100			350			925	
Relative weight		100			103			106	

FIGURE 8.5
Comparison of solid and sandwich laminates for out-of-plane loads. (From Greene, E., Marine composites, *ACMA Composites 2004 Conference*, Tampa, FL, October 7, 2004.)

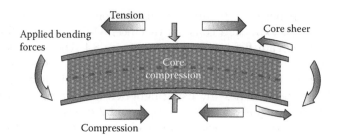

FIGURE 8.6
Bending force works on a sandwich structure. (From http://oneoceankayaks.com.)

The advantages of foam-cored sandwich construction are as follows:

- Consisting of a low-to-moderate stiffness core that is connected with two stiff exterior face sheets, the sandwich composite has a considerably higher shear stiffness/weight ratio than an equivalent beam made of only the core material or the face-sheet material.
- The high stiffness of the face sheet leads to a high bending stiffness/ weight ratio for the composite.
- The concept of sandwich construction gives more flexibility to designers by the selection of various facing materials, core types, core density, core finishing format configuration, lamination methods, and combinations of all of the above, based on the particular application.
- As an added benefit, foam-cored sandwich panels offer excellent heat insulation, sound abatement, fire proofing. and vibration damping, with the appropriate selection of the core material.

Some downsides to sandwich construction include the cost of the core material and the employee learning curve toward proper installation. Rigid structural core materials are invariably more expensive than the resin and glass that they are replacing by weight, and, in some cases, the labor savings will not offset the cored laminate price. More care and attention need to be taken when processing cored laminates. Employees need to be aware of all the possible problems that could occur if the core materials are not handled or bonded to the skins correctly.[2]

8.2 Rigid Foams and Properties as Core Materials

8.2.1 Requirements of Foam Core Materials

As a structural sandwich core material, many unique properties are required; thus, a prudent selection of the polymer from which the foam is produced is most important. However, not every polymer can be foamed commercially, and this is why not many polymers are available as core materials. First of all, the foam should be sufficiently stiff. As the core in a structural sandwich, its shear modulus is a representative indication of sandwich stiffness. A PVC foam core with a density of 40 kg/m^3, a general-purpose core material, has a shear modulus of 12 MPa, whereas a PMI foam core, a high-performance core material, with a lower density of 30 kg/m^3, has a shear modulus of 13 MPa. However, a rigid PUR foam with a density of 70 kg/m^3 only has a shear modulus of 7.6 MPa, which is why PUR foam is a lower-performance core material.

The second important property is toughness or *ductility*, an indication of the impact resistance of a core material; it is reflected by the value of the foam's shear elongation at break (SEB). A rigid PVC core with a density of 40 kg/m^3 has an SEB of approximately 9%, which increases with increasing density. PMI has an SEB of 3% because of its high rigidity.

Equally important are the surface properties of the foam core. Not only should it have a sufficiently high surface energy so that it can be wetted out (compatible) with the all adhesives and liquid thermosetting resins that are used in composite industries, but it also should have good resistance against solvent attack so that it cannot be softened by any solvents in the resin or adhesive. The closed-cell content of the foam core is also a critical property that should be, generally, more than 80%. If the open-cell content is too high, the liquid resin will migrate into the core during laminating, increasing the weight of the sandwich laminate. In summary, the foam should have a high-enough surface energy to bond to the adhesive or resin for laminating the facing but cannot be attacked by them physically or chemically. The bond strength must be high enough to withstand the constant bending, compressive, or tensile forces of dynamic loading, such as the forces on a boat hull or a rotating turbine blade. The coefficient of thermal expansion of the core, the laminate material, and the resin or adhesive must be compatible to ensure that thermal cycling does not cause unequal expansions and contractions, leading to debonding.

The hot-and-cold temperature resistance is another important property of the structural core materials. During lamination, the core will endure a heat history because of the need to cure the resin at elevated temperatures and by the heat released by the adhesive or resin curing (exothermic). The bond layer between the facing and the foam surface will shrink or swell if the foam core does not have a sufficiently high heat resistance, reducing the bond strength and possibly causing a laminate to fail. For the specific laminate application or service, the sandwich structure could be exposed to extremes in temperature. The foam core should maintain mechanical properties within an acceptable design range to ensure that the composite sandwich product performs the intended function.

The foam core also needs to be fatigue resistant. A product made by using the foam-cored sandwich construction, such as a boat or a wind turbine blade, will endure cyclical forces. For example, a water or wind blade may be designed for over 10 million cycles in its service life. If the core cannot maintain sufficient mechanical properties for its design life in a dynamically loaded environment, the product will prematurely fail.

The additional properties to be considered for the foam core include fire resistance, water resistance, and appropriate cell structure, depending on the application. For applications in building construction and public transportation, the fire/smoke/toxic gases (FST) behavior is critical and usually must pass special tests, as required by the prevailing authority.

8.2.2 Important Foam Cores Used in the Composite Industries

8.2.2.1 PVC Foam

Of the various structural foam cores, the most commonly used is PVC foam, which is actually a hybrid of PVC, PUR, and polyurea.[1,9,10] Commercially, two types of PVC foam are available. Cross-linked, or semirigid, foams are relatively stiff and strong, perform well at temperatures up to 120°C/250°F, and are resistant to styrene, so they can be used with unsaturated polyester and vinyl ester resins. Linear or ductile PVC foams, made with a different polymer formulation, are more elastic than cross-linked varieties and are widely used in marine applications, where they offer a high deformation before failure and excellent impact resistance. While linear PVC is easier to thermoform around complex curves, the trade-off is somewhat lower mechanical properties and resistance to temperature and styrene compared to the cross-linked version. Both offer good properties toward fatigue resistance.

Most of the cross-linked PVCs are currently used in applications of the wind energy and marine industries. The flow chart in Figure 8.7 briefly shows the process of making rigid PVC foam. PVC and other materials, which include one or two different isocyanates, one or two different anhydrides, blowing agents, and surfactants, are mixed thoroughly into a plastisol at room temperature, under conditions to avoid contact with moisture. The plastisol is injected into aluminum or steel molds, and the molds are heated in a multiple opening press at 170°–175°C under a pressure of 200–400 bar for approximately 20 min. The partially cured but unfoamed products, called embryos, are cooled and released from molds. They are then put into a chamber for expanding into a foam plank with steam heating at approximately 90°C–100°C for 24 h. Finally, the foam planks are kept in a steam heating room at 40°C–70°C for several days to several weeks, depending on the target density, for finishing the expansion and curing reactions. After cutting off the outer skin and trimming off edges, the planks, called the foam buns, are ready for slicing into foam sheets that can be processed into the different core products.

The typical mechanical properties of PVC foam cores are listed in Table 8.1.[11] Several excellent properties of the PVC foam cores are (a) good fatigue resistance, (b) high ratio of stiffness and strength to weight, and (c) resistance to styrene and other chemicals, making them suitable for processing with

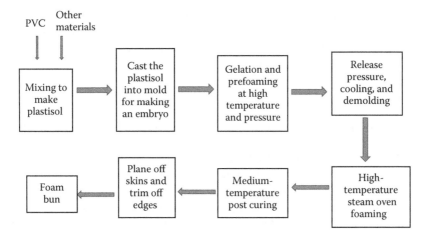

FIGURE 8.7
Production flow chart of rigid PVC foam.

TABLE 8.1

Typical Mechanical Properties of Rigid PVC Foam Cores

Apparent	ISO	kg/m³	48	60	80	100	130	200
density	845	lb./ft.³	3.0	3.8	5.0	6.3	8.1	12.5
Compressive	ASTM	Mpa	0.65	0.85	1.35	2.21	3.02	4.81
strength	1621B	psi	94	123	196	321	438	698
Compressive	ASTM	Mpa	52	71	92	136	172	242
modulus	1621B	psi	7542	10,298	13,343	19,725	24,947	35,099
Tensile	ASTM	Mpa	1.42	1.61	2.21	3.52	4.78	7.15
strength	C297	psi	206	234	321	511	693	1037
Tensile	ASTM	Mpa	55	75	95	132	175	253
modulus	C297	psi	7975	10,875	13,775	19,145	25,375	36,695
Shear	ASTM	MPa	0.55	0.70	1.11	1.63	2.19	3.52
strength	C273	psi	80	102	161	236	318	511
Shear	ASTM	MPa	16	22	31	36	51	86
modulus	C273	psi	2321	3191	4496	5221	7397	12,473
Shear	ASTM	%	12	20	30	40	40	40
elongation	C273							
Water	ISO	%	1.9	1.8	1.6	1.5	1.4	1.2
absoption	2896							

Source: Strucell, P., Specification. Available at http://strucell.com.

unsaturated polyester, vinyl ester, and epoxy resins. The limitations of PVC foam cores are (a) their modest heat resistance and (b) the toxic gases that are liberated during burning, related to the PVC resin. Also, PVC foam contains carbon dioxide gas under pressure, so outgassing can occur at elevated temperatures over time—that is, the gas diffuses from the closed cells and migrates to voids or unbounded areas in the laminate. In some instances, outgassing has been blamed for delamination and blistering in the sandwich construction, especially in parts that are made at elevated cure temperatures or finished in dark colors and in direct exposure to the sun.

8.2.2.2 PET Foam

After foamed PVC, PET foam core is the second type of thermoplastic structural foam core that is made by a continuous extrusion process. It is also the second-largest volume foam that is currently used in the sandwich composite industries. Because of being totally made by thermoplastics, the foam's production waste, scrap, and products at the end of service life can be reused by conventional extrusion melting and repelletizing processes. Currently, the wastes of PVC, PUR, and other semi- and fully thermosetting structural foam impose a large negative impact to the environment. PET foam is one of the most ecological solutions for reducing composites' environmental footprint.

The production route of PET foam is shown in Figure 8.8.[12] Unlike PS, PET foam is not easily processed into thick foam planks by using a slot die extrusion method. The commercially successful process uses a multiorifice die and a simple device to shape the multistrands together into a thick foam plank. So, PET foam currently in the market is also called a tPET strand foam core. The process of using a slot die continues in the industry, but the resulting planks are of limited thickness.

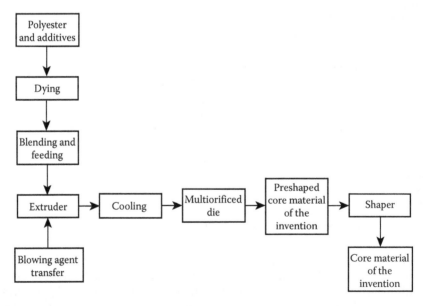

FIGURE 8.8
Production flow chart of PET foam processing.

Table 8.2 lists the typical mechanical properties of PET foam cores.[13] Because of its behavior as a semicrystalline resin with an approximately 250°C melting temperature, PET foam can be used with almost all thermosetting resins that are employed in the composite industries and can be processed at a relatively high temperature without any outgassing.

8.2.2.3 PUR Foams

Rigid PUR foams have been used as structural foam cores for many years in the marine industry. There are two types of rigid PUR foam: (1) polyisocyanurate (PIR) formulations and (2) PUR. These designations are misleading, however, because both are PURs—that is, they are polymers that are formed by reacting a monomer containing at least two isocyanate functional groups with another monomer having at least two hydroxyl groups. The resulting polymer is a chain of organic, monomeric units that are joined by urethane (carbamate) linkages. In PIR foam, the proportion of methylene diphenyl diisocyanate (MDI) is greater than for PUR. The PIR reaction also

TABLE 8.2

Typical Mechanical Properties of PET Strand Foam

Density	ISO 845	kg/m³	**Average**	**65**	**110**	**145**	**210**
			Typical range	*60–70*	*105–115*	*140–150*	*200–220*
Compressive strength perpendicular to the plane	ISO 844	N/mm²	**Average**	**0 80**	**1.4**	**2.2**	**3.5**
			Minimum	*0.7*	*1.2*	*2.0*	*3.2*
Compressive modulus perpendicular to the plane	DIN 53421	N/mm²	**Average**	**50**	**85**	**115**	**170**
			Minimum	*35*	*75*	*100*	*145*
Tensile strength perpendicular to the plane	ASTM C297	N/mm²	**Average**	**1.5**	**2.2**	**2.7**	**3.0**
			Minimum	*1.2*	*1.6*	*2.2*	*2.4*
Tensile modulus perpendicular to the plane	ASTM C297	N/mm²	**Average**	**85**	**120**	**170**	**225**
			Minimum	*70*	*90*	*140*	*180*
Shear strength	ISO 1922	N/mm²	**Average**	**0.46**	**0.8**	**1.2**	**1.85**
			Minimum	*0.4*	*0.7*	*1.1*	*1.5*
Shear modulus	ISO 1922	N/mm²	**Average**	**12**	**20**	**30**	**50**
			Minimum	*10.5*	*18*	*26*	*44*
Shear elongation at break	ISO 1922	%	**Average**	**25**	**10**	**8**	**5**
			Minimum	*15*	*5*	*4*	*3*
Thermal conductivity at room temperature	ISO 8301	W/m.K	**Average**	**0.033**	**0.033**	**0.036**	**0.041**

uses a polyester-derived polyol in place of a polyether polyol. PIR foam typically has an MDI/polyol ratio (also called its index) of between 200 and 500, whereas PUR indices are normally approximately 100. The net result is that PIR foams are more cross-linked and, therefore, are stiffer but more friable than PUR foams.[3]

PIR foams, sometimes called *trimmer* foams, generally have low density (1.8–6 lb./ft.[3]) with high insulating values and good compressive strength. Friability, however, limits their utility in sandwich panel applications because this lack of toughness can result in failures at the foam-to-facing bond under dynamic loads. As a result, the structural use of these foams is usually limited to two roles: (1) as core-carrier material for glass–fiber reinforcement or (2) to provide an internal mold shape for laminate overlays.

Considerably tougher (less friable) than PIR foams, PUR foams are produced in densities ranging from 2 to 50 lb./ft.[3], depending on the formulation, and can retain a substantial strength and toughness up to 275°F/135°C, which allows their use with high-temperature curing prepreg in ovens and autoclaves. Rigid PUR foam is also made using polyester-derived polyols for increased rigidity. PUR foam is widely used in structural sandwich shapes, such as transoms and bulkheads, in foam-cored RTM parts and composite tooling, and as an edge closeout in honeycomb-cored aircraft sandwich interior panels.

A third category, a blend of PIR and PUR foams, attempts to get the best of both worlds. These foams provide some improvement in strength compared to PUR foams, with reduced friability relative to PIR foams.

PUR foams can be produced by batch and continuous process to produce foam buns that will be sliced into any thickness. The continuous process can also be used for making sandwich panels by using the coiled sheets of the skin materials. A batch process is shown in Figure 8.9 that uses hand mixing and a mold with the lid.[14] The mold may also have an open top, and the walls of mold can be hinged together, or the hinges are released for allowing the foam to expand freely in three directions. If a completely closed mold is used, a rectangular-shaped foam bun will be produced. On closed molds, adjusting the pressure applied to the mold controls the density of the foam.

The principle of the continuous mixing and production of foam slabstock is shown in Figure 8.10, which is the most economical method to produce large sections of foam. The side walls are necessary for controlling expansion and ensuring uniform density and cell structure. To avoid crown formation and reduce cutting losses, the flat-topped slab method is mainly used for continuous production. With this process, a disposable paper web is fed above the rising foam zone, and it serves as a separator to slide the continuous bun along the immobile upper platen.

The typical mechanical properties of rigid PUR foam are listed in Table 8.3.[15] PUR foam is widely used in structural shapes, such as transoms

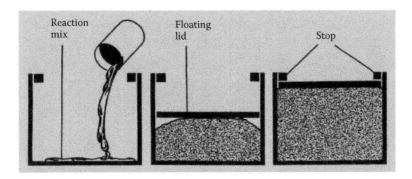

FIGURE 8.9
Batch process for making PUR bun by using a close-lid mold. (From The production of rigid polyurethane foam, available at http://plastics.bayer.com.)

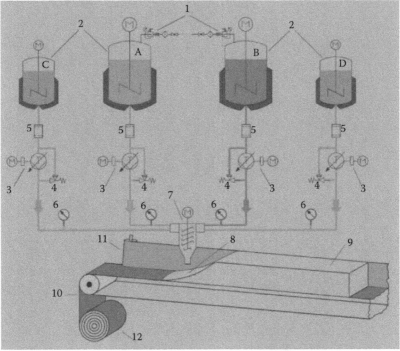

A: Polyol mix	1: Pressure relief valve	7: Agitator mixing head
B: Polyisocyanate	2: Agitator tank	8: Reaction mix
C: Blowing agent	3: Metering pump	9: Reacted foam
D: Activator	4: Safety valve	10: Lower paper
	5: Filter	11: Side paper
	6: Manometer	12: Paper roll

FIGURE 8.10
Diagram of continuous-producing PUR rigid foam slabstock. (From The production of rigid polyurethane foam, available at http://plastics.bayer.com.)

TABLE 8.3

Typical Mechanical Properties of PET PUR Rigid Foam

Density ASTM D1622	lb./ft.^{3}	5	6	8	10	15	18	20	24	29
Compressive Strength ASTM D1621	psi	118	173	257	369	690	914	1241	1755	2018
Compressive Modulus ASTM D1621-B	psi	3266	5349	6942	9427	17,349	23,101	28,674	36,577	49,112
Tensile Strength ASTM D1623	psi	191	287	329	421	654	801	989	1149	1489
Shear Strength ASTM C273	psi	90	127	179	268	498	678	771	1029	1353
Shear Modulus ASTM C273	psi	1386	2380	3754	4588	6241	8880	10,843	13,123	20,741
Flexural Strength ASTM D790	psi	172	259	369	478	839	1013	1304	1501	2105
Water Absorption ASTM D2842	lb./ft.^{2}	0.063	0.063	0.058	0.053	0.040	0.029	0.027	0.017	0.007

Source: http://www.polyumac.com/upload/files/Air-Cell+T_Info_and_Data_Sheet_English_Jan_20_2013.pdf.

and bulkheads, in sandwich RTM parts and composite tooling, and as an edge closeout in honeycomb aircraft interior panels. However, because the PUR foam has a relatively low fatigue resistance, especially in the lower-density range, only the higher-density PUR foams are used for structural application, such boat transoms and stringers.

For improving certain properties of PUR foam, some modifications have created a few new generations of core materials by introducing fiberglass into the foam. One product of the fiber-reinforced composite core, shown in Figure 8.11, is made by assembling PUR foam strips into a plank with heat-activated glass scrim, after individually wrapping the strips with biaxial glass or even glass roving. The glass reinforcement–filled gap between the strips will be filled with liquid resin during closed molding, and the glass reinforcement in the gap will reinforce the foam in the thickness direction. This core technology combines fiberglass and closed-cell foam in an engineered architecture to create a very efficient sandwich core solution. Specifically designed for resin infusion processes, infusion grooves allow the resin to move quickly and efficiently throughout each panel ensuring a high-quality infusion of the sandwich panel's internal structure.

8.2.2.4 Poly(styrene-co-acrylonitrile) Foam

Poly(styrene-co-acrylonitrile) (SAN) foam core was first invented by Rohm GmbH and later licensed for commercial production in the 1990s by Canadian

FIGURE 8.11
PUR foam reinforced by assembling foam strips wrapped with the glass roving. (From http://tycor.milliken.com.)

firm ATC Chemicals Inc. under the trade name Corecell®. The linear closed-cell thermoplastic foam combines good static mechanical properties with high elongation, resulting in impact toughness and fatigue resistance—in essence, delivering the best properties of both cross-linked and linear PVC. Due to its high acrylonitrile content, Corecell has quite good chemical resistance and thus can be used with most of the resin systems that are used in composite construction.[1]

SAN foam can be processed at temperatures up to 85°C/185°F, which accommodates lower temperature curing epoxy prepreg systems. SAN foam also has excellent strength, buoyancy, and insulating values in subsea structural laminates that are subjected to hydrostatic loads, such as remotely operated vehicles for the offshore oil industry. The high-temperature-resistant version of SAN foam offers high rigidity and thermal stability for process temperatures as high as 110°C/230°F.

SAN foam is made by first polymerizing between two tempered glass sheets a blend of styrene and acrylonitrile with PUR chemistries and blowing agent shown in Figure 8.12.[16] After the removal of the glass sheets, the clear, unfoamed sheet is expanded at a temperature in the range of 50°C–70°C/120°F–160°F. The typical properties of SAN foam are listed in Table 8.4. Sandwiches cored with SAN foam give the highest heat resistance when used with epoxy prepreg and vacuum infusion lamination. However, since the foam is a copolymer of a styrene monomer, it has a limitation to use with the unsaturated polyester and vinyl ester at elevated temperatures, such as for pultrusion and heated RTM processing. Due to its partial PUR structure, SAN foam is not recyclable by traditional extrusion melting processes into postconsumption products.

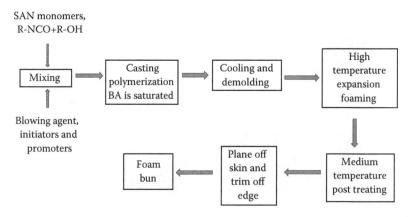

FIGURE 8.12
SAN foam manufacturing processes.

TABLE 8.4

Typical Mechanical Properties of SAN Foam

Nominal density		kg/m³	71	94	104	115	143
		lb./ft.³	4.4	5.9	6.5	7.2	8.9
Density range		kg/m³	66–76	89–99	100–107	108–122	133–153
		lb./ft.³	4.1–4.7	5.6–6.2	6.2–6.7	6.7–7.6	8.3–9.6
Compressive strength	ASTM D1621	MPa	0.88	1.41	1.67	1.98	2.85
		psi	128	205	242	287	413
Compressive modulus	ASTM D1621 1973	MPa	62	101	120	143	209
		psi	8992	14,649	17,405	20,740	30,313
	ASTM D1621 2004	MPa	45	69	79	90	119
		psi	6527	10,008	11,458	13,053	17,259
Shear strength	ASTM C273	MPa	0.81	1.15	1.30	1.47	1.93
		psi	117	167	189	213	280
Shear modulus	ASTM C273	MPa	28	40	46	52	70
		psi	4061	5802	6672	7542	10,153
Shear elongation at break	ASTM C273	%	24%	17%	15%	13%	10%
Tensile strength	ASTM D1623	MPa	1.30	1.72	1.91	2.11	2.62
		psi	189	249	277	306	380
Tensile modulus	ASTM D1623	MPa	85	118	134	151	196
		psi	12,328	17,114	19,435	21,901	28,427
Thermal conductivity	ASTM C518	W/m.K	0.03	0.04	0.04	0 04	0.04
Heat deflection temperature (HDT)	DIN 53424	°C	100	100	100	100	100
		°F	212	212	212	212	212

Source: http://www.gurit.com/gurit-corecell-t.aspx.

8.2.2.5 PMI Foam

PMI foam cores, commercially available in 11 different grades, exhibit the highest mechanical properties of any foam core type, at a comparable density. For example, the shear and tensile modulus are 20% higher for PMI foam than PVC foam. HDT ranges from 177°C to 235°C, the highest for any foam core product, which makes PMI foam suitable for laminates with high-cure temperatures as well as high, sustained temperature applications.[1]

Because of their high-temperature performance, low-smoke evolution during burning, and halogen-free chemistry, PMI foams have been used in public transportation applications such as high-speed marine vessels and trains, where fire performance is regulated by the authorities. These foams have been incorporated into a number of intercity trams and newer high-speed trains, in both interior and structural exterior applications. PMI foam can be thermoformed to conform to the complex part shapes and can be used with epoxy prepreg that cures up to 125°C–175°C for the train's construction. The entire thermoforming and layup can be vacuum-bagged and cocured in concert.

The excellent resistance of PMI foam to compression creep, plus its ability to fully support the prepreg during the curing resulting in a high dimensional stability, are responsible for its long and successful application for manufacturing aerospace sandwich components. This includes secondary structures, as well as primary structures, for state-of-the-art aircraft. At present, PMI foams are listed in more than 170 aerospace specifications worldwide, including helicopter, military, and civil airplanes.[17]

PMI foam is produced by thermal expansion of a copolymer sheet of methacrylic acid and methacrylonitrile or acrylonitrile. During the foaming process, the copolymer sheet is converted to PMI. The principle of PMI manufacturing is shown in Figure 8.13. By controlling the formulation and the processing conditions, the resulting PMI foam has a very homogeneous cell structure, small cell size, and isotropic properties. The cell size of some special grades of PMI foam is below 100 μm.

The PMI foam's typical properties are listed in Table 8.5. As mentioned in the last section, PMI foam can offer very low density foam but still has good mechanical properties. However, because of its high price and low productivity, PMI foam applications are limited to those requiring a very low weight, high mechanical properties, high temperature resistance and low-cost sensitivity.

8.2.2.6 PEI Foam

PEI foam is a rigid lightweight, low flame, smoke, and toxicity (FST) foam core based on the PEI polymer. The material combines a high strength/weight ratio with low moisture absorption. The foam also possesses excellent dielectric properties. The foam is targeted at applications where structural

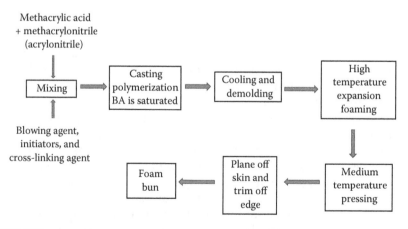

FIGURE 8.13
PMI foam manufacturing processes.

TABLE 8.5

Typical Mechanical Properties of PMI Foam

Density	kg/m³	32	52	75	110	ISO 845
	lb./ft.³	2.00	3.25	4.68	6.87	ASTM D1622
Compressive strength	MPa	0.4	0.9	1.5	3.0	ISO 844
	psi	58	130	217	435	ASTM D1621
Tensile strength	MPa	1.0	1.9	2.8	3.5	ISO 527-2
	psi	145	275	406	507	ASTM D638
Shear strength	MPa	0.4	0.8	1.3	2.4	DIN 53294
	psi	58	116	188	348	ASTM C273
Elastic modulus	MPa	36	70	92	160	ISO 527-2
	psi	5220	10,150	13,340	23,200	ASTM D638
Shear modulus	MPa	13	19	29	50	DIN 53294
	psi	1885	2755	4205	7250	ASTM C273
Strain at break	%	3	3	3	3	ISO 527-2
						ASTM D638

Source: http://www.rohacell.com/sites/lists/PP-HP/Documents/ROHACELL-IG-IG-F-mechanical-properties-EN.pdf.

fire performance, radar transparency, or performance under extreme hot or cold environments is required. PEI foam is thermoformable and compatible with phenolic and bismalimide prepregs. These properties make it an excellent candidate as a sandwich core for many of the current high-performance composite applications, including aerospace/transportation interiors, military radomes, telecommunication systems, composite tooling, and medical imaging equipment.[6]

PEI foam is a resilient foam for use in fire and structural applications within an operating temperature range of −194°C (−317°F) to +160°C (+320°F) and has many unique characteristics—fulfills most stringent fire requirements, high-impact resistance (nonbrittle failure mode), three-dimensionally (3D) thermoformable, good radar transparency, and can be used for prepreg processing up to 180°C, adhesive bonding, hand layup, fiber spraying, thermoforming method, etc.

Commercially, PEI foams have been made by two methods. One is by expanding the hot-pressed embryo that is made from a mixture of PEI powder and organic solvent blowing agent. The other is by an extrusion foaming process. The typical properties of PEI foam are listed in Table 8.6.[18] PMI foam's outstanding low moisture absorption gives it a decisive advantage compared to PMI foam during high-temperature processing. Often, PMI foam boards must be conditioned (dried and/or stored in a special area) before they can be machined, compression molded, or thermoformed. This extra step adds time, costs, and overhead to the process. PEI foam avoids this scenario. Further, PMI foam may require a multistep annealing process. As one of its detractors, the closed-cell percentage is quite low for PEI foam,

TABLE 8.6

Typical Mechanical Properties of PEI Foam

Density	ISO 845	kg/m^3	**Average**	**60**	**80**	**110**
			Typical range	*54–69*	*72–95*	*99–126*
Compressive strength perpendicular to the plane	ISO 844	N/mm^2	**Average**	**0.70**	**1.1**	**1.4**
			Minimum	*0. 60*	*0.9*	*1.2*
Compressive modulus perpendicular to the plane	DIN 53421	N/mm^2	**Average**	**46**	**62**	**83**
			Minimum	*40*	*56*	*60*
Tensile strength in the plane	ISO 527 1-2	N/mm^2	**Average**	**1.7**	**2.0**	**2.2**
			Minimum	*1.2*	*1.7*	*1.9*
Tensile modulus in the plane	ISO 527 1-2	N/mm^2	**Average**	**45**	**54**	**64**
			Minimum	*35*	*50*	*54*
Shear strength	ISO 1922	N/mm^2	**Average**	**0.80**	**1.1**	**1.4**
			Minimum	*0.65*	*0.9*	*1.15*
Shear modulus	ASTM C393	N/mm^2	**Average**	**18**	**23**	**30**
			Minimum	*15*	*20*	*25*
SEB	ISO 1922	%	**Average**	**25**	**23**	**18**
			Minimum	*15*	*15*	*10*
Impact strength	DIN 53453	kJ/m^2	**Average**	**1.0**	**1.3**	**1.4**
Thermal conductivity at room temperature	ISO 8301	W/m.K	**Average**	**0.036**	**0.037**	**0.040**

rendering it unsuitable for vacuum infusion processing (VIP) unless it is first sealed with a coating.

The low moisture absorption combined with the proven, excellent FST, dielectric, acoustic, and thermal performance of PEI foam makes an ideal thermoplastic foam solution for the aircraft industry. Applications include luggage bins, galleys, and lower wall panels.

In addition, moisture absorption itself can have a disruptive effect on electronics (interference) and may cause condensation on sensitive areas of the interior. The cycle of moisture absorption and drying that occurs as the aircraft travels through different environmental conditions (altitudes) also has the potential to cause the delamination of a composite structure and can distort the dimensions of a part. Such results can lead to more frequent repairs and downtime.

8.2.3 Special Properties and Testing Methods of Foam Cores

8.2.3.1 Specialties of Sandwich Core Materials

As mentioned in Section 8.1.2, sandwich construction is a term that is used to describe a composite consisting of two, relatively thin, high tensile and compressive strength facings laminated around a lightweight, high-shear-strength core. A sandwich laminate is composed of a total of five structural elements. The two facings are generally materials exhibiting a high strength and modulus of elasticity, including reinforced plastics, a single or multiple plies of wood, or metals such as aluminum or steel. As with a steel I-beam, when the panel is subjected to flexural loads, the loaded facing experiences compressive stresses, whereas the opposite facing is in tension. When greater core thicknesses are utilized, the flexural rigidity is increased by much more than a linear relationship with only a minor increase in weight.

The adhesion between each facing and the core constitutes the third and fourth structural elements of this assembly. This interface may be a film adhesive, as is the case for metal-faced laminations. Or, in the case of wet lamination, the unsaturated polyester or epoxy resin used as a matrix for the reinforcement would serve as the adhesive layer. Depending on the nature of the load imposed on the sandwich panel, the adhesive layer may be stressed in tension, shear, compression, or peel.[19]

The last and important structural element, the core, serves a far-greater demand than simply separating the two facings. The prime requirements for structural core panels include high shear, compressive, and tensile strengths and moduli at densities much lower than the facings. Most sandwich constructions are large flat or contoured structures with a high ratio of length or width to thickness, such as a boat hull, truck side wall, rail vehicle floor, and wind turbine blade.

The core is used as a flat sheet with large area but with small thickness. Most external forces, such as water and wind, uniformly load the construction so that the flat sandwich structure usually reacts with a flexural response. When the sandwich structure experiences tensile stresses on one facing, and compressive stresses on the other facing, the core must transfer the force from one facing to the other in shear, and that is the origin of the shear stress. When the sandwich panel serves as a floor, the core experiences the normal, flatwise pressure. When the panel is used as a truck sidewall, a vertical force works on the top of the panel, so the facings experience a buckling force. If the facing is buckling is outward, the core with adjacent adhesive attempts to hold the facing and core together, and it experiences a flatwise tensile stress, as shown in Figure 8.14. In conclusion, the core in the sandwich structure suffers the shear, compression, and tension stress during the different time and in the different location in most applications. The shear stress on the core is parallel to the face, but the compression and tension stress are perpendicular to the faces. In other words, the flat shear, flatwise tension, and compression properties are important behaviors of the foam core materials.

The essential core static performance characteristics and test methods were identified and documented by the US Department of Defense in the form of MIL-STD-401B.[20] Consisting predominantly of the American Society for Testing and Materials (ASTM) standard methods, MIL-STD-401B is used for specifying structural core materials. The International Organization for

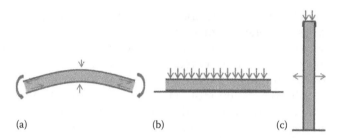

FIGURE 8.14
Sandwich panel in three different loading cases: (a) bending, (b) compression, and (c) vertical compression.

Standardization (ISO) also has similar test methods corresponding to the appropriate ASTM methods. The testing principles, fixtures, and important results for characterizing the core materials will be presented in Sections 8.2.3.2 through 8.2.3.5

8.2.3.2 Flatwise Compressive Properties

As mentioned in the last section, the sandwich structure is often used as a floor, and in the bending load case, the core will be subject to localized compression. As well, highly localized dead loads on the surface of a sandwich panel will be transmitted through the facing to the core. Thus, the flatwise compressive properties of the core are important performance characteristics for the applications of sandwich construction.

ASTM C365, ISO844, and GB/T8813 are standard test methods of compressive strength and modulus of core materials, as tested in the plane of the core surface. The test setup is shown in Figure 8.15. The compressive loads between two flat anvils are measured by means of a load cell. Strain

FIGURE 8.15
Flatwise compressive test setup.

is determined by the precise measurements of the foreshortening of a gauge length, measured at the center or on both sides of the sample. From the cross-sectional area and the load-versus-strain data, the core compressive strength and modulus may be determined.

Test specimens should be of square or circular cross section having areas not exceeding 100 cm² (16 in.²) but not less than 6.25 cm² (1 in.²). The height of the specimen should be equal to the thickness of the sandwich construction and thus changes with the requirements of the application.[21]

Calculations for compressive strength and modulus are accomplished according to the standard procedures, using the load at failure and the slope of the load/strain response. For reference, the appropriate equations are provided as follows:

$$\text{Foam core compressive strength: } \sigma = \frac{P}{A}$$

where
σ = core compressive strength, MPa (psi);
P = ultimate load, N (lb.); and
A = cross-section area, mm² (in.²).

$$\text{Foam core compressive modulus: } E = \frac{St}{A}$$

where
E = core compressive modulus, MPa (psi);
S = ($\Delta P / \Delta \mu$) slope of initial linear portion of load-deflection curve, N/mm (lb./in.);
μ = displacement with respect to the two loading anvils, mm (in.); and
t = core thickness, mm (in.).

A high modulus and high strength are required for a good structural foam core material.

8.2.3.3 Flatwise Tensile Properties

Flatwise tensile strength is a valuable indication of the inherent bonding strength of the foam core material. When the bending, edgewise compression, or normal pulling load works on a sandwich construction, the foam core will experience tensile stress in certain locations and during certain times. For evaluating the tensile properties of the foam core, both edgewise and flatwise tests are commonly used. Especially for a low-density and low-strength foam, the tensile grips used for conventional tensile *dog-bone* specimens will crush the end of the foam sample. For that reason, the flatwise tensile test is used and accepted by more industries and researchers.

ASTM C297, ASTM D1623, and GB/T 1542 are used as standard test methods for evaluating the flatwise, or normal to facing, tensile properties of the foam core material. Two heavy steel or aluminum blocks, each having a through hole perpendicular to the axial pulling direction, are bonded to each facing surface of a core sample, as illustrated in Figure 8.16. Double gimbals are used for pinning the blocks to the load cell and base of the test machine, to ensure than only axial forces are imposed on the specimen. The test method consists of subjecting a foam-cored sandwich specimen to a uniaxial tensile force normal to the plane of the foam sheet. The force is transmitted to the foam through thick loading blocks, which are bonded to the core.[22] The ultimate failure load is normalized for the sample face area to obtain strength. Modulus is determined in the same manner as for compression but with forces acting in the opposite direction:

$$\text{Foam core flatwise tensile strength: } \sigma = \frac{P}{A}$$

where
  σ = core tensile strength, MPa (psi);
  P = ultimate load, N (lb.); and
  A = cross-section area, mm^2 (in.^{2}).

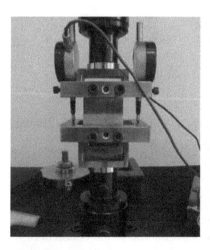 

FIGURE 8.16
Flatwise tensile test setup.

$$\text{Foam core tensile modulus: } E = \frac{St}{A}$$

where
   E = core tensile modulus, MPa (psi);
   S = ($\Delta P / \Delta \mu$) slope of initial linear portion of load-deflection curve, N/mm (lb./in.);
   μ = displacement with respect to the two loading blocks, mm (in.); and
   t = core thickness, mm (in.).

$$\text{Foam core tensile elongation at ultimate load: } \varepsilon = \left(\frac{\mu}{t} \right) 100\%$$

where
   ε = elongation at ultimate load, %;
   μ = displacement with respect to the two loading blocks, mm (in.); and
   t = core thickness, mm (in.).

Aside from exhibiting a high tensile modulus and high strength, an excellent foam core material should possess a large ultimate elongation that is representative of the toughness of the core material.

8.2.3.4 Plate Shear Properties

Shear is defined as the force that causes two contiguous parts of the same body to slide relative to each other in a direction parallel to their plane of contact. Shear strength is the stress that is required to yield or fracture the material in the plane of the core material.

When large sandwich structures, such as airplane fuselages and wings, boat hulls, truck sidewalls and wind-energy turbine blades, experience bending loads, the core materials function as a transition layer to transfer the force from one facing to the other and thus experience shear stresses more commonly than other stress. Shear stress also results when the sandwich structure simultaneously experiences compressive and tensile stresses, so when the core experiences any of these two stresses, it will transform them into shear stress within the core material. For these reasons, the shear properties of the foam core are the most important of all the mechanical properties.

Two prominent standard methods have been developed for evaluating the shear properties of sandwich foam core materials. The most commonly used is ASTM C273 *Standard Test Method for Shear Properties of Sandwich Core Materials.*[23] Samples are bonded between two steel plates by structural adhesive and tested by forcing the plates to translate with respect to each other, as shown in Figure 8.17. The plates can move in either a tensile or compressive mode. From the load versus plate-to-plate displacement response, the shear modulus may be calculated. By normalizing the ultimate or breaking load for the sample dimensions, a core shear strength may be calculated. For reference, the appropriated equations are provided as follows:

$$\text{Foam core shear strength: } \tau = \frac{P}{Lb}$$

where
   τ = core shear strength, MPa (psi);
   P = load on specimen, N (lb.);
   L = length of specimen, mm (in.); and
   b = width of specimen, mm (in.).

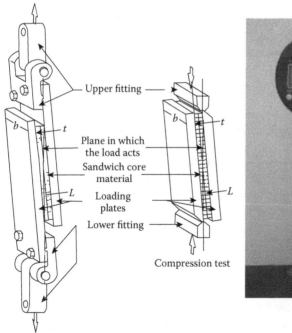
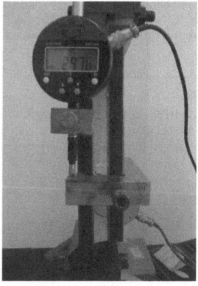

FIGURE 8.17
Plate shear test setup.

$$\text{Foam core shear modulus: } G = \frac{St}{Lb}$$

where

G = core shear modulus, MPa (psi);
S = $(\Delta P/\Delta \mu)$ slope of initial linear portion of load-deflection curve, N/mm (lb./in.);
μ = displacement with respect to the two loading plates, mm (in.); and
t = thickness of core, mm (in.).

$$\text{Foam core shear ultimate elongation: } \gamma = \left(\frac{\mu}{t}\right)100\%$$

where

γ = core engineering shear elongation at specimen break, %;
μ = displacement with respect to the two loading plates, mm (in.); and
t = thickness of core, mm (in.).

In addition to ASTM C273, the ASTM C393 *Core Shear Properties Testing of Sandwich Constructions by Beam Flexure* procedure can also be used as a test method for obtaining the shear strength of a foam core material. For conducting the ASTM C393 test, the foam core must be laminated with facings that are sufficiently strong, such that failures are exclusively within the core, in shear.[24]

Another test method, ASTM C394 *Standard Test Method for Shear Fatigue of Sandwich Core Materials*, can be used for evaluating the fatigue resistance of the foam core, by repeatedly imposing shear stresses, at levels below the core's static shear strength.[25]

8.2.3.5 Other Important Properties

Even though the most common design criteria of structural core materials are their mechanical properties, in many cases, it is a combination of a core material's structural property and other characteristics that are of interest, for example, thermal insulation, sound dampening, and dielectric properties. The following list will give a brief description for each property[26]:

- *Thermal conductivity* is the property or capability of a material to conduct heat or the inverse value of thermal insulation capacity for a given material if it has a low thermal conductivity. Thermal conductivity is not a constant among the various core materials and is affected by temperature, foam density, service history, and moisture content. For a low-density material with air or a gas inside of the foam's cells, it generally provides for a good insulating capacity, which explains the use of foam-cored sandwich structures as thermal insulation components where both insulation and load-bearing requirements are expected.

- *Heat deflection temperature* (HDT) is the temperature at which the foam deforms under a specified stress. The value of the HDT can be determined by using a thermomechanical analyzer or by testing the deformation of the sample at different temperatures in response to a specified stress. The HDT value varies with the density of the foam core, with the type of polymer used to produce the foam, and is a basic property for establishing the maximum processing and operating temperatures of a foam core material.

- *Maximum continuous operating temperature* is the temperature at which the core, with or without facings, can be safely used without being reduced below the minimum mechanical properties required by the particular application and imposed loads.

- *Maximum process temperature* is the temperature at which the core may be laminated with facing materials by thermal curing and consolidation processing without losing physical and mechanical properties. It is dependent on the time, pressure, and processing conditions and is usually lower than the HDT.

- *Dielectric constant* refers to the relative dielectric constant, that is, a ratio of capacitance for the material compared to capacitance for vacuum as the dielectric, when a high-frequency potential is imposed upon it. It represents how readily electromagnetic signals (radio and radar waves) pass through the material. Dielectric constants vary with the density of the foam, the polymer from which the foam was produced, and the frequency of the signal. A low value of dielectric constant means that electromagnetic signals travel through the material with only a small loss.

- *FST behavior* is a requirement in most public transportation and building construction applications, and the authority's specific requirements vary in different regions of the world. The foam core can be an inherent fire-retardant material or be modified into a fire-retardant grade by the use of an appropriate additive. The FST behavior is defined by the aspects of flame spreading, dripping during burning, smoke density, heat release, toxic fumes, limiting oxygen index, etc.

8.2.4 Formats of Foam Core

For use as the core of a sandwich composite, the foam sheet or block needs to be processed into different formats based on the requirements of the application and laminating method. The different formats of the foam core are shown in Table 8.7.[27] Initially, the foam block needs to be cut into a *plain* or

TABLE 8.7

Different Formats of Foam Core Materials

Surface Finishing	Figure	Remarks
Plain foam		After the sheets are cut, no extra surface treatment steps are needed. For a contoured surface, the sheet can be adapted by thermoforming the plain sheets.
Perforated foam		Cylindrical through holes are perforated in a regular pattern distributed over the sheet area. The diameter and distance between holes can be adjusted based on the application's requirements. By using this format core, the air under the sheet can be expelled out, and the resin will flow to the other side through the holes.
Grooved foam		The sheet surface is grooved to provide resin flow channels. The grooves can be made on one side or both sides, in one direction or two crossing directions.
Grooved and perforated foam		Combination of the above two formats, providing both flow channel and air-expelling functions.
Double-cut foam		Cut to a depth of 55%–60% of the core thickness. The cut location of both sides is staged. It is used on slightly contoured surfaces.

(Continued)

TABLE 8.7 (CONTINUED)

Different Formats of Foam Core Materials

Surface Finishing	Figure	Remarks
Contoured foam	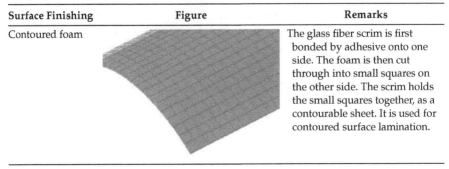	The glass fiber scrim is first bonded by adhesive onto one side. The foam is then cut through into small squares on the other side. The scrim holds the small squares together, as a contourable sheet. It is used for contoured surface lamination.

rigid sheet with required thickness. Following that, different formats will be processed according to the specific processing method and contours of the mold or layup surface.

When multipiece foam core sheets are used for making a large construction, such as a boat hull and a wind turbine blade, they need to be cut into different thicknesses, different shapes, and even different 3D formats for meeting the requirements of the mechanical, geometric, and aesthetic designs of the end product. One set of these core sheets is called a *kit* of cores. The kit cores for one shell of a wind turbine blade is shown in Figure 8.18. Many specialized machines are designed and used for making core kits.

FIGURE 8.18
Core kits for a wind turbine blade.

8.3 Laminating Processes and Application of Foam-Cored Sandwich Composites

A structural sandwich is a layered composite that is formed by bonding two relatively thin facings to a relatively thick core. The core increases the spacing between the two facings and transmits forces that are imposed on one facing to the other in shear so that they are effective about a common neutral axis. The facings resist nearly all of the applied edgewise (in-plane) loads and flatwise bending moments. For low-density cores, the thin, separated facings provide nearly all of the bending rigidity to the construction. The core also provides most of the shear rigidity of the sandwich construction. By proper choice of materials for facings and core, constructions with high ratios of stiffness to weight can be achieved.

The facing materials can be almost any structural materials that are available in the form of thin sheet and can be used to form the facings of a sandwich panel. The properties of primary interest for the facings are (a) high axial stiffness giving high flexural rigidity, (b) high tensile and compressive strength, (c) impact and temperature resistance, (d) environmental resistance, (e) wear resistance, and optionally, (f) an attractive surface finish.

The commonly used facing materials can be divided into two main groups: (1) metallic and (2) nonmetallic materials. The metallic group is composed primarily of steel, stainless steel, and aluminum alloys. Within each type pf metal, there are a vast variety of alloys with different strength properties, whereas the stiffness variation is very limited. The second group is the larger in the two groups, including plywood, cement, veneer, reinforced plastic, and fiber-reinforced composites.[28]

The facing can also be divided into another two groups: (1) rigid sheets before laminating onto the cores, which will be constructed with the cores by using adhesives and by a dry consolidation process, and (2) facings formed during the laminating process on to the cores by a wet process. Most large sandwich products and most fiber-reinforced sandwich composites belong to the second group that are made by using thermosetting liquid resin, the dry fiber reinforcement materials and the different wet laminating processes.

The dry fabrication process is completed by using three separate materials: (1) face sheets, (2) core, and (3) adhesive. The faces are metallic or nonmetallic rigid sheets; the adhesives could be liquid resins or thermosetting or thermoplastic adhesive films. The fabrication processes are hot pressing, vacuum bagging, pinch roller pressing, and continuous pressing, with many variations for all of these.

Wet lamination is commonly used for making fiber-reinforcing composites in which the facing is fabricated during the production process. The facing consists of dry, fibrous materials, such as glass, carbon, synthetic, metallic, and natural fibers. The matrix materials are primarily the thermosetting resins and, to a lesser extent, thermoplastic resins that are gaining attention because of their recyclable properties. The wet processes can be completed by using liquid resin and different fabrication methods, such as vacuum infusion, RTM, chop-and-spray lamination, hand layup, hot press, and pultrusion. The facing is simultaneously created and bonded to the core during wet lamination processes.

8.3.1 Dry Laminating Process

8.3.1.1 Facing Materials

Metal sheets are the primary facing materials for producing sandwich panels that are used in the building construction market. The Young's (elastic) modulus, thickness, surface treatment for bonding, cosmetic requirements, and environmental protection are the most important factors. Stainless steel and aluminum sheets, with a thickness of 0.4–0.7 mm, are primarily laminated using a continuous process.

Thermosetting fiber-reinforced plastics sheets, referred to as FRP in the composites industry, are mainly used as facings for making sandwich wall panels for recreational vehicles, delivery trucks, and semitrailer trucks. FRP sheets, approximately 1–3-mm thick, are generally made with unsaturated polyester or vinyl ester as a matrix, chopped glass fiber or fabric as reinforcing materials, and laminated in a continuous process. The surface of the FRP is coated with a gelcoat that can be a variety of colors and glossy and has the function of weather and impact resistance (Figure 8.19).

A continuous fiber-reinforced thermoplastic sheet (CFRTP) is made by impregnating nonwoven continuous fibers with thermoplastics. The fibers can be glass, carbon and, aramid materials, and the plastics can be polypropylene, polyethylene, PET, polyamide 6, and polyvinylidene fluoride.

FIGURE 8.19
FRP and RV with the wall made with FRP.

For oriented unidirectional or biaxial CFRTPs, the unidirectional 0° direction tape is produced first, and then optionally slit into ribbons, woven, and heat laminated together to create a 0°/90° bidirectional structure that has balanced (or intentionally unbalanced) physical and mechanical properties. The CFRTP product resists rot, corrosion, and mildew. Some CFRTP sheets with a thin scrim liner material on one side promote adhesion for sandwich composite laminating (Figure 8.20). The scrap, waste, and recovered product, after serving its useful life, can be recycled into new products by traditional hot-melting processes because the matrix is thermoplastic. Also, the sheets can be laminated by applying heat without adhesives and emit no volatile organic compounds (VOCs). The products are primarily used for producing sandwich panels in such applications as truck bodies, shipping containers, and building construction wall panels (Figure 8.21).

8.3.1.2 Adhesives

There exists a variety of adhesives for dry laminating sandwich composites. The adhesive must meet the mechanical requirements of the structure: (a) providing a good adhesive bond between the facing and core components in the environment that the structure is to work and (b) performing satisfactory

FIGURE 8.20
CFRTP sheet and sandwich lamination. (From http://www.polystrand.com.)

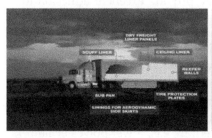

FIGURE 8.21
Applications of CFRTP. (From http://uslco.com.)

against such considerations as static strength and fatigue, heat resistance, aging, and creep.[28] The following are the most frequently used adhesives:

- Hot-melt adhesives provide optimal adhesion and superior manufacturing flexibility compared with waterborne or solvent-based adhesives and are widely used in a variety of applications, including product assembly, woodworking, and pressure-sensitive tapes and labels. The products are solvent-free, thermoplastic-based materials that are specially formulated and are applied in the molten state at temperatures varying from 120°C to 180°C. Most hot-melt adhesives are composed of three main components: (1) high-molecular-weight polymer (e.g., EVA or synthetic rubber and PUR), which acts as a backbone and provides the primary mechanical properties of the adhesive, (2) tackifying resin, which provides wetting and adhesion properties, and (3) plasticizer, such as an oil or wax, which controls the viscosity of the blend and enables the adhesive to be handled by simple machinery.

- There are film, rod, and pellet types of hot-melt adhesives that can be dispensed by directly laminating and melting using a hot press, and by glue gun, or by roller coating and spraying after melting. The hot-melt adhesives can be solidified at short time and with very little VOCs released.

- Epoxy resins and adhesives can be formulated to cure at a wide range of temperatures, but for most consumer sandwich applications are low-temperature curing resins, normally between 20°C and 90°C. For aerospace and other demanding applications, some epoxy formulations are formulated for high-temperature curing (130°C–220°C). In general, epoxy resins and adhesives have the advantage of being used without solvents, cure without creating volatile by-products, and have a relatively low volumetric shrinkage rate. The absence of solvents makes epoxies usable with almost every type of core material. Epoxy is available as pastes, films, powders, or as solid adhesives. The typical shear strength of epoxy is approximately 20–25 MPa.

- PUR adhesives are likely the most widely used for bonding sandwich elements. This is because they provide excellent adhesion to most materials. They can be used as a paste or liquid in a wide range of viscosities, may have long or short cure times, and can be made fire retardant and water resistant. Most PUR adhesives contain virtually no solvents and are thus environmentally friendly and, once cured, are the least toxic of all the adhesives.

- Urethane acrylate adhesives are resins that are compatible with polyesters and vinyl esters. In fact, urethane acrylates are so compatible that they can be incorporated, in, e.g., a wet polyester laminate. Urethane acrylates are very tough and exhibit almost no curing volumetric shrinkage.

- Polyesters and vinyl esters are low-cost, commonly used adhesives or resins for dry lamination. Prefabricated laminates can be bonded to foam, balsa, or honeycomb cores using the same resin as in the laminate. One problem with these resins is that their curing volumetric shrinkage often creates very high interfacial shear stresses. One way of decreasing the effect of shrinkage is to add a low-profile additive or some fillers to the resin for reducing the shrinkage.

8.3.1.3 Laminating Processes

Dry laminating is an assembling process for bonding the core and facings together by using adhesives and specially engineered equipment. The process can be used for producing small and simple products but can also be used for making large and complicated products. Some lamination processes may just employ a hot press or a vacuum bag; whereas others require a large and complicated system. In the process, adhesive layers are interleaved between the faces and the core, and the whole stack is subjected to elevated temperature and pressure as required by the adhesive or resin until it cures; then, the sandwich is cooled. For high-performance applications, the bonding process likely takes place using a vacuum bag and an autoclave, whereas for less demanding applications, it may be sufficient use a vacuum bag and/ or weights or a hydraulic press. Since, ideally, there should be little or no resin bleeding if the bonding is correctly performed, the vacuum-bagging arrangement is the simplest when compared to wet laminate manufacture.

It is normally necessary to prepare the surfaces to be bonded in order to achieve a high-quality adhesion. Unless already done, foam cores are typically vacuum cleaned to remove all loose particles, and they may also be primed if necessary.

The typical processing sequence for bonding of composite laminates to a foam core using an epoxy or PUR adhesive is shown in Figure 8.22.

The dry lamination process has been used in automatic production for making large-scale sandwich panels for the past 20 years in various industries. Several technologies have been developed for continuous or semicontinuous production. A pinch roller laminating system is a rapid fabrication process using hot-melt PUR adhesive. The adhesive is initially melted and then coated to the foam core panel using a specially designed rubber roller. The face sheets are laid below and above the foam panel, and then the

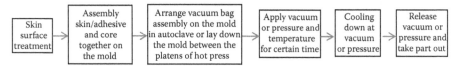

FIGURE 8.22
Dry lamination procedures by using heat and pressure.

assembly is sent into the pinch roller system for consolidation, as shown in Figure 8.23. The adhesive can cure and then cool down to give enough strength (green strength) for holding the three-layer assembly together after pinch rolling and will have a maximum strength after just 24 h.

A PUR moisture-curing adhesive is used in many large-scale lamination processes for making RV and truck sandwich wall, floor, and roof panels. A multiopening vacuum-powered press for producing a sandwich panel with FRP facings up to 15-m long and 3-m wide is shown in Figure 8.24. Each platen opening can operate by drawing vacuum independently, whereas other platens open for moving their laminate in or out.

Adhesive films and a double-belt press are commonly used in continuous laminating processes for making sheet metal-faced sandwich panels. Hot-melt adhesive films are composed mainly of polyethylene copolymer or thermoplastic PUR. The double-belt press may be set up for heating and cooling in different zones so that the laminate is heated first under pressure, then cooled under pressure until leaving the press. Two laminating lines shown in Figures 8.25 and 8.26 are used for making metal-faced sandwich panels

FIGURE 8.23
Pinch roller lamination process. (From http://blackbrothers.com.)

FIGURE 8.24
Four-level vacuum-powered press for making sandwich panel. (From Bonding System International.)

FIGURE 8.25
Metal skin sandwich panel laminating system. (From Wabash National.)

FIGURE 8.26
Sandwich panel continuous processing. (From steelformer.com.)

for semitrailer truck sidewalls and for building construction wall panels, respectively.

8.3.2 Wet Laminating Process

8.3.2.1 Reinforcement Materials

In the wet laminating process, three main materials are used for making the sandwich composite products: (1) core, (2) reinforcement material, and (3) liquid resin. Fabrication of the facing and bonding the facing to the core are completed in a one-shot process. The fiber-reinforcing fabrics are the major materials that are used in the wet laminating process because they have a high Young's modulus and thus high stiffness, are flexible, and are air permeable, so the liquid resin can penetrate them and bond them to the core after curing to make sandwich composite products with a complicated shape, high stiffness, and unique mechanical properties.

The most commonly used fiber-reinforcing materials are glass, carbon, mineral, synthetic and natural fibers. Of these, *E-glass*, or electrical glass, fibers enjoy the greatest utilization as a reinforcement material in modern composite industries. Fiberglass in its current style was first made in the middle of the 1930s and was used as a reinforcement material with unsaturated polyester for making composites in the early 1940s. Following that, many modifications have been accomplished for making new grades of glass fibers with a higher modulus, a wide variety of construction formats, and many more applications. Currently, fiberglass is classified by chemical composition as E-glass, S-glass, A-glass, R-glass, and others.

Carbon fiber having a high tensile strength and modulus was first discovered in the late 1950s and is produced by slowly carbonizing synthetic fiber filaments, such polyacrylonitrile fiber, at a high temperature in a controlled atmosphere. Carbon fibers have a much higher tensile modulus and much lower density than fiberglass. Due to their premium cost, carbon fiber reinforcements are mainly used for making high-performance composite products in applications for aerospace, sport goods, military, and medical equipment applications.

The greatest usages of synthetic fibers in the composite industry are aramid fibers, made from aromatic polyamide, and Spectra® fiber, made from ultraoriented, ultrahigh-molecular-weight polyethylene. These have properties of light weight, high strength, and high toughness. The key properties of fibrous aramid make it an important material in many different composite markets, such as bulletproof vests, aircraft body parts, fibrous optic and electromechanical cables, and friction linings for clutch plates and brake pads.[29]

Natural fibers are enjoying greater usage recently for reducing energy consumption and applications desiring the use of sustainable natural materials in the composite industries. Natural fibers include flax, hemp, jute, and kenaf, which are well adapted for making door panels, backseats, and trunk liners of automobiles.

Other fibers used as reinforcement materials are mineral fibers, such as basalt fiber, ceramic fiber, and boron fiber, all of which have a higher tensile modulus and strength, and compressive strength, than fiberglass. Metal fibers are also used as reinforcement materials, such as fibers that are made by ferrous and nonferrous alloys including steel, stainless steel, bronze, aluminum, brass, and copper. Applications for metal fibers in composites include those requiring electrical conductivity and providing electromagnetic interference shielding.

Fiber materials have different formats for various processing methods and applications. The basic format of continuous fibers is roving, as shown in Figure 8.27, which can be used directly for filament winding and pultrusion processes, and can be chopped into short-length fibers for spray lamination, and for producing sheet molding composite prepreg, chopped strand mats, and surface finishing veil, as illustrated in Figure 8.28. Roving is the basic starting material for making a variety of different fabric formats for wet laminations.

FIGURE 8.27
Roving of different fiber materials.

FIGURE 8.28
Different fiber mats.

One widely used fabric is woven roving, which is produced by interlacing continuous fiber roving into relatively heavyweight fabrics. Woven roving is generally compatible with most resin systems and used primarily to increase the flexural and impact strength of FRP laminates. It is ideal for multilayer hand layup applications where increased material strength is required, along with good drapeability, ease of wet out, and cost efficiency. Woven roving is a *plain-weave* style, available in a variety of weights, widths, and finishes to suit a wide range of applications.

Multiaxial nonwoven stitched fabrics are ideally suited for selective reinforcement. Fibers are bundled and stitched together, rather than woven, so there is no crimp in the yarns. As a result, these reinforcements offer optimized directional strength for their finished composite parts. For this reason, they are popular in the wind energy industry. These fabrics are ideal for wet layup, pultrusion, and VIP, to create strong, stiff composite parts, and are compatible with polyester, vinyl ester, and epoxy resins (Figure 8.29).

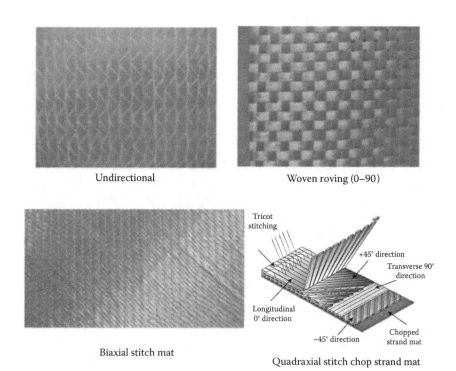

Undirectional   Woven roving (0–90)

Biaxial stitch mat

Tricot stitching

+45° direction

Transverse 90° direction

Longitudinal 0° direction

−45° direction

Chopped strand mat

Quadraxial stitch chop strand mat

FIGURE 8.29
Different fibrous fabrics. (From http://vectorply.com/stitch-bonded-reinforcements/.)

8.3.2.2 Resin Systems

The purpose of the resin system, or *matrix*, is to bind the dry fibers of the reinforcement together into integrated composite facings, to transfer load from fiber to fiber, and to carry the load from the composite facings to the core. The choice of resin will depend upon many factors: (a) the structural requirements of the part; (b) the lamination method; (c) the cost limitations; (d) the mixing, handling, and mold-curing facilities; (e) the service environment expected; (f) the cosmetic requirements; and (g) the length of service life expected or desired.[30]

Almost all of the resins used in composites are a combination of several resins and various additives. Hence, the term resin system is often used instead of resin. Such additives can raise or lower the viscosity, change the resistance to ultraviolet radiation, improve the interlaminar shear strength, as well as the toughness, increase the strength of resin-rich areas in the laminate, increase or reduce translucence, or add a core material to the laminate, and change the surface tension and or the wettability during the low-viscosity period just before cure.

Unsaturated polyester resins are the most widely used in sandwich composite wet lamination, thanks to their ease of handling; a good balance of mechanical, electrical, and chemical properties; and relatively low cost. Typically coupled with E-glass fiber reinforcements, polyesters adapt well to a range of fabrication processes and are most commonly used in open-mold spray-up, compression molding, RTM, vacuum infusion, and pultrusion.[31]

Unsaturated polyester properties can be modified to meet specific performance criteria, based on the selection of glycol and acid elements and reactive monomers (most commonly, styrene monomer). Polyester resins are often differentiated in terms of their base diacid ingredients: (a) orthopolyesters, (b) isopolyesters, and (c) terephthalic resins. Dicyclopentadiene (DCPD) can be added to polyester resins to lower viscosity, improve cosmetics, and provide sufficient mechanical properties at a lower styrene content.

Styrene enables polyester resins to cure from a liquid to a solid by *cross-linking* the molecular chains. Catalysts (actually are initiators) are added to the resin prior to molding, to activate cross-linking, but do not take part in the polymerization reaction. Catalysts used with polyester are organic peroxides or modified peroxides. A promoter, usually a cobalt *soap*, also needs to be added to the catalyzed resin to enable the reaction to proceed at prevailing temperatures and at an increased rate. Additives, also called modifiers, offer embellished processing or performance attributes, with the most common including pigments, fillers, and flame or fire retardants.

Vinyl ester is a resin that is produced by the esterification of an epoxy resin with an unsaturated monocarboxylic acid, most commonly methacrylic acid. The reaction product is then dissolved in a reactive solvent, such as styrene, to 35%–45% content by weight, offering a bridge between lower-cost, rapid-curing, and easily processed polyesters and higher-performance epoxy resins. Their molecular structure is very similar to polyester's but with reactive sites only at the ends of the molecular chains and with fewer ester groups. Since ester groups are susceptible to hydrolysis, less of these increase vinyl esters' resistance to water and chemically corrosive environments. Vinyl esters are favored in making the chemical tanks and the corrosion-resistance equipment and add value in structural laminates that require a high degree of moisture resistance (such as boat hulls and decks). They are processed and cured very similarly to polyesters, with the potential to offer improved toughness, although, just as for orthopolyester resins, this usually requires an elevated temperature postcure.

Epoxy resins have a chemical structure based on diglycidyl ether of bisphenol, creosol novolacs, or phenolic novolacs. Epoxies are not cured with a catalyst, like polyester resins, but instead use a hardener (also called a curing agent). The hardener (part B) and the base resin (part A) coreact in an *addition reaction* according to a fixed ratio. Epoxy resins contribute strength, durability, and chemical resistance to a composite. They offer high performance at elevated temperatures, with hot/wet service temperatures up to 121°C/250°F. Epoxy resin is known in the marine and wind energy industry for its incredible toughness and bonding strength. Quality epoxy resins adhere to other materials with a shear strength of 2000 psi versus only 500 psi for vinyl ester resins and less for polyesters. In areas that must be able to flex and strain with the fibers without microfracturing, epoxy resins offer much greater capability. Epoxy effectively bonds to all sorts of fibers very well and offers excellent results in repairability when it is used to bond two different materials together.

Phenolic resins are based on a combination of phenol and formaldehyde or resorcinol. They find application in flame-resistant aircraft interior panels and in commercial markets that require low-cost, flame-resistant, low-smoke products; excellent char yield; and heat-absorbing characteristics. They have proven to be successful in aerospace applications, in components for offshore oil and gas platforms, and in mass transit and electronics applications. Phenolic resins polymerize via a condensation reaction, releasing water vapor and formaldehyde during cure, which can produce voids in the composite. As a result, their mechanical properties are somewhat lower than those of epoxies and most other high-performance resins. Multiple *degassing* or venting steps during lamination can assist to remove the water and formaldehyde that are liberated and improve the physical properties.

Cyanate esters are chemical substances generally based on a bisphenol or novolac derivative in which the hydrogen atom of the phenolic OH group is substituted by a nitrile or cyanide group. The resulting product with a—OCN group is called a cyanate ester. These are versatile matrices that provide excellent strength and toughness, allow very low moisture absorption, and possess superior electrical properties compared to other polymer matrices, although at a higher cost. Cyanate esters feature hot/wet service temperatures to 149°C/300°F and are usually toughened with thermoplastics or spherical rubber particles. They process similarly to epoxies, but their curing process is simpler because of their viscosity profile and nominal volatiles. Current applications in high-performance composites range from radomes, antennae, missiles, and nosecones to microelectronics and microwave products.[32]

8.3.2.3 Laminating Processes

There are numerous methods for fabricating sandwich composite products. Some methods have been borrowed (injection molding, for example), but many were developed to meet specific sandwich design or manufacturing challenges. Selection of a method for a particular part, therefore, will depend on the materials, the part design, and end use or application.[8]

In a wet laminating process, the material-selection criteria for core, resin, and reinforcement are workability, cost, static and dynamic strength, and elevated temperature performance.[5] The wet laminating processes involve some form of molding, to shape the core, resin, and reinforcement into a specially designed product prior to and during cure. After wet laminating, the resin in the facings needs to cure in order to bond the fibers together and bond the facing to the core. Several curing methods are available. The most basic is to simply allow the cure to occur at room temperature. Cure time can be accelerated, however, by applying heat, typically within an oven, and pressure, by means of a vacuum bagging that consolidates the plies of material and significantly reduces voids due to incomplete penetration and wetting of the matrix through the reinforcements and core surface.

8.3.2.3.1 Hand Layup Process

The most basic fabrication method for making sandwich composites is the hand layup, also called the wet layup, which typically consists of wetting dry fabric layers with liquid resin, and prewetting the core surfaces with liquid resin, while assembling these onto a mold to form a laminate stack by hand. The trapped air in the fabric is rolled out by a hand roller, as shown in Figure 8.30. The laminate can also be consolidated by vacuum bagging, air pressure, or by the force of a platen press. Following the deairing, the resin is allowed to cure either in an open condition or under a vacuum bag or press, at room temperature. The curing time may be accelerated by curing in a heated oven or using a hot platen press.

Hand layup has been widely used for many years; is easy to learn and practice using a low-cost mold if curing is accomplished at room temperature; and can accommodate a wide choice of cores, resins, and fiber materials. This method is often used for producing small-size components, such as a prototype, a sample for testing, and certain products having a small part count. The disadvantages of a hand layup are that the resulting quality strongly depends on the skill of the operator, that it results in a lower fiber/resin ratio because the hand roller cannot squeeze excess resin out, and that the volatile components of the resin system more readily evaporate into the air.

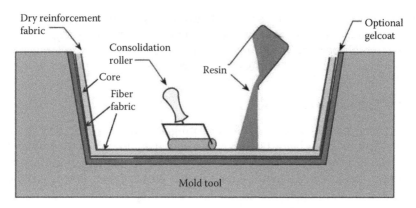

FIGURE 8.30
Hand layup laminating. (From http://fiber-reinforced-plastic.com.)

8.3.2.3.2 Spray Layup Lamination

Spray layup processing is a method to apply the liquid resin by the use of a spray gun instead of by hand. The fiber is simultaneously applied with sprayed resin and broadcast onto the mold by the use of a chopper gun and mixing process, or separately applied to the mold as a fiber fabric before spraying resin, as shown in Figure 8.31. For laminating a sandwich structure, the first facing must be prepared by spray layup processing, rolled out to remove trapped air, and then the prewetted core material rolled over that. After the mold-side facing cures, a second facing is spray layup processed over the back side of the core and rolled out to remove trapped air. Vacuum bagging may be used for consolidating the laminate and drawing out the air after finishing the spray layup processing but before the resin cures. The product usually cures at room temperature or alternatively cures in a heated room.

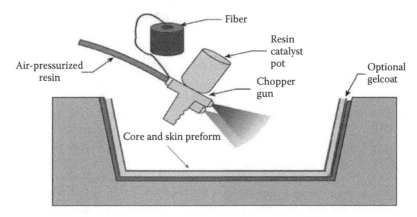

FIGURE 8.31
Spray layup process. (From http://www.netcomposites.com.)

The resins used in the spray layup process are primarily unsaturated poly-ester and vinyl ester. All types of foam cores can be used in this process, pro-vided that they are compatible with the resin. Chopped strand mat and all woven and knitted fabrics are suitable for this process. Spray layup processing can be performed in a continuous process for making flat sandwich or solid FRP panels, but mostly, it is used in the intermittent process for making cus-tom parts in low- to medium-volume quantities, such as bathtubs, swimming pools, boat hulls, storage tanks, duct and air-handling equipment, and the like.

The advantages of the spray layup process include a very economical pro-cess for making small to large parts, using a low-cost mold, as well as low-cost materials. The limitations of the process are that it is not suitable in making components with a high glass/resin ratio; it is difficult to control the fiber volume fraction, as well as the thickness; the quality of the component highly depends on operator skill; often, there is a high volatile chemical emission; and it cannot produce a product with a good finish on both surfaces.

8.3.2.3.3 Resin-Infused Closed-Mold Lamination

The closed-mold processes are opposite in their boundary conditions to open-mold processes, such as hand layup and spray layup processes, because resin is drawn into the component by tubing and not directly exposed to the workshop air. Resin infusion is only processed after the mold is tightly closed, and thus no VOCs such as styrene monomer are released to the air. Closed-mold processes include VIP and RTM, the latter optionally vacuum-assisted. If the resin is being *pushed* into the mold or tool, this is considered RTM as shown in Figure 8.32, while if the resin is *pulled* into the mold, this is considered VIP as shown in Figure 8.33. RTM uses an above atmospheric pressure to send the resin into the composite tool, as shown in Figure 8.32. VIP uses a lower than atmospheric pressure to suck the resin into the com-posite tool.

For performing closed molding, three main materials of a sandwich struc-ture have their individual requirements. The fiber reinforcement materials are usually fabrics and mats that can be dry laid down on the mold and allow the resin to flow through them under pressure or vacuum force. The resins should be low in viscosity so that they can easily flow and saturate all dry fiber fabrics and the surfaces of the core. They should have the catalyst or hardener added and uniformly mixed, such as for unsaturated polyester and epoxy, respectively, before being introduced into the mold, and then they should react or cross-link without releasing any low-molecular by-products. The core should be finished into a particular shape to meet the design requirements and fit of the mold, and cut with flow channels or perforated that allow the infused liquid resin to readily flow horizontally and vertically.

Closed-mold resin infusion also requires a few auxiliary materials for completing the process, such as flow media, release film, peel ply, spiral resin flow hose, vacuum tape and film, vacuum system or resin pressure injection equipment, and the like.

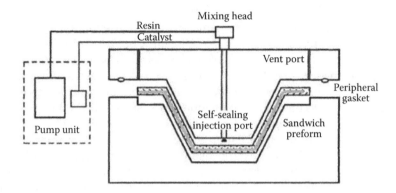

FIGURE 8.32
RTM setup. (From http://www.fiber-reinforced-plastic.com.)

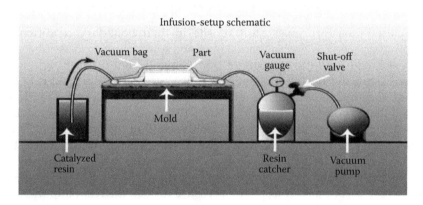

FIGURE 8.33
Vacuum infusion setup. (From http://www.mid-composties.com.)

8.3.2.3.4 Vacuum Infusion

Vacuum infusion is a variation of vacuum bagging where the resin is introduced into the mold after the vacuum has pulled the vacuum bag down and compacted the still-dry laminate. The method is defined as having a lower pressure than atmospheric pressure in the mold cavity. The reinforcement and core material are placed dry in the mold by hand or using machine assistance, which provides the opportunity to precisely position the core and reinforcements. The mold may optionally be coated with a gelcoat before the placement of reinforcements and core. A peel-ply fabric or a perforated release film is then placed over the dry reinforcement. Next, a flow media consisting of a coarse *fishnet* mesh or a *crinkle* ply is positioned, and perforated tubing is positioned as a manifold to distribute resin across the laminate. The vacuum bag is then positioned and sealed around the mold perimeter. A tube is connected between the vacuum bag and the resin container. Initially, the vacuum pump draws air out from under the vacuum bag until the lowest level of vacuum is achieved, and any leaks are then sealed. Finally, the valve between the bag and the resin container is opened, and resin enters the bag and wets out all dry reinforcements and core surfaces. The vacuum is maintained until the resin has wetted out the entire laminate, and then the resin valve is closed to stop drawing resin in. After that, the laminate is kept under vacuum until the resin cures. For making a large part, multiple hoses are installed to draw air out and resin into the bag, as shown in Figure 8.34, an example of a vacuum infusion process for making a wind turbine blade.

By using VIP, one can make products with the highest glass/resin ratio, a minimum amount of resin, thus lowest in weight, and with superior mechanical properties. This is because when resin is pulled into the mold, the laminate is already compacted; therefore, there is no room for extra resin. Vacuum infusion is suitable for molding very large structures. This process uses the same low-cost tooling as open molding and requires a modest amount of additional equipment. Vacuum infusion offers the potential for a substantial reduction in emissions compared to either open molding or wet-layup vacuum bagging processes. The limitations are a greater level of skill for producing quality parts, the risk of air leaks, the extra expense on auxiliary disposable materials, and the capability for producing only one cosmetic side to the part. This process is widely used for producing wind turbine blades and nacelles, boat hulls, aircraft components, and industrial parts.

8.3.2.3.5 RTM

RTM is an intermediate volume-molding process for producing composites. The RTM process involves injecting resin under pressure into a mold cavity. RTM can use a wide variety of tooling, ranging from low-cost, composite molds to temperature-controlled metal tooling. This process can be

FIGURE 8.34
Making a wind turbine blade by vacuum infusion. (From http://www.mathfem.it.)

automated and is capable of producing rapid cycle times. Vacuum assist can be used to enhance resin flow in the mold cavity.

The mold set is gel coated conventionally, if required. The reinforcement (and core material) is positioned in the mold, and the mold is closed and clamped. The resin is injected under pressure, using mix/meter injection equipment, and the part is cured in the mold. The reinforcement can be either a preform or pattern-cut roll stock material. Preforms are reinforcements that are preformed in a separate process and can be quickly positioned in the mold. RTM can be done at room temperature. However, heated molds are required to achieve faster cycle times and greater product consistency. Clamping can be accomplished with perimeter clamping or clamping in a press.

The process can give a high rate of production, is easy to control, ideal for medium-to-large series of small-to-medium parts, has a high level of consistency, and provides good dimensional stability. However, the cost of the equipment is high, requires high-precision molds, and requires time to learn. Applications are abundant in the production of trucks, buses and cars, and structural and industrial products.

8.3.2.3.6 Pultrusion

Pultrusion is a continuous process for the manufacture of products having a constant cross section, such as rod stock, structural shapes, and beams and channels. After modifying the die and installing a core-aligning and feeding system, sandwich composite panels can be processed by the pultrusion process, as shown in Figure 8.35. Pultrusion produces a sandwich panel with extremely high fiber content and the high structural properties.

For producing sandwich panel by pultrusion, a continuous glass strand mat, cloth, or surfacing veil is impregnated in a resin bath, and then the

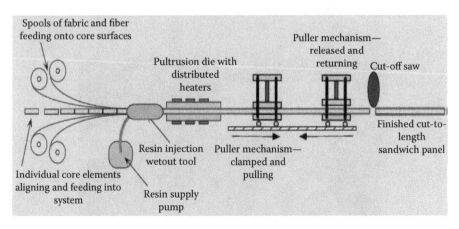

FIGURE 8.35
Sandwich panel pultrusion process. (From http://www.fiber-reinforced-plastic.com.)

core is fed between two sets of the impregnated reinforcements and pulled through a steel die by a powerful tractor mechanism or a continuous puller. The steel die consolidates the saturated reinforcement, sets the shape of the stock, and controls the fiber/resin ratio. The resin for pultrusion is specially formulated to cure at a high temperature and thus at a fast speed. The die's platen is designed with multiheating zones that are controlled by a computer. The reinforcements are positioned into the die by tensioning devices and guides. The hardened steel die system is machined, chromium coated, and includes a preform area to do the initial shaping of the resin-saturated reinforcements. The latest pultrusion technology uses direct injection dies in which the resin is injected into the die rather than applied through an external resin bath.

The process is a continuous operation that can be readily automated with high efficiency and produces products with consistent properties. The composite skins of a sandwich panel are cured under tensile stress and at a high temperature, so the panel has greater flexural properties and a high heat resistance. The pultrusion process has been used for producing sandwich panels from 5- to 100-mm thick, and up to 3-m width, with different foam cores and facing materials.

8.3.3 Applications of Sandwich Composites

Sandwich structural composites (SSCs) have been used for building aircraft fuselage and wings in WWII, which was the first composite application for a balsa core sandwich. At present, SSC is still one of the major materials for the aerospace industry. After WWII, foam-cored sandwiches had been used for building power and sailboats, yachts, and other marine vessels. After the 1990s, the wind energy industry employed SSC to fabricate large wind turbine blades, which are now the largest market application of sandwich constructions. Foam core sandwich composites have also been used in applications of manufacturing rail vehicles, trucks, trailers, buses, sporting goods, construction buildings, military battle ships and vehicles, antenna radomes, medical equipment, and industrial infrastructure.

8.3.3.1 Marine Industry

For over 60 years, foam cores have been utilized in marine applications to produce boat and yacht hulls, decks, and fly bridges, providing for a light, stiff, and strong construction. A variety of marine vessels have been produced using foam, end-grain balsa wood, and other core materials, along with glass reinforcements and liquid thermosetting resins. Foam-cored sandwich composite structures can be employed for producing the entire boat hull by laminating within a female mold, as shown in Figure 8.36. Primarily, PVC, PET, and SAN foam cores, along with glass roving and fabric as the reinforcements, and thermosetting liquid resins, such as polyester,

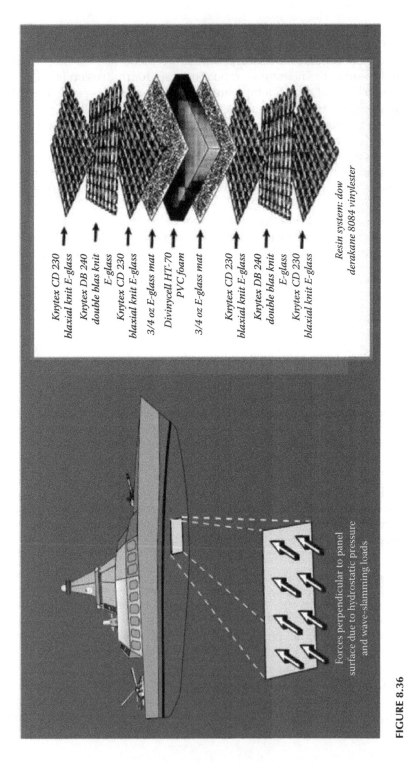

Knytex CD 230 blaxial knit E-glass

Knytex DB 240 double blas knit E-glass

Knytex CD 230 blaxial knit E-glass

3/4 oz E-glass mat

Divinycell HT-70 PVC foam

3/4 oz E-glass mat

Knytex CD 230 blaxial knit E-glass

Knytex DB 240 double blas knit E-glass

Knytex CD 230 blaxial knit E-glass

Resin system: dow
derakane 8084 vinylester

Forces perpendicular to panel
surface due to hydrostatic pressure
and wave-slamming loads

FIGURE 8.36

Sandwich composite used for building boat hulls. (From https://en.wikipedia.org/wiki/De_Havilland_Mosquito.)

vinyl ester, and epoxy, are used for marine structural sandwich applications. The lamination process utilized depends upon the strategy of the boat yard and the skills of the workers. A female mold is necessary for building a boat. The gelcoat must first be coated on the mold surface, and then composites construction using hand layup and spray layup and open-air curing is the simplest method for producing a boat. Currently, more boat builders use vacuum bagging for assisting in laminate consolidation and the reduction of voids. However, for environmental and cost reasons, a growing number of boat hulls and other components are fabricated by using RTM or vacuum infusion.

Marine products made by sandwich composites include hulls, stringers, bulkheads, flat deck panels, containers and furniture, for racing boats, lifeboats, sailing boats, and leisure yachts, as shown in Figure 8.37. Some military ships and submarines also use foam-cored sandwich construction for components such as topside superstructures, tanks, doors, electrical enclosures, tables and worktops, shower units, and other installations.

8.3.3.2 Wind Energy Industry

Sandwich construction has been used in wind energy since the early 1990s. Wind blade manufacturers typically make blades in half-shells, which are secondarily assembled. Each blade shell is a sandwich structure, often featuring a combination of different types and thicknesses of core materials, each used in a particular location, as shown in Figure 8.38. The wind energy market is currently the number one consumer of core materials, reinforcements, and resins in the composites industry, and as the needs of this market evolve, so will the demands on technology for all of these materials. Blade manufacturers are, like all composites processors, balancing many variables against each other, each picking and choosing different combinations of materials to meet cost and performance goals. Core selection is influenced by several metrics: (a) thickness of the structural sandwich, (b) mechanical strength and stiffness, (c) blade durability, and (d) cost. Some core types offer better mechanical properties but are more costly. Others are economical, but lack mechanical strength, in which case, a thicker sandwich might be needed to meet a specific requirement.[4]

To accommodate the complex and unique curvature of the blade at each point along its length, foam and balsa wood sheet stock is cut to specified dimensions, scored or kerfed, and then labeled and arranged by core suppliers into kits. For producing wind blade and nacelle housings, the reinforcement materials are predominately E-glass construction, but for some designs, and in critical areas, carbon fiber reinforcements are included. End-grain balsa, PVC, and PET foam are the primary types of core materials. Liquid thermosetting resins, such as vinyl ester and epoxy, are most often

FIGURE 8.37
Marine vessels made by sandwich composites.

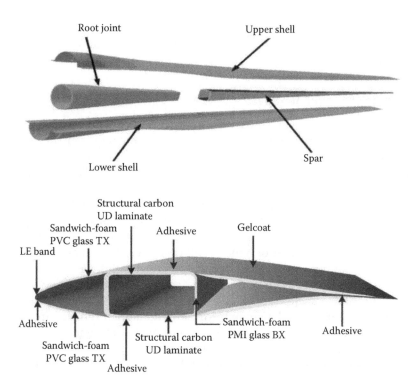

FIGURE 8.38
Wind blade structure. (From http://www.compositeworld.com.)

used. VIP is the primary process for producing blades and shear webs, along with a small number of prepreg applications for wind energy.

It is estimated that, in 2017, global wind blade production will be 140,000 units for turbines that range from 0.75-MW to more than 3-MW capacity.[7] An average of 10 m³ of core materials will be consumed for every three blades

FIGURE 8.39
Wind blade in manufacturing and in application. (From http://technologyreview.com.)

of each turbine tower, and a total of 1.4 million m³ of core materials will be used each year. Wind turbine blades, in manufacturing and in application, are shown in Figure 8.39.

8.3.3.3 Airplane and Aerospace

One of the key features of sandwich construction is a high ratio of stiffness to weight. Aircraft and aerospace vehicles all have the requirements for light weight and stiff and strong construction. Usage of sandwich structures for fabricating aircraft started in the early 1940s during WWII. The de Havilland *Mosquito*, as shown in Figure 8.40, had a sandwich structure of flat-grain balsa that is sandwiched between veneers of Canadian Birch for its fuselage and wings.[33] Currently, aluminum and Nomex®-manufactured honeycomb cores are most commonly used as core materials for aircraft construction. Because of the greater cost of honeycomb cores, foam cores are used wherever possible for building aircraft and aerospace vehicles.

PEI, PMI, and other high-performance cores are widely recognized for their performance in the aerospace sector where light, strong, and resilient sandwich structures are demanded. Aircraft and launch vehicles are typical target platforms. Flight control surfaces, rotor blades, radomes, and satellite containment fairings are other common applications. Their facings are composed of fibers that include glass, carbon, Kevlar® and other reinforcements that are impregnated with a number of resin systems, including epoxy, phenolic, bismaleimide (BMI), cyanate ester, and others. Producing the finished sandwich structure requires the use of typical processes to cure the resin in the facings and cobond it to the core. The most common are vacuum infusion for liquid thermosetting resins or, in the case of the use of a prepreg, autoclave high-temperature curing. Critical to curing the resin is applying the precise amount of pressure at the right temperature environment.

FIGURE 8.40
Aircraft built by sandwich composites in WWII.

(Approximately 180°C is most frequently used in the aircraft industry.) The end result is an extremely light but rigid and robust structure. Typical applications include landing gear doors, belly fairings, flight control surfaces, radomes, and nacelles.

Foam-cored sandwich structures can be used for producing the fuselage and wings for small, private aircraft, as shown in Figure 8.41. Most components of large aircraft, such as the floor, interior ceiling, bulkheads and sidewall paneling, window sets, luggage stow bins, seat sets, serving carts, and landing gear doors, also are fabricated of sandwich structures (Figure 8.42).

8.3.3.4 Transportation and Other Industries

Sandwich composites have been used in the transportation industry for a long time, including applications for rail vehicles, buses, cars, and trucks, and for a variety of industrial equipment and components for different products. For these applications, the core, reinforcement, and resin should be selected based on the requirements, as well as the processing method.

Fast train vehicles are typically constructed of foam core sandwiches, for structural components such as the floor, sidewall panels, doors, roofs, and

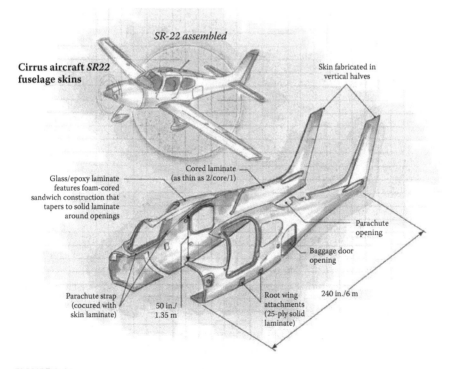

FIGURE 8.41
Airplane fuselage and wing made by using foam core sandwich. (From http://composites world.com.)

FIGURE 8.42
Commercial airplane components made by foam core sandwich composites. (From Sloan, J., Structural polyurethanes: Bearing bigger loads, *Composites Technology*, March 10, 2010.)

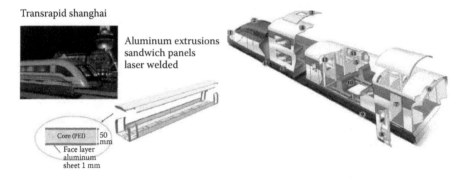

FIGURE 8.43
Fast train vehicle made by foam-cored sandwich construction.

head caps, to the internal components, such as engine cab, dividing walls, luggage bins, and window sets, as shown in Figure 8.43. Some of the more critical applications are made using metal (typically aluminum)-faced sandwiches, but most are made of fiberglass composite facings with various foam core materials. The composite-faced sandwich parts are primarily fabricated by wet lamination.

Bus and car modules, as shown in Figures 8.44 and 8.45, can be made using a *one-shot* vacuum infusion process by employing PVC, PET, or PMI foam cores, fiberglass or carbon fiber renforcements, and liquid epoxy or vinyl resins. The resulting light weight and stiff structure reduces gas consumption and environmental footprint.

FIGURE 8.44
Bus module made by foam core sandwich structure.

FIGURE 8.45
Carbon fiber foam sandwich car module. (From http://operationsroom.files.wordpress.com; http://extremtech.com.)

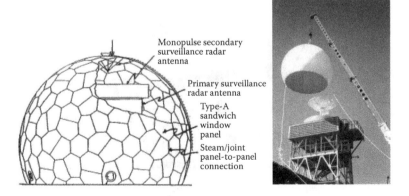

Monopulse secondary surveillance radar antenna

Primary surveillance radar antenna

Type-A sandwich window panel

Steam/joint panel-to-panel connection

FIGURE 8.46
Sandwich radome geometry and installation. (From http://www.L-3com.com; http://radome.net.)

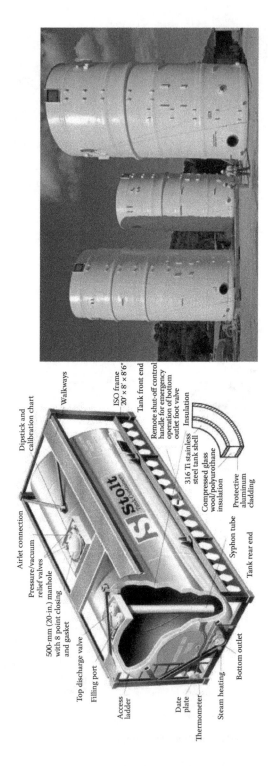

FIGURE 8.47

Chemicals tank made by sandwich construction. (From http://shipping container 24.com; http://compositesworld.com.)

Radomes (domes covering radar antennae) are fabricated by assembling multipiece sandwich panels that are laminated by RTM or other processing methods, as shown in Figure 8.46. All materials in a radome must be as radar frequency transparent as possible, since the radar equipment must transmit and recieve signals with very little attenuation.

Chemical storage tanks, as shown in Figure 8.47, are made of composites for their primary structural requirements and their corrosion resistance. The sandwich structure is employed for increasing stiffness in comparison to a solid FRP design and providing thermal insulation. Most commonly, wet lamination using glass reinforcements and vinyl ester resins are employed for chemical storage tanks and chemical processing ductwork. Usually, the bottom and top caps of storage tanks are made by the vacuum infusion process, and the central cylinder is made by a filament-winding process.

References

1. Black, S. 2003. Getting to the core of composite laminates. *Composites Technology*, October 1.
2. Gundberg, T. Foam core materials in the marine industry. Available at http://boatdesign.net.
3. Sloan, J. 2010. Structural polyurethanes: Bearing bigger loads. *Composites Technology*, March 10.
4. Sloan, J. 2010. Wind foam sources: PET, SAN & PVC. *Composites Technology*, June 1.
5. Material Notes, Evonik Corporation Rohacell® 110 WF High Heat Grade Polymethacrylimide (PMI) Foam. Available at http://matweb.com.
6. Sabic. I.P. 2011. Ultem foam for high-performance applications. http://www.plasticwire.com, October.
7. Sloan, J. 2010. Core for composites: Winds of change. *Composites Technology*, June 10.
8. Greene, E. 2004. Marine composites. *ACMA Composites 2004 Conference*, Tampa, FL, October 7.
9. Lauri, L., Ang, S.S., Stigsson, J.J.C., and Bressan, R. 2012. US Patent 8,168,293 assigned to Diab International AB.
10. Lauri, L. 1994. US Patent 5,352,710 assigned to Prima S.p.A.
11. Strucell, P. Specification. Available at http://strucell.com.
12. Ma, W. and Feichtinger, F. 2011. US Patent 7,951,449 assigned to Baltek Corporation.
13. The fire resistance structural foam. Available at http://3accorematerials.com.
14. The production of rigid polyurethane foam. Available at http://plastics.bayer.com.
15. Air-cell polyester foam core. Available at http://polyumac.com.
16. Ang, S. 2012. Rigid polymer foams for wind blade applications. *Polymer Foam Conf.* spon. by AMI, Iselin, NJ, October.
17. Seibert, H. 2006. Applications for PMI foams in aerospace sandwich structures. *Reinforced Plastics*, 50.1: 44–48.

18. High performance structure foam. Available at http://3accorematerials.com.
19. Feichtinger, K. 1988. Test methods and performance of structural core materials—I. Static properties. *44th Annual ASM International/Engineering Society of Detroit,* September 13–15, pp. 1–11.
20. US Military Standard. 1967. Sandwich constructions and core materials: General test methods. *Mil-STD-401B.*
21. ASTM C365/C365M—11a. 2003. Standard Test Method for Flatwise Compressive Properties of Sandwich Cores. *ASTM International,* West Conshohocken, PA. Available at http://astm.org.
22. ASTM C297/C297M—15. 2003. Standard Test Method for Flatwise Tensile Strength of Sandwich Constructions. *ASTM International,* West Conshohocken, PA. Available at http://astm.org.
23. ASTM C273/C273M—11. 2003. Standard Test Method for Shear Properties of Sandwich Core Materials. *ASTM International,* West Conshohocken, PA. Available at http://astm.org.
24. ASTM C393/C393M—11e1. 2003. Standard Test Method for Core Shear Properties of Sandwich Constructions by Beam Flexure. *ASTM International,* West Conshohocken, PA. Available at http://astm.org.
25. ASTM C394/C394M—13. 2003. Standard Test Method for Shear Fatigue of Sandwich Core Materials. *ASTM International,* West Conshohocken, PA. Available at http://astm.org.
26. DIAB. Guide to core and sandwich. Available at http://diabgroup.com.
27. CTNM. Foam core surface finishing. Available at http://strucell.com.
28. Manufacturing of sandwich composites. Available at http://angelfire.com.
29. M2 basic constituent materials in composite. Available at http://coursehero.com.
30. Taylor, D. and Bader, S. Composites manufacturing techniques. Available at http://textilearchitecture.polimi.it.
31. 2014. The composites matrix. *Composites World,* January 1.
32. Hamerton, I. (Ed.). 1994. *Chemistry and Technology of Cyanate Ester Resins.* Blackie Academic and Professional: Glasgow, Scotland.
33. Composites fabrication methods. Available at https://en.wikipedia.org/wiki/De_Havilland_Mosquito.

9

Microcellular Polyimide Foams: Fabrication and Characterization

Yang Li, Wentao Zhai, and Wenge Zheng

CONTENTS

9.1 Introduction

Polyimide (PI) foam belongs to a class of high-performance cellular plastics compared to traditional polyurethane and polystyrene (PS) foams, due to their superior properties such as low dielectric constant, high thermal stability, excellent flame retardancy, less smoke generation and radiation, etc. [1–3]. Because of this, PI foams are employed widely in high-tech applications of aerospace, submarine, and high-speed vehicles.

PI foams were primarily developed by Monsanto and DuPont in the late 1960s [4]. To date, a variety of PI foams have been fabricated. According to the morphology of PI foams, we can roughly divide them into three categories: (1) nanoporous, (2) bicontinuous, and (3) microcellular PI foams. The nanoporous PI foams are frequently referred to as PI foams with pore sizes on the nanometer scale and have received considerable attention in the microelectronics industry as they may be used as low-dielectric-constant materials to protect microelectronic devices from signal delays and cross talk [1].

Typically, PI nanofoams are fabricated by a *template* method, where the raw materials (block copolymers or composite polymers) consist generally of both a thermally stable continuous phase and a thermally labile dispersed phase [5–8]. Upon heating, the thermally labile phase undergoes pyrolysis, leaving nanopores inside the thermally stable matrix.

With the development of the foaming technologies, compressed gases like supercritical carbon dioxide are gradually employed to the solid-state foaming of thermoplastic PI [9]. Actually, it is the usage of CO_2 that brings the discovery of bicontinuous PI foams. When Krause et al. [10,11] systematically investigated the foaming behaviors of poly(ether imide) (PEI), poly(ether sulfone) (PES), and polysulfone/PEI blends, they found that a cellular-to-bicontinuous transition would occur for these polymers at a certain saturation level of CO_2. For a PEI film with a thickness of 75 μm, a foam with opened nanoporous bicontinuous structure will be produced when the CO_2 saturation level is above 72–75 mg CO_2/g PEI. While below this level, closed microcellular PEI foam will be obtained. Based on the report of Krause et al. [10], Miller et al. [12] further investigated the solid-state batch foaming of PEI sheets via subcritical CO_2 and found that the CO_2 saturation level for the micro-/nanotransition was up to 94–110 mg CO_2/g PEI when the sample thickness increased to 1.5 mm. Additionally, a third hierarchical structure with a microcellular structure as the first-order structure and a nanocellular structure as the second-order structure was also produced. A similar phenomenon was also reported by Zhou et al. [13].

Further increasing the cell sizes of PI foams to the micrometer scale, we may obtain microcellular PI foams. The primary motivation for fabricating microcellular foam is to reduce the consumption of polymers in the packaging industry. The common polymer matrices to fabricate microcellular foams are PS, polyvinyl chloride, and polylactic acid [14–19]. The invention of microcellular PI foam extensively exploits the applications of microcellular foams in high-tech fields. Previous work on the fabrication of microcellular PI foam has been contributed by a few groups [10,12,13,20,21]. However, there are still technical challenges in the fabrication of microcellular PI and PI nanocomposite foams with high expansion ratios. This chapter mainly summarizes the current progress of our group in the preparation of high-expansion-ratio microcellular PEI and PEI/graphene foams and their functional property characterizations.

9.2 Fabrication of Microcellular Thermoplastic PI Foams

9.2.1 Blowing with Compressed Gas

There are plenty of strategies for fabricating microcellular PI foam, including a template method and chemical and physical foaming [5,10,22–24]. Among them, physical foaming is the most favorable method for the large-scale fabrication of PI foams, as the foaming agents like compressed CO_2 are nonpolluting and can be used immediately without further purification. With compressed gas, microcellular foams can be processed not only by extrusion and injection molding but also by solid-state foaming [25,26]. Compared to extrusion and injection molding, solid-state foaming is more suitable for the foaming process of PI, for the plasticization effect of compressed gas could reduce the energy barrier for the movement of PI chains and, consequently, decrease the T_g of PI/gas system that was related to the cell-nucleation and growth process [10,12,27]. For example, Miller et al. [12] found that the effective T_g of the PEI/CO_2 mixture saturated at 5 MPa was only 95°C, much lower than the 215°C for neat PEI. This phenomenon helps to reduce the foaming temperature of PI, improving the processibility of PI foaming.

For solid-state foaming, the typically process is composed mainly of three stages (Figure 9.1), including (1) gas saturation, (2) foaming procedure, and (3) cell stabilization [21,28,29]. Gas saturation is usually conducted under high pressure at a certain temperature to facilitate the diffusion of gas into a polymer matrix, and the gas solubility can be adjusted in terms of sample dimensions by parameters like saturation temperature, pressure, and time. While the foaming procedure is performed by creating a thermodynamic instability like a sudden pressure drop or a temperature increase to induce cell nucleation and growth, the morphologies of the resultant foams can be controlled by parameters including foaming temperature and time. For convenience of reference, the experimental parameters associated with the solid-state foaming process have been collected in Table 9.1. As to the cell stabilization, it is commonly conducted by cooling to room temperature.

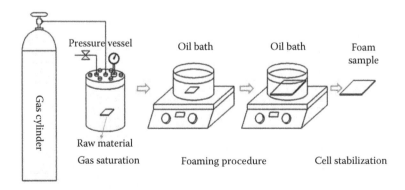

FIGURE 9.1
Schematic illustration for typical solid-state foaming process.

TABLE 9.1

Comparison of PI Foams under Different Foaming Conditions

Polymer	Sample Thickness (mm)	Foaming Agent	Saturation Temperature (°C)	Saturation Pressure (MPa)	Saturation Time (h)	Foaming Temperature (°C)	Expansion Ratio [Ref.]
PEI[a]	0.075	CO_2	25	4.6	2	200	2.6 [10]
PEI[a]	1.5	CO_2	21	1	520	210	3.5 [12]
PEI[a]	1.5	CO_2	21	4	325	190	2.6 [12]
PEI[a]	1.5	CO_2	21	5	280	200	2.5 [12]
PEI[a]	0.5	CO_2	50	6.5	48	210	5.0 [21]
PEI[a]	0.76	CO_2	35	8	98	155	1.2 [13]
PEI[a]	0.76	CO_2	35	8	98	202	2.6 [13]
PEI[a]	0.5	CO_2	25	5	24	205	2.6[g]
PI[b]	0.1	CO_2	25	5	2	290	1.6 [9]
PI[c]	0.1	CO_2	0	5.5	2	240	1.7 [9]
PI[d]	0.075	CO_2	25	5	2	300	1.3 [9]
PI[e]	1.2–1.4[f]	CO_2	80	5	504	270	2.9 [30]
PI[e]	1.2–1.4[f]	CO_2/THF	80	5	168	270	15.7 [30]

[a] PEI resin, ULTEM-1000.
[b] PI resin, Matrimid-5218.
[c] PI resin, LaRC-CP1.
[d] PI resin, Kapton-HN.
[e] PI resin, YZPI-JL10.
[f] Diameter of PI beads.
[g] The data obtained from our study.

9.2.2 Fabricated by Mixed Foaming Agents

In our previous work, we used subcritical CO_2 as a blowing agent to investigate the foaming behaviors of PEI sheets (ULTEM-1000, 0.5 mm in thickness) and PI beads (YZPI-JL10, 1.2–1.4 mm in diameter) via solid-state foaming. The results revealed that the highest expansion ratio for the PEI and PI foams reached 3.2 and 2.9 [30], respectively, which were comparable to the values achieved by other groups (Table 9.1). In fact, most microcellular PI foams are prepared using CO_2 as a blowing agent, and their expansion ratios are less than 5. The low expansion ratio can be assigned to the high stiffness of the PI matrix and their low gas solubility, which is a challenge in the preparation of lightweight, high-performance polymeric foams [31].

It has been said that organic solvent owns much higher solubility in polymers than compressed gas and is usually employed to prepare polymeric foams with high expansion ratios and low foam densities, due to its low supersaturation and poor cell nucleation [32,33]. To further improve the expansion ratio of microcellular PI foam, our group used tetrahydrofuran (THF) as a coblowing agent to work with CO_2 and prepared microcellular PI foam with high expansion ratio of 15.7 (Figure 9.2) [30]. As shown in Figure 9.3, the PI foam prepared by the mixed blowing agent (CO_2/THF) exhibited cell sizes of 6.8–14.5 μm and a cell density of 4.2×10^9 cell cm^{-3}. The control, however, just had an average cell size of 1.1 μm and a cell density of 1.7×10^{12} cell cm^{-3}, indicating that the introduction of THF could improve the expansion ratio of PI foams. The synergetic effect of compressed CO_2 and THF facilitated both the cell nucleation and coalescence process, leading to the high expansion ratio of the PI foams. Furthermore, the introduction of THF significantly reduced the saturation time of PI beads from 504 h for the control to 168 h.

The PI resin used in the above experiment was YZPI-JL10 and can be noted as YZ-1. To further improve the foaming performance of YZ-1, our group investigated another kind of PI, which was derived from a structure modification of the YZ-1 and could be noted as YZ-2. Compared with the T_g of YZ-1 (257°C), YZ-2 possessed a much lower T_g of 221°C. Using the same mixed foaming agent as the YZ-1, our group prepared the YZ-2-type bead foam with an ultrahigh expansion ratio over 50 (Figure 9.4a) and extensively investigated the effect of saturation press, foaming temperature, and desorption time on the final volume expansion ratio of the YZ-2 beads (Figure 9.4b through d). As shown in Figure 9.4d, the highest expansion ratio of the YZ-2 bead foam reached 57 under a saturation pressure of 3 MPa and desorption time of 0 days. Further extending the desorption time would decrease the content of the foaming agent in the PI matrix, leading to the decrement in the expansion ratio. When the desorption time reached 18 days, the corresponding expansion ratio of the YZ-2 was still 30. The structure evolution of the YZ-2 bead foams could be observed by their scanning electron microscopy

FIGURE 9.2
Optical micrographs of PI beads and the foamed PI beads. (a) PI pellets and (b, c) the foamed PI pellets. (Reproduced from W. T. Zhai et al., *Industrial & Engineering Chemistry Research*, **51**, 12827–12834, 2012. With permission.)

(SEM) images shown in Figure 9.5. This PI foam with an ultrahigh volume expansion ratio will be very promising in the large-scale commercial fabrication of PI foams.

9.2.3 Fabricated by a Nonsolvent-Induced Phase-Separation Foaming Process

Nonsolvent-induced phase separation (NIPS) has widely been used in the fabrication of membranes [34–36]. In 1992, it was introduced by Hatori et al. [37–39] to foam Kapton-type PI. The NIPS process is usually performed by the exchange of a nonsolvent with a solvent. When the nonsolvent for a polymer is introduced to its homogeneous solution, the polymer will precipitate, and two phases will be generated in the solution: (1) polymer-rich phase and (2) polymer-lean phase. After the removal of the solvent, the former turns into the matrix of membrane, while the latter forms the pores inside. Generally, the phase separation with respect to the fabrication of MIPs can be roughly divided into the dry method and the wet method. The dry method is referred to as water vapor–induced phase separation (VIPS), where a polymer solution

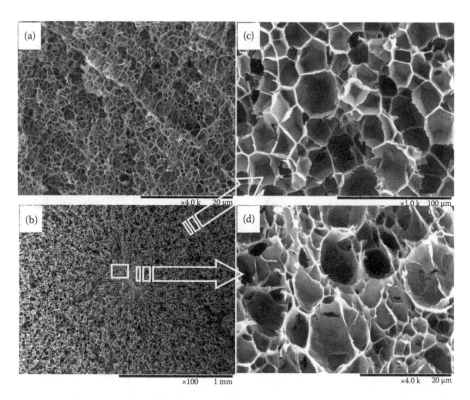

FIGURE 9.3
SEM micrographs of PI foams with various magnifications: (a) PI foam was prepared using CO_2; (b) PI foam was prepared using mixed blowing agents; (c, d) amplified SEM micrographs of (b). (From W. T. Zhai et al., *Industrial & Engineering Chemistry Research*, **51**, 12827–12834, 2012.)

needs to be exposed in a humid atmosphere, and the water vapor from ambient air is normally the nonsolvent of polymer (Figure 9.6). The wet method is usually conducted by immersing the polymer solution in a coagulant bath containing nonsolvent.

Ren et al. [40] found that the microcellular PI foam fabricated with the VIPS did not have the dense skins that are produced in the solid-state foaming process, which promotes the improvement of the dielectric properties of microcellular PI foam. In the VIPS process, the morphology of microcellular PI foam can be strongly affected by the relative humidity of the solution-casting atmosphere, as well as the concentration of the polymer solution. Park et al. [41], for example, used N-methyl-2-pyrrolidone as the solvent of polysulfone (PSU) and fabricated PSU membranes via the VIPS. The results revealed that the phase separation occurred when the relative humidity was higher than 65% and that the pore size of PSU membranes increased with both the relative humidity

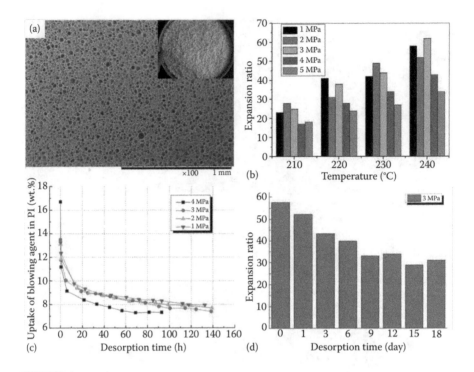

FIGURE 9.4

(a) SEM image of the as-prepared YZ-2 bead foam with ultrahigh volume expansion, and the inset was the optical picture of the corresponding YZ-2 foams; (b) expansion ratio of YZ-2 foams under different saturation pressure versus foaming temperature; (c) uptake of blowing agent in YZ-2 resin versus desorption time; (d) expansion ratio of YZ-2 foams at the saturation pressure of 3 MPa versus desorption time.

and the polymer concentration decreasing. To improve the foaming efficiency of PEI (ULTEM-1000), our group used *N,N'*-dimethyl formamide (DMF) as the solvent of PEI and fabricated microcellular PEI/graphene composite foams under a controlled humidity of 75% and a PEI solid content of 8 wt% via the VIPS [42]. As shown in Figure 9.7, microcellular PEI/graphene foams exhibited a decreasing tendency in pore size, and this phenomenon was more obvious when the graphene loading was higher than 1 wt%. The reason was due to the two fold role of graphene in the phase-separation process. On one hand, the presence of graphene provided a great number of nucleation sites for the cell nucleation. On the other hand, too much graphene increased the viscosity of the casting solutions, causing a physical barrier to the cell coalescence. As a result, the pore size decreased with the graphene content increasing. The same method was conducted as well in our other reports [43,44].

By the NIPS method, we also synthesized Kapton-type PI/graphene composite foams [45]. Due to the hard processibility of Kapton-type PI, we used its precursor, poly(amic acid) (PAA), as the raw material of the NIPS. In this process, *N,N'*-DMF was the solvent of PAA, whereas the mixture of alcohol

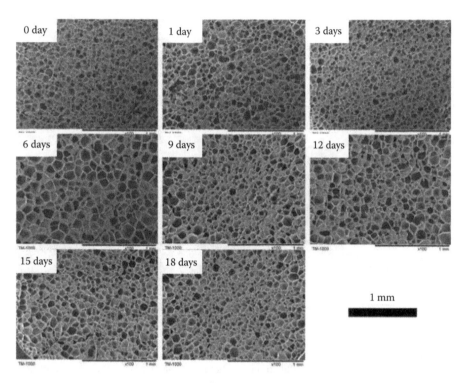

FIGURE 9.5
Cell-structure evolution of YZ-2 bead foams as the function of desorption time.

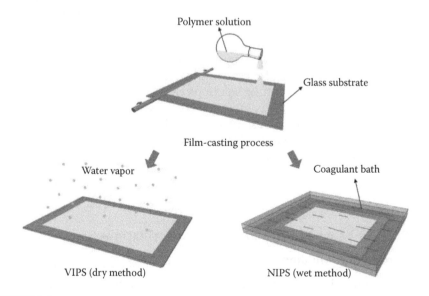

FIGURE 9.6
Scheme of VIPS and NIPS process.

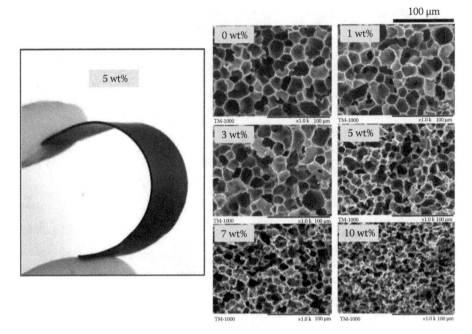

FIGURE 9.7
Optical and SEM micrographs of PEI/graphene nanocomposite foams as the function of graphene content. (From J. Ling et al., *ACS Applied Materials & Interfaces*, 5, 2677–2684, 2013.)

and distilled water was its nonsolvent (noted as coagulant bath). Because the morphology of membranes had a close relationship with the components of the coagulant bath and the concentration of the polymer solution [34], the optimum parameters for this study were determined as 1:1 for the volume ratio of alcohol to water in the coagulant bath and 10 wt% for the PI solid content. The whole foaming process continued only 30 min at room temperature after PAA-containing casting solutions were immersed in the coagulant bath. It is obvious that NIPS is more effective than solid-state foaming, as the latter always needs a long time for gas solubility [12,30]. The resultant PI composite foams exhibited a similar foam morphology with that of our previous PEI composite foams [42,43] and an expansion ratio of 2.4–5.2.

9.3 PI Bead Foam Molding and Characterizations

9.3.1 Bead Foam Molding

Assembling our microcellular PI bead foam into three-dimensional integration material means a lot for its practical applications. Though traditional steam-chest molding has been widely applied in the fabrication of expanded polypropylene

(EPP) bead foam [46,47], it is unsuitable for the molding process of PI bead foam. The reason is that the PI bead foam tends to shrink instead of surface fuse and expand in hot air, which makes the steam-chest molding unavailable.

To address this issue, our group used PEI resin (ULTEM) as a gel to form effective bonding among foamed YZ-1 beads (Figure 9.8) [30]. Specifically, PEI resin is firstly dissolved in chloroform, forming a homogeneous PEI solution. And then the PEI solution is uniformly coated onto the surface of the YZ-1 bead foams. After the compression molding, the solvent is evaporated, and the PEI resin is solidified, generating strong interbead bonding among the YZ-1 bead foams. Figure 9.9a and c shows the uniform PEI coating, indicating the good infiltration character of PEI solution on the surface of PI foams. Moreover, the crack path (Figure 9.9a) almost 100% went through the cross section of the foams rather than the bead boundaries, suggesting the strong interfacial interactions between the PEI resin and the YZ-1 bead foams.

With the same molding method, we also prepared YZ-2 foam products (Figure 9.10a). The difference was that the gel and corresponding solvent we used in the YZ-2 foam products was epoxy and acetone, respectively. The reason was that the YZ-2 resin with the structure modification from YZ-1 can be dissolved by THF, and the porous structure inside would be destroyed. After the compression molding process, the YZ-2 foam product exhibited

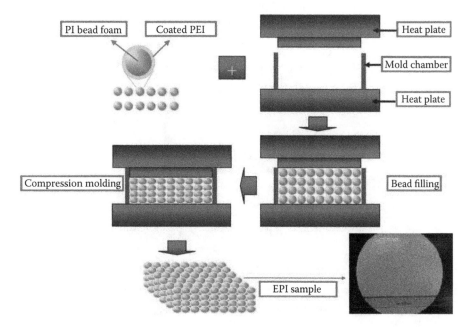

FIGURE 9.8
Schematic illustration of the compression molding process of the foamed PI beads. The optical micrograph was the resultant molded PI bead foam product. (From W. T. Zhai et al., *Industrial & Engineering Chemistry Research*, **51**, 12827–12834, 2012.)

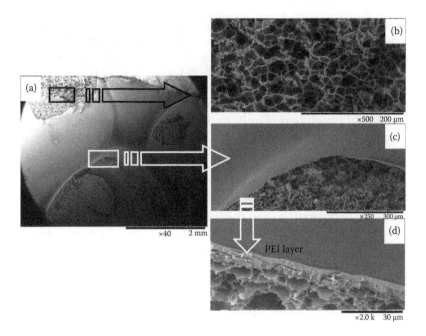

FIGURE 9.9
SEM micrographs of the resultant MPI product: (a, c, d) fracture surface of MPI foam with various magnifications, respectively; (b) cell morphology of MPI foam. (From W. T. Zhai et al., *Industrial & Engineering Chemistry Research*, **51**, 12827–12834, 2012.)

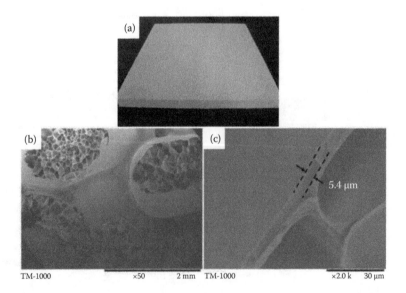

FIGURE 9.10
(a) Digital picture of molding YZ-2 bead foam product and (b, c) typical SEM images of the products.

a typical compacted cross-section morphology like the YZ-1 foam product (Figure 9.10b) and an epoxy coating layer in thickness of 5.4 μm (Figure 9.10c).

9.3.2 Mechanical Properties

The mechanical properties of the resultant YZ-1 foam products are affected significantly by the concentration of PEI solution and the applied compression force during the molding process [30]. The effect of PEI solution concentration was investigated using two different PEI concentrations: (1) 0.08 and (2) 0.13 g mL^{-1}, and the results revealed that the thickness of the PEI layer increased from 1.2 to 2.7 μm with the PEI concentration (Figure 9.11a and b). Moreover, the thickness of the PEI layer would be further improved by the compression force that is acted on the PEI solution during compression molding (Figure 9.11c). Mechanical properties testing revealed that both the beading strength and the compression strength of the YZ-1 foam products increased with increasing PEI concentration (Figure 9.12a). As to the effect of the compression force, three different values were researched. The SEM images in Figure 9.11d through f show that with the compression force increasing, the interbead void fraction in the molded YZ-1 products reduced significantly, and the bending strength for the samples increased gradually from 0.34 to 1.27 MPa. When the PEI concentration was kept at a constant of 0.08 g mL^{-1}, the compression strength of the resultant foam products would present an increasing tendency with compression force (Figure 9.12b). Here,

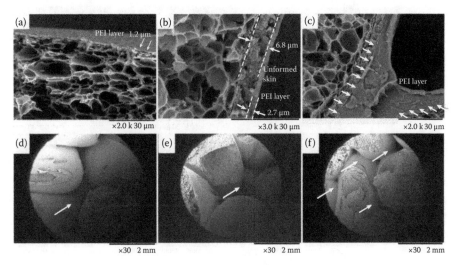

FIGURE 9.11
SEM micrographs illustrating the thickness of the PEI layer on PI foams: the PEI solution concentrations were (a) 0.08 g/mL and (b, c) 0.13 g/mL, respectively; SEM micrographs illustrating the formation of void areas with the application of various compression forces: the compression force increased gradually from sample (d) to (f). (From W. T. Zhai et al., *Industrial & Engineering Chemistry Research*, **51**, 12827–12834, 2012.)

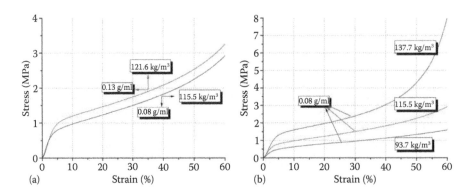

FIGURE 9.12
Compressive stress–strain behavior of YZ-1 foam products (a) with various PEI solution concentrations and (b) with the application of various compression forces. (From W. T. Zhai et al., *Industrial & Engineering Chemistry Research*, **51**, 12827–12834, 2012.)

we need to mention that the highest compression stress of the YZ-1 foam products at 50% strain reached 4.31 MPa, which was much higher than that for the TEEK PI foam [48].

The compression molding process of YZ-2 bead foams was similar to that of YZ-1 bead foams. The mechanical properties of the resultant foam products were adjusted by both the epoxy concentration and the molding

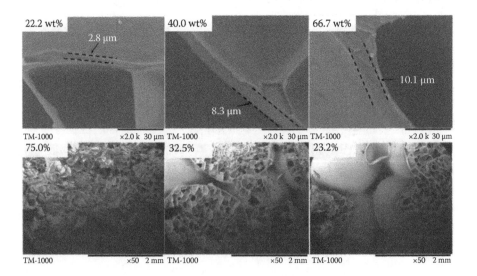

FIGURE 9.13
SEM images of PI foams coated with epoxy layers: (the upper three) the epoxy solution concentrations were 22.2, 40.0, and 66.7 wt% (the mass fraction of epoxy resin in YZ-2 foam products); (the bottom three) the compression force during compression molding process decreased gradually from left to right (the compression force during the molding process was expressed in forms of the deformation degree of the volume change in YZ-2 bead foams).

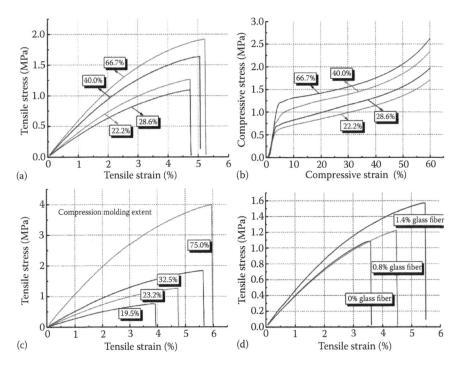

FIGURE 9.14

(a) Tensile and (b) compressive stress–strain curves of YZ-2 foam products with various epoxy concentrations. (c) Tensile stress–strain curves of YZ-2 foam products molded with various compression forces. (d) Compressive stress–strain curves of YZ-2 foam products reinforced with various glass fiber concentrations.

compression force. Generally, the higher the epoxy concentration, the thicker the epoxy coating (Figure 9.13). The mechanical properties, including tensile and compressive strength, also exhibited an increasing tendency with epoxy concentration (Figure 9.14a and b). The foam products prepared under higher compression force also exhibited a much more compacted structure (Figure 9.13) and higher tensile strength value (Figure 9.14c). The addition of glass fiber into the epoxy resin could further help to improve the tensile strength of YZ-2 foam products (Figure 9.14d).

9.4 Fabrication and Characterization of PI/Graphene Composite Foams for Electromagnetic Interference Shielding

With the rapid development of high-frequency electronic devices these days, electromagnetic interference (EMI) is becoming increasingly acute. Too much

EMI can exert a negative influence on the normal work of electronics and threaten the physical health of human beings. So the fabrication of materials to shield EMI is necessary. With the development of EMI shielding materials, traditional metals are gradually replaced by conductive polymer composites (CPCs) due to their low density, erosion resistance, and good processibility [49–51].

When CPCs are foamed, great progress will be made in their EMI shielding effectiveness (SE), mechanical properties, and mass density. Thomassin et al. [52] found that the SE of polycaprolactone (PCL)/multiwalled carbon nanotube (MWCNT) composite foams was commonly higher than that of their solid counterparts. A similar phenomenon was also observed by Ameli et al. [53,54] in polypropylene (PP)/MWCNTs and PP/stainless-steel fiber composite foams. The difference between conductive polymeric composite foams (CPFs) and CPCs in SE was mainly concentrated on the absorption loss for electromagnetic waves. It has been found that CPFs generally contribute much more absorption loss (SE_A) to their total SE (SE_{total}) than CPCs do, which results from the higher impedance match of CPFs with ambient atmosphere and the enhanced multireflection loss of CPFs [52–54]. Additionally, a porous structure can help to improve the ductility and toughness of CPCs and reduce their mass density simultaneously [55].

Most CPFs are fabricated by filling conductive fillers into a polymer matrix and a variety of polymers, such as PS, polymethylmethacrylate (PMMA), poly(dimethyl siloxane) (PDMS), PCL, polyvinylidene fluoride (PVDF), and PP, have been included in the fabrication of CPFs [52,54–60]. However, these foams can hardly meet the harsh requirements of the aerospace industry, where high thermal stability, extraordinary flame retardancy, and mechanical properties are required. Therefore, Zhan et al. [61,62] prepared silver-coated PI foams and PI/silver nanowire (AgNWs) hybrid foams by combining the liquid foaming process with a physical spraying method for EMI shielding. But there is still a great need to fabricate PI foams with effective EMI SEs over a much higher frequency range.

As a crucial conductive filler, graphene has a great potential in EMI shielding due to its excellent electrical conductivity, high aspect ratio, and huge surface area [63–68]. Based on our previous work on graphene preparation [69,70], our group investigated the EMI shielding performance of PEI/graphene composites and microcellular composite foams [42]. The results revealed that microcellular PEI/graphene composite foams showed cell sizes of 9.0–15.3 μm and a similar mass density of 0.28 g cm^{-3}, corresponding to an expansion ratio of 4.6. Though the SE of our PEI composite foams was lower than that of the solid counterparts (Figure 9.15a and b), their specific SE was much higher due to their low density (Figure 9.15c). For example, when graphene loading was 10 wt%, the specific SE of the PEI foam was 44 dB cm^3 g^{-1}, approximately 2.5 times higher than the unfoamed counterparts. More importantly, the contribution of SE_A to SE_{total} of the foams with graphene loading of 3–10 wt% was 90.6%–98.9%, much higher than the 76.2%–90.8% for the solid counterparts (Figure 9.15d). The reason was assigned to the enhanced

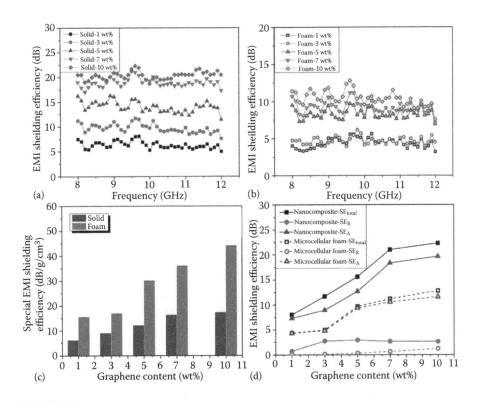

FIGURE 9.15
EMI SE of (a) PEI/graphene composite and (b) microcellular foams at different frequency. (c) The specific EMI SE of PEI/graphene nanocomposite and microcellular foams at 9.6 GHz. (d) SE_{total}, SE_R, and SE_A of PEI/graphene nanocomposite and microcellular foams at 9.6 GHz. (From J. Ling et al., *ACS Applied Materials & Interfaces*, 5, 2677–2684, 2013.)

multireflection loss of the PEI composite foams. It was believed that when an electromagnetic wave propagated in PEI/graphene foams, multiple reflections would occur among the cell walls full of graphene sheets (Figure 9.16). Because of the dielectric loss of graphene, the incident electromagnetic power would be converted into heat and dissipate before escaping from the PEI composite foams. As a result, the SE_A to SE_{total} of the microcellular PEI composite foams was improved. However, the highest SE of our PEI composite foams was only 11 dB over the frequency range of 8–12 GHz (X band), which was lower than the standard value (20 dB) for commercial applications [42].

To further improve the SE of PEI composite foams, our group functionalized graphene sheets with ferromagnetic Fe_3O_4 nanoparticles ($G@Fe_3O_4$) through the coprecipitation method and then incorporated the $G@Fe_3O_4$ into PEI foams [43]. From the perspective of electromagnetic theory, there are two means for a shield to attenuate electromagnetic waves: (1) dielectric loss and (2) magnetic loss [71–73]. Pristine graphene is nonmagnetic and contributes to SE mainly through the dielectric loss. Thereby, we endowed PEI composite

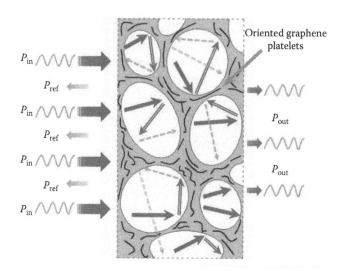

FIGURE 9.16

Schematic description of the microwave transfer across PEI/graphene nanocomposite foam. (From J. Ling et al., *ACS Applied Materials & Interfaces*, **5**, 2677–2684, 2013.)

foams with ferromagnetism by filling Fe_3O_4-decorated graphene sheets (Figure 9.17a and b), hoping to improve the magnetic loss of PEI composite foams for electromagnetic waves. As expected, the PEI composite foam with 10 wt% $G@Fe_3O_4$ exhibited a higher SE value of 14–18 dB in the X band (Figure 9.17c) than our previous PEI foam with the same content of graphene did (11 dB in Figure 9.15b) [42]. Furthermore, the contribution of SE_A to SE_{total} of the PEI composite foams with 7–10 wt% $G@Fe_3O_4$ was varied from 96.9% to 97.2% at 9.6 GHz (Figure 9.17d), pretty much higher than that of the foam with 10 wt% graphene (90.6%) (Figure 9.15d).

In another report, we fabricated PEI/nickel composite foams in which ferromagnetic nickel nanoparticles are gradiently distributed throughout the foam cross section [44]. The composite foam with 70 phr (parts per hundred of resins) nickel exhibited a high SE of 86.7–106.5 dB over the frequency range of 50–3000 MHz, slightly higher than the 61.6–95.6 dB for the silver-coated PI foam [61]. Nonetheless, there are still many problems in the PEI/graphene composite foams: (a) the nonuniform dispersion of graphene in PEI matrix, (b) the brittle feature of the foams at high graphene loading, and (c) the high sample thickness (>2.3 mm) needed to acquire sufficient SE. With electronics getting smaller, faster, and thinner, it has become a challenge to fabricate EMI shields with low thickness, high thermostability, and effective SE.

For these challenges, our group fabricated Kapton-type PI/reduced graphene oxide (rGO) foam sheets via a three-step method: (1) *in situ* polymerization, (2) NIPS, and (3) thermal imidization (Figure 9.18) [45]. Meanwhile, the dispersibility of rGO in PI was improved through the nucleophilic substitution reaction between the residual epoxy groups in rGO and the amine groups

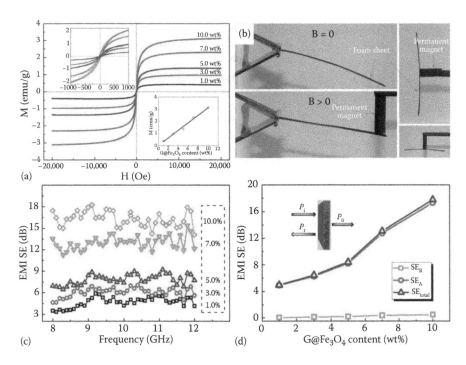

FIGURE 9.17
Superparamagnetic microcellular PEI/G@Fe3O4 foam sheets: (a) magnetization of the composite foams with different G@Fe$_3$O$_4$ loading, (b) response of free-standing composite foam sheet to external magnetic field, (c) EMI SE of PEI/G@Fe$_3$O$_4$ foams in X band, and (d) SE$_{total}$, SE$_R$, and SE$_A$ of microcellular foams at 9.6 GHz. (From B. Shen et al., *ACS Applied Materials & Interfaces*, **5**, 11383–11391, 2013.)

in organic diamine with the assistance of ultrasonication. The as-obtained PI composite foams had mass densities of 0.28–0.60 g cm^{-3} and pore sizes of 0.6–1.9 μm. As a result, the PI/16 wt% rGO composite foam with the sample thickness of only 0.8 mm exhibited a high SE of 17–21 dB in X band (Figure 9.19a), which resulted from the high dielectric loss at high rGO content and the multireflection loss inside the cell structure. The shielding mechanism for our PI composite foams with rGO loading of 8–16 wt% is dominated by absorption, and the contribution of SE$_A$ to SE$_{total}$ showed a decreasing tendency in the range of 88.0%–95.2% with rGO content increasing (Figure 9.19b), probably because of the lower multireflection loss from the weakened cell structure at higher rGO loading. TGA cures indicated that the addition of rGO significantly improved the thermal degradation temperature of PI resin (Figure 9.19c). Considering the fact that Kapton-type PI possesses much higher T_g (approximately 380°C) than that of PEI resin (217°C for ULTEM-1000) [9,30], we believed that the PI composite foams possess more excellent thermostability than our previous PEI composite foams do [42,43]. In addition, the PI foam was very flexible even when rGO loading was up to 16 wt% (Figure 9.19d). Tensile testing suggested

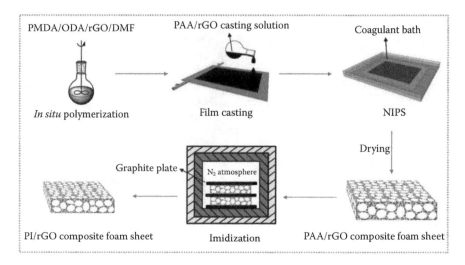

FIGURE 9.18
Schematic illustration for the fabrication process of PI/rGO composite foam sheets.

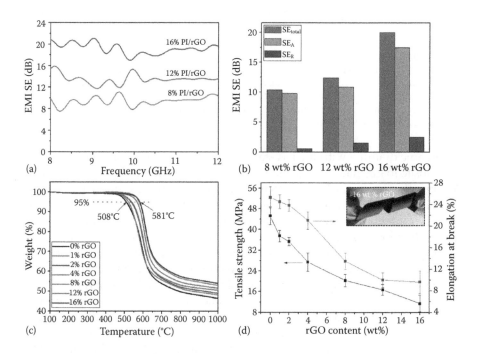

FIGURE 9.19
(a) EMI SE of PI/rGO foams with different rGO content in X band and (b) their SE_{total}, SE_A, and SE_R at 9.6 GHz; (c) TGA curves and (d) tensile properties of PI/rGO foams as a function of rGO content. The inset in (d) is the PI foam containing 16 wt% rGO. (From Y. Li et al., *RSC Advances*, **5**, 24342–24351, 2015.)

that the PI/16 wt% rGO foam had a tensile strength of 11.4 MPa, much higher than the 3.5 MPa for our previous PEI/7 wt% graphene foam [42].

Recently, our group prepared ultrathin flexible graphene films via the direct evaporation of graphene oxide (GO) suspension and graphitization postprocessing (Figure 9.20a) [74]. The resultant graphene film thermally treated at 2000°C, together with its precursor GO film, exhibited a smooth surface and well-defined layer-by-layer structure throughout the entire cross section (Figure 9.20b through f). The EMI SE of the graphene film tested on a polyolefin elastomer (POE) substrate (Figure 9.21a) was approximately 9.1 dB in the X band when the sample thickness was 2.7 μm (Figure 9.21b). Further

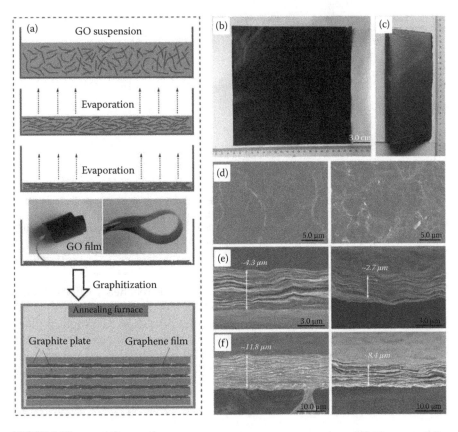

FIGURE 9.20

(a) Schematic representation of a proposed self-assembly process of GO film by the evaporation of GO suspension, as well as the followed graphitization process. The insert were photographs of a flexible GO film. (b) Photograph of a free-standing and dark brown GO film (approximately 20 cm × 20 cm) flattened on a paper. (c) Photograph of a free-standing and shiny metallic GF-2000 (approximately 20 cm × 20 cm) folded on a paper. (d) SEM images showing the surface morphology of GO film (left) and GF-2000 (right). (e and f) SEM images showing a layer-by-layer nanostructure in the cross section of GO film and GF-2000 (with different thickness). (From B. Shen et al., *Advanced Functional Materials*, **24**, 4542–4548, 2014.)

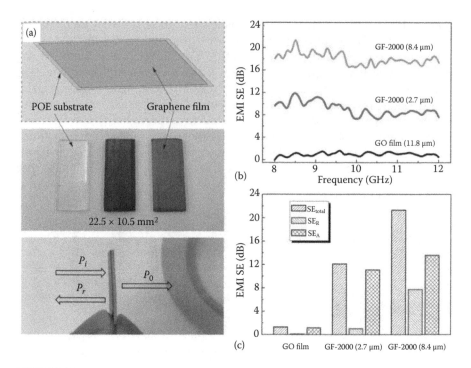

FIGURE 9.21
(a) Schematic representation of the structure of POE/GF-2000 (or GO film) sample. The EMI SE is defined as the logarithmic ratio of incoming (P_i) to outgoing power (P_o) of radiation. (b) The EMI SE as a function of frequency measured in 8–12 GHz range of GO film and GF-2000. (c) SE_{total}, SE_R, and SE_A of GO film and GF-2000 at 8.5 GHz. (From B. Shen et al., *Advanced Functional Materials*, **24**, 4542–4548, 2014.)

increasing the sample thickness to 8.4 µm, the SE reached to 19.1 dB. And the shielding mechanism of the graphene film was dominated by absorption (Figure 9.21c). What is more, the graphene film also showed a high in-plane thermal conductivity of ca. 1100 W m⁻¹ K⁻¹ and excellent mechanical flexibility, which makes it very promising in the fields of microelectronic packaging to solve the heat dissipation and EMI problems simultaneously.

9.5 Conclusion

This chapter has reviewed the fabrication and property characterizations of microcellular PI foam. With the low expansion ratio of traditional microcellular PI foam, our group employed a mixed foaming agent (CO_2/THF) to improve the expansion ratio of microcellular PI foam from *less than 5* to *higher than 10*. This result is the synergetic effect of CO_2 inducing cell nucleation and THF promoting cell growth. For the resultant microcellular PI bead foams, we proposed a novel molding method and thus used PEI resin as a gel to coat outside the PI bead foams. The as-prepared PI foam products exhibited excellent mechanical property and retardancy. To fully exploit the functionality of microcellular PI foam, our group prepared a variety of microcellular PI/graphene composite foams and ultrathin graphene films for EMI shielding.

References

1. J. L. Hedrick, K. R. Carter, J. W. Labadie, R. D. Miller, W. Volksen, C. J. Hawker, D. Y. Yoon, T. P. Russell, J. E. McGrath and R. M. Briber. 1999. Nanoporous polyimides. *Progress in Polyimide Chemistry*, **141**, 1–43.
2. Y. N. Sazanov. 2001. Applied significance of polyimides. *Russian Journal of Applied Chemistry*, **74**, 1253–1269.
3. W. J. Farrissey, J. S. Rose and P. S. Carleton. 1970. Preparation of a polyimide foam. *Journal of Applied Polymer Science*, **14**, 1093–1101.
4. E. S. Weiser, T. F. Johnson, T. L. St. Clair, Y. Echigo, H. Kaneshiro and B. W. Grimsley. 2000. Polyimide foams for aerospace vehicles. *High Performance Polymers*, **12**, 1–12.
5. J. L. Hedrick, R. D. Miller, C. J. Hawker, K. R. Carter, W. Volksen, D. Y. Yoon and M. Trollsas. 1998. Templating nanoporosity in thin-film dielectric insulators. *Advanced Materials*, **10**, 1049–1053.
6. J. L. Hedrick, K. R. Carter, H. J. Cha, C. J. Hawker, R. A. DiPietro, J. W. Labadie, R. D. Miller et al. 1996. High-temperature polyimide nanofoams for microelectronic applications. *Reactive and Functional Polymers*, **30**, 43–53.
7. J. S. Fodor, R. M. Briber, T. P. Russell, K. R. Carter, J. L. Hedrick and R. D. Miller. 1997. Transmission electron microscopy of 3F/PMDA-polypropylene oxide triblock copolymer based nanofoams. *Journal of Polymer Science Part B-Polymer Physics*, **35**, 1067–1076.
8. K. R. Carter, R. A. DiPietro, M. I. Sanchez, T. P. Russell, P. Lakshmanan and J. E. McGrath. 1997. Polyimide nanofoams based on ordered polyimides derived from poly(amic alkyl esters): PMDA/4-BDAF. *Chemistry of Materials*, **9**, 105–118.
9. B. Krause, G. H. Koops, N. F. A. van der Vegt, M. Wessling, M. Wubbenhorst and J. van Turnhout. 2002. Ultralow-k dielectrics made by supercritical foaming of thin polymer films. *Advanced Materials*, **14**, 1041–1046.
10. B. Krause, H. J. P. Sijbesma, P. Munuklu, N. F. A. van der Vegt and M. Wessling. 2001. Bicontinuous nanoporous polymers by carbon dioxide foaming. *Macromolecules*, **34**, 8792–8801.
11. B. Krause, K. Diekmann, N. F. A. van der Vegt and M. Wessling. 2002. Open nanoporous morphologies from polymeric blends by carbon dioxide foaming. *Macromolecules*, **35**, 1738–1745.

12. D. Miller, P. Chatchaisucha and V. Kumar. 2009. Microcellular and nanocellular solid-state polyetherimide (PEI) foams using sub-critical carbon dioxide I. Processing and structure. *Polymer*, **50**, 5576–5584.

13. C. Zhou, N. Vaccaro, S. S. Sundarram and W. Li. 2012. Fabrication and characterization of polyetherimide nanofoams using supercritical CO_2. *Journal of Cellular Plastics*, **48**, 239–255.

14. G. Rizvi, L. M. Matuana and C. B. Park. 2000. Foaming of PS/wood fiber composites using moisture as a blowing agent. *Polymer Engineering and Science*, **40**, 2124–2132.

15. W. Zhai, J. Yu and J. He. 2008. Ultrasonic irradiation enhanced cell nucleation: An effective approach to microcellular foams of both high cell density and expansion ratio. *Polymer*, **49**, 2430–2434.

16. R. P. Juntunen, V. Kumar and J. E. Weller. 2000. Impact strength of high density microcellular poly(vinyl chloride) foams. *Journal of Vinyl & Additive Technology*, **6**, 93–99.

17. G. Ji, W. Zhai, D. Lin, Q. Ren, W. Zheng and D. W. Jung. 2013. Microcellular foaming of poly(lactic acid)/silica nanocomposites in compressed CO_2: Critical influence of crystallite size on cell morphology and foam expansion. *Industrial & Engineering Chemistry Research*, **52**, 6390–6398.

18. W. Zhai, J. Yu, L. Wu, W. Ma and J. He. 2006. Heterogeneous nucleation uniformizing cell size distribution in microcellular nanocomposites foams. *Polymer*, **47**, 7580–7589.

19. X. Lan, W. Zhai and W. Zheng. 2013. Critical effects of polyethylene addition on the autoclave foaming behavior of polypropylene and the melting behavior of polypropylene foams blown with *n*-pentane and CO_2. *Industrial & Engineering Chemistry Research*, **52**, 5655–5665.

20. D. Miller and V. Kumar. 2011. Microcellular and nanocellular solid-state polyetherimide (PEI) foams using sub-critical carbon dioxide II. Tensile and impact properties. *Polymer*, **52**, 2910–2919.

21. L. Sorrentino, M. Aurilia and S. Iannace. 2011. Polymeric foams from high-performance thermoplastics. *Advances in Polymer Technology*, **30**, 234–243.

22. M. Inagaki, T. Morishita, A. Kuno, T. Kito, M. Hirano, T. Suwa and K. Kusakawa. 2004. Carbon foams prepared from polyimide using urethane foam template. *Carbon*, **42**, 497–502.

23. L.-Y. Pan, M.-S. Zhan and K. Wang. 2010. Preparation and characterization of high-temperature resistance polyimide foams. *Polymer Engineering and Science*, **50**, 1261–1267.

24. X.-Y. Liu, M.-S. Zhan and K. Wang. 2012. Thermal properties of the polyimide foam prepared from aromatic dianhydride and isocyanate. *High Performance Polymers*, **24**, 373–378.

25. K. Wang, F. Wu, W. Zhai and W. Zheng. 2013. Effect of polytetrafluoroethylene on the foaming behaviors of linear polypropylene in continuous extrusion. *Journal of Applied Polymer Science*, **129**, 2253–2260.

26. Q. Ren, J. Wang, W. Zhai and S. Su. 2013. Solid state foaming of poly(lactic acid) blown with compressed CO_2: Influences of long chain branching and induced crystallization on foam expansion and cell morphology. *Industrial & Engineering Chemistry Research*, **52**, 13411–13421.

27. W. Zhai, J. Yu, W. Ma and J. He. 2007. Influence of long-chain branching on the crystallization and melting behavior of polycarbonates in supercritical CO_2. *Macromolecules*, **40**, 73–80.

28. G. Ji, J. Wang, W. Zhai, D. Lin and W. Zheng. 2013. Tensile properties of microcellular poly(lactic acid) foams blown by compressed CO_2. *Journal of Cellular Plastics*, **49**, 101–117.

29. P. Jia, J. Hu, W. Zhai, Y. Duan, J. Zhang and C. Han. 2015. Cell morphology and improved heat resistance of microcellular poly(L-lactide) foam via introducing stereocomplex crystallites of PLA. *Industrial & Engineering Chemistry Research*, **54**, 2476–2488.

30. W. T. Zhai, W. W. Feng, J. Q. Ling and W. G. Zheng. 2012. Fabrication of lightweight microcellular polyimide foams with three-dimensional shape by CO_2 foaming and compression molding. *Industrial & Engineering Chemistry Research*, **51**, 12827–12834.

31. W. Jin, C. Xingguo, Y. Mingjun and H. Jiasong. 2001. An investigation on the microcellular structure of polystyrene/LCP blends prepared by using supercritical carbon dioxide. *Polymer*, **42**, 8265–8275.

32. W. Zhai, J. Yu, W. Ma and J. He. 2007. Cosolvent effect of water in supercritical carbon dioxide facilitating induced crystallization of polycarbonate. *Polymer Engineering and Science*, **47**, 1338–1343.

33. J. S. Colton and N. P. Suh. 1987. The nucleation of microcellular thermoplastic foam with additives: Part I: Theoretical considerations. *Polymer Engineering & Science*, **27**, 485–492.

34. G. R. Guillen, Y. Pan, M. Li and E. M. V. Hoek. 2011. Preparation and characterization of membranes formed by nonsolvent induced phase separation: A review. *Industrial & Engineering Chemistry Research*, **50**, 3798–3817.

35. G. R. Guillen, G. Z. Ramon, H. P. Kavehpour, R. B. Kaner and E. M. V. Hoek. 2013. Direct microscopic observation of membrane formation by nonsolvent induced phase separation. *Journal of Membrane Science*, **431**, 212–220.

36. M. Sadrzadeh and S. Bhattacharjee. 2013. Rational design of phase inversion membranes by tailoring thermodynamics and kinetics of casting solution using polymer additives. *Journal of Membrane Science*, **441**, 31–44.

37. H. Hatori, Y. Yamada and M. Shiraishi. 1992. Preparation of macroporous carbon-films from polyimide by phase inversion method. *Carbon*, **30**, 303–304.

38. H. Hatori, Y. Yamada and M. Shiraishi. 1995. Preparation of macroporous carbons from phase-inversion membranes. *Journal of Applied Polymer Science*, **57**, 871–876.

39. H. Hatori, T. Kobayashi, Y. Hanzawa, Y. Yamada, Y. Iimura, T. Kimura and M. Shiraishi. 2001. Mesoporous carbon membranes from polyimide blended with poly(ethylene glycol). *Journal of Applied Polymer Science*, **79**, 836–841.

40. Y. Ren and D. C. C. Lam. 2008. Properties and microstructures of low-temperature-processable ultralow-dielectric porous polyimide films. *Journal of Electronic Materials*, **37**, 955–961.

41. H. C. Park, Y. P. Kim, H. Y. Kim and Y. S. Kang. 1999. Membrane formation by water vapor induced phase inversion. *Journal of Membrane Science*, **156**, 169–178.

42. J. Ling, W. Zhai, W. Feng, B. Shen, J. Zhang and W. G. Zheng. 2013. Facile preparation of lightweight microcellular polyetherimide/graphene composite foams for electromagnetic interference shielding. *ACS Applied Materials & Interfaces*, **5**, 2677–2684.

43. B. Shen, W. Zhai, M. Tao, J. Ling and W. Zheng. 2013. Lightweight, multifunctional polyetherimide/graphene@Fe_3O_4 composite foams for shielding of electromagnetic pollution. *ACS Applied Materials & Interfaces*, **5**, 11383–11391.

44. W. Zhai, Y. Chen, J. Ling, B. Wen and Y.-W. Kim. 2014. Fabrication of light-weight, flexible polyetherimide/nickel composite foam with electromagnetic interference shielding effectiveness reaching 103 dB. *Journal of Cellular Plastics*, **50**, 537–550.

45. Y. Li, X. Pei, B. Shen, W. Zhai, L. Zhang and W. Zheng. 2015. Polyimide/graphene composite foam sheets with ultrahigh thermostability for electromagnetic interference shielding. *RSC Advances*, **5**, 24342–24351.

46. W. Zhai, Y.-W. Kim and C. B. Park. 2010. Steam-chest molding of expanded polypropylene foams. 1. DSC simulation of bead foam processing. *Industrial & Engineering Chemistry Research*, **49**, 9822–9829.

47. W. Zhai, Y.-W. Kim, D. W. Jung and C. B. Park. 2011. Steam-chest molding of expanded polypropylene foams. 2. Mechanism of interbead bonding. *Industrial & Engineering Chemistry Research*, **50**, 5523–5531.

48. A. Kuwabara, M. Ozasa, T. Shimokawa, N. Watanabe and K. Nomoto. 2005. Basic mechanical properties of balloon-type TEEK-L polyimide-foam and TEEK-L filled aramid-honeycomb core materials for sandwich structures. *Advanced Composite Materials*, **14**, 343–363.

49. N. Li, Y. Huang, F. Du, X. B. He, X. Lin, H. J. Gao, Y. F. Ma, F. F. Li, Y. S. Chen and P. C. Eklund. 2006. Electromagnetic interference (EMI) shielding of single-walled carbon nanotube epoxy composites. *Nano Letters*, **6**, 1141–1145.

50. J. J. Liang, Y. Wang, Y. Huang, Y. F. Ma, Z. F. Liu, F. M. Cai, C. D. Zhang, H. J. Gao and Y. S. Chen. 2009. Electromagnetic interference shielding of graphene/epoxy composites. *Carbon*, **47**, 922–925.

51. B. Yuan, C. Bao, X. Qian, L. Song, Q. Tai, K. M. Liew and Y. Hu. 2014. Design of artificial nacre-like hybrid films as shielding to mitigate electromagnetic pollution. *Carbon*, **75**, 178–189.

52. J.-M. Thomassin, C. Pagnoulle, L. Bednarz, I. Huynen, R. Jerome and C. Detrembleur. 2008. Foams of polycaprolactone/MWNT nanocomposites for efficient EMI reduction. *Journal of Materials Chemistry*, **18**, 792–796.

53. A. Ameli, P. U. Jung and C. B. Park. 2013. Electrical properties and electromagnetic interference shielding effectiveness of polypropylene/carbon fiber composite foams. *Carbon*, **60**, 379–391.

54. A. Ameli, M. Nofar, S. Wang and C. B. Park. 2014. Lightweight polypropylene/stainless-steel fiber composite foams with low percolation for efficient electromagnetic interference shielding. *ACS Applied Materials & Interfaces*, **6**, 11091–11100.

55. H. B. Zhang, Q. Yan, W. G. Zheng, Z. X. He and Z. Z. Yu. 2011. Tough graphene-polymer microcellular foams for electromagnetic interference shielding. *ACS Applied Materials & Interfaces*, **3**, 918–924.

56. Y. L. Yang, M. C. Gupta, K. L. Dudley and R. W. Lawrence. 2005. Conductive carbon nanoriber-polymer foam structures. *Advanced Materials*, **17**, 1999–2003.

57. Y. Yang, M. C. Gupta, K. L. Dudley and R. W. Lawrence. 2005. Novel carbon nanotube-polystyrene foam composites for electromagnetic interference shielding. *Nano Letters*, **5**, 2131–2134.

58. D. X. Yan, P. G. Ren, H. Pang, Q. Fu, M. B. Yang and Z. M. Li. 2012. Efficient electromagnetic interference shielding of lightweight graphene/polystyrene composite. *Journal of Materials Chemistry*, **22**, 18772–18774.

59. Z. Chen, C. Xu, C. Ma, W. Ren and H.-M. Cheng. 2013. Lightweight and flexible graphene foam composites for high-performance electromagnetic interference shielding. *Advanced Materials*, **25**, 1296–1300.

60. V. Eswaraiah, V. Sankaranarayanan and S. Ramaprabhu. 2011. Functionalized graphene-PVDF foam composites for EMI shielding. *Macromolecular Materials and Engineering*, **296**, 894–898.
61. X.-Y. Liu, M.-S. Zhan and K. Wang. 2013. Preparation and characterization of electromagnetic interference shielding polyimide foam. *Journal of Applied Polymer Science*, **127**, 4129–4137.
62. J. Ma, M. Zhan and K. Wang. 2015. Ultralightweight silver nanowires hybrid polyimide composite foams for high-performance electromagnetic interference shielding. *ACS Applied Materials & Interfaces*, **7**, 563–576.
63. J.-M. Thomassin, C. Jérôme, T. Pardoen, C. Bailly, I. Huynen and C. Detrembleur. 2013. Polymer/carbon based composites as electromagnetic interference (EMI) shielding materials. *Materials Science and Engineering: R: Reports*, **74**, 211–232.
64. B. Shen, W. Zhai, D. Lu, W. Zheng and Q. Yan. 2012. Fabrication of microcellular polymer/graphene nanocomposite foams. *Polymer International*, **61**, 1693–1702.
65. Y. Chen, Y. Wang, H.-B. Zhang, X. Li, C.-X. Gui and Z.-Z. Yu. 2015. Enhanced electromagnetic interference shielding efficiency of polystyrene/graphene composites with magnetic Fe_3O_4 nanoparticles. *Carbon*, **82**, 67–76.
66. S. T. Hsiao, C. C. M. Ma, H. W. Tien, W. H. Liao, Y. S. Wang, S. M. Li and Y. C. Huang. 2013. Using a non-covalent modification to prepare a high electromagnetic interference shielding performance graphene nanosheet/water-borne polyurethane composite. *Carbon*, **60**, 57–66.
67. N. Yousefi, X. Sun, X. Lin, X. Shen, J. Jia, B. Zhang, B. Tang, M. Chan and J.-K. Kim. 2014. Highly aligned graphene/polymer nanocomposites with excellent dielectric properties for high-performance electromagnetic interference shielding. *Advanced Materials*, **26**, 5480–5487.
68. H. B. Zhang, W. G. Zheng, Q. Yan, Z. G. Jiang and Z. Z. Yu. 2012. The effect of surface chemistry of graphene on rheological and electrical properties of polymethylmethacrylate composites. *Carbon*, **50**, 5117–5125.
69. B. Shen, D. Lu, W. Zhai and W. Zheng. 2013. Synthesis of graphene by low-temperature exfoliation and reduction of graphite oxide under ambient atmosphere. *Journal of Materials Chemistry C*, **1**, 50–53.
70. H.-B. Zhang, J.-W. Wang, Q. Yan, W.-G. Zheng, C. Chen and Z.-Z. Yu. 2011. Vacuum-assisted synthesis of graphene from thermal exfoliation and reduction of graphite oxide. *Journal of Materials Chemistry*, **21**, 5392–5397.
71. T. Chen, F. Deng, J. Zhu, C. Chen, G. Sun, S. Ma and X. Yang. 2012. Hexagonal and cubic Ni nanocrystals grown on graphene: Phase-controlled synthesis, characterization and their enhanced microwave absorption properties. *Journal of Materials Chemistry*, **22**, 15190–15197.
72. Y.-J. Chen, G. Xiao, T.-S. Wang, Q.-Y. Ouyang, L.-H. Qi, Y. Ma, P. Gao, C.-L. Zhu, M.-S. Cao and H.-B. Jin. 2011. Porous Fe_3O_4/carbon core/shell nanorods: Synthesis and electromagnetic properties. *Journal of Physical Chemistry C*, **115**, 13603–13608.
73. C. Hu, Z. Mou, G. Lu, N. Chen, Z. Dong, M. Hu and L. Qu. 2013. 3D graphene-Fe_3O_4 nanocomposites with high-performance microwave absorption. *Physical Chemistry Chemical Physics*, **15**, 13038–13043.
74. B. Shen, W. Zhai and W. Zheng. 2014. Ultrathin flexible graphene film: An excellent thermal conducting material with efficient EMI shielding. *Advanced Functional Materials*, **24**, 4542–4548.

10

Recent Innovations in Thermoplastic Foams

Tomoo Tokiwa and Tatsuyuki Ishikawa

CONTENTS

10.1 Part I: Blow Molding FOAMCORE™ Technology and Its Application

Tomoo Tokiwa

10.1.1 Introduction

JSP started foam extrusion production in the early 1960s for packaging applications, then into molded beads in the early 1980s for food and automotive parts protection [1]. JSP is not only active in new foam technology development but also very conscientious of its environmental and social responsibilities [2].

FOAMCORE, which consists of a solid outer layer and foamed core material, is produced in a single mold produced by blow molding (Figure 10.1). The molded product is superior to the lamination process in terms of surface performance, durability and impact resistance since the outer layer and the foamed core part adhere to each other when the same resin type is used [3].

It is possible to control the rigidity and weight of the products freely depending on the materials used for the outer layer and the expanded beads used for the core material (Figure 10.2).

A 1–10-mm thickness for the outer layer and 5–60 times expansion ratio of the expanded beads for the core material are available, as shown in Figure 10.3. It has high flexural strength due to its sandwich structure. Moreover, it is possible to add features such as weatherability, flame retardancy, and antibacterial properties in the outer layer.

10.1.2 Manufacturing Process Details

Figure 10.4 illustrates the basic steps of the FOAMCORE process.

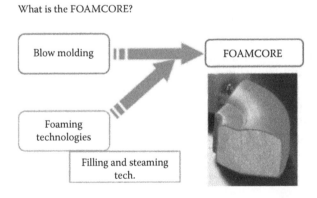

FIGURE 10.1
FOAMCORE.

In one manufacturing step,

form a shell by blow molding,

fill the core by expanded bead, and heat fuse them.

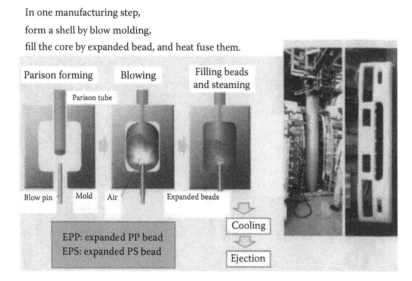

FIGURE 10.2
Production method.

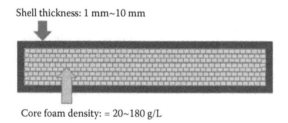

FIGURE 10.3
Technical range.

First, blow molding is used to form the shell.

As a second step, the beads are injected into the shell through inserted nozzles.

After that, the beads are fused by steam that is supplied from the inserted pins through the shell.

Finally, the molded part is cooled and ejected from the mold.

All is done within a regular blow-molding cycle.

The FOAMCORE molding cycle time is similar to that of standard blow-molded products since the beads are molded while the shell is cooled.

10.1.3 Benefits of FOAMCORE

FOAMCORE has high flexural strength due to its sandwich structure.

It also has excellent adhesion between the shell and foam core.

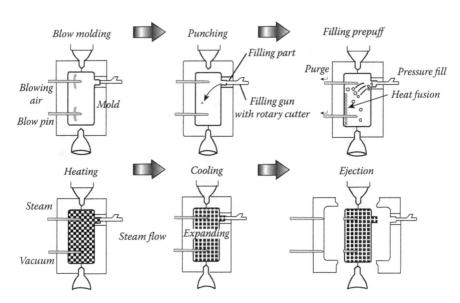

FIGURE 10.4
Basic step of FOAMCORE process.

It is lightweight, has shock-absorbing properties, and has superior surface appearance. There is no need for postprocessing. It provides thermal insulation and sound insulation and is fully recyclable.

Figure 10.5 shows the characteristics of FOAMCORE products.

Figure 10.6 illustrates the possible combinations of FOAMCORE.

To achieve high flexural strength, the combination of resin type is important to achieve optimal adhesion between the shell and the foamcore.

In some cases, the adhesion between the shell and the core is not sufficient.

Polypropylene–expanded polypropylene (PP-EPP) and polystyrene–expanded polystyrene (PS-EPS) are commonly selected due to their optimal

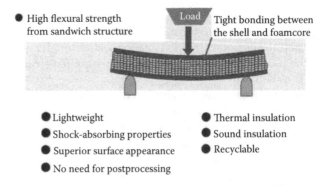

FIGURE 10.5
Characteristics of FOAMCORE products.

Combinations		Characteristics
Shell	Core	
PP	EPP	High bending stiffness
TPO	EPP or EPS	Soft touch
PE	EPE or EPS	Weatherability
PS	EPS	Thermal insulation
ABS	E-AS or EPS	Flame retardant
PLA	E-PLA	Bio-base

FIGURE 10.6
Resin combinations of FOAMCORE.

resin compatibility to achieve high flexural strength due to good adhesion between the shell and the core.

JSP has produced products using the following resin types:

- PP—polypropylene
- TPO—thermoplastic polyolefin
- PE—polyethylene
- PS—polystyrene
- ABS—acrylonitrile butadiene styrene
- PLA—polylactic acid

Figure 10.7 shows mold density versus flexural modulus.

The flexural modulus of FOAMCORE is 2.6 times that of EPP without the shell. Due to the sandwich structure, it is possible to obtain three to four times more flexural strength with weight reduction.

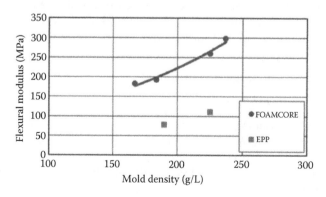

FIGURE 10.7
Flexural modulus of FOAMCORE.

	Thermal conductivity (W/mK)
FOAMCORE (EPS core)	0.040
EPS (50x)	0.040
FOAMCORE (EPP core)	0.044
EPP (45x)	0.044
Conventional blow molding (PP)	0.170

FIGURE 10.8
Thermal insulation of FOAMCORE.

Items	Conventional blow molding	FOAMCORE
Flexural strength	1	3~4
Thermal insulating property	1	5
Part weight	1	0.7~0.9
Molding cycle time	1	1
Cost	1	0.9~1.2

FIGURE 10.9
Cost performance of FOAMCORE.

Figure 10.8 shows the thermal insulation of FOAMCORE. The thermal conductivity is nearly equal to that of foam by itself. Conventional blow molding is not very good.

Figure 10.9 shows the cost performance of FOAMCORE. The cost–benefit performance of FOAMCORE will be significantly higher than that of conventional blow molding since there is no significant difference between the production costs. By reduction of shell thickness, cost reduction is possible. The final cost depends on the core volume and part design.

10.1.4 Recent Technology Advancement

10.1.4.1 Thinner FOAMCORE

As an updated technology, a thinner FOAMCORE with a minimum thickness of 15 mm in total, which has been considered to be conventionally difficult to produce, was successfully developed.

FOAMCORE with 15 mm thickness will comply with the requirement of auto makers with a weight requirement of less than 3 kg/m^2. This product was successfully produced using JSP's original process.

Figure 10.10 shows the 15 mm thickness of the FOAMCORE part.

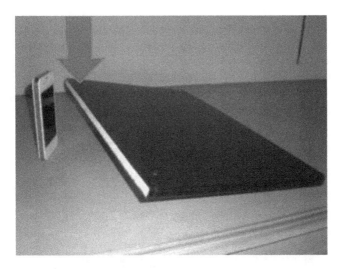

FIGURE 10.10
15 mm thickness of the FOAMCORE part.

10.1.4.2 FOAMCORE with Excellent Surface

The excellent surface quality was achieved by modifying the outer layer and the molding conditions, although it was generally difficult for the blow molding to have optimal surface flatness compared with other technologies. Figure 10.11 shows a painted FOAMCORE product.

Such technical innovation is what contributed to developing this new application.

Injection-molded article

FOAMCORE

FIGURE 10.11
Painted FOAMCORE product.

10.1.5 Patents

This JSP technology is protected by patents. JSP has numerous registered patents throughout the world. The oldest ones were registered in 1992. The most recent one was in 2014, and more are in progress. JSP is continuing research and development to improve the FOAMCORE technology [4]. The related patent numbers are shown below.

Patent No.	Constitution
JP2014-156044	Equipment
CN103372981 A, EP2656995 A1 US2013/0280468 A1, JP2013-223974	PS
CN103203878 A, EP2614942 A1 US2013/0175725 A1, JP2013-141780	Equipment
EP1987934 A1, US8221876 B2 JP5161481	PP
JP5437017	Application
JP4322019	PP

10.1.6 Recent Application Development

New applications of FOAMCORE are generated with advanced technology.

Examples of current applications include bathroom ceilings, tractor roofs, cart panels, pallets for clean environment, floats, and emergency lifeboats.

10.1.6.1 Bathroom Ceilings and Shower Booths

JSP provides insulation boards for prefabricated bathrooms (Figure 10.12) and shower booths (Figure 10.13). FOAMCORE products provide superior

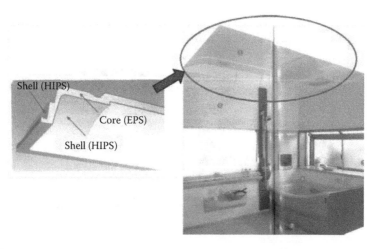

FIGURE 10.12
Bathroom ceiling.

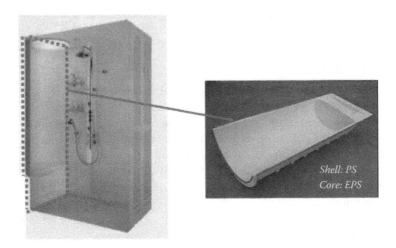

FIGURE 10.13
Shower booth.

performance for thermal insulation properties to the users of these products. For this product application, high-impact polystyrene is used for the shell and expanded polystyrene for the core material.

10.1.6.2 Tractor Roof

FOAMCORE is applied to the roofs of agricultural tractors (Figure 10.14). Polypropylene is used for the shell and expanded polypropylene for the core material in this application. In the example in Figure 10.14, the high strength

FIGURE 10.14
Tractor roof.

and thermal-insulation properties of FOAMCORE were evaluated by the customers and found to be most beneficial.

10.1.6.3 Bumper

Figure 10.15 shows an application where FOAMCORE technology is used for a truck bumper.

FOAMCORE technology makes it possible to produce an integrated bumper system where polypropylene is used for the shell, and expanded polypropylene is used for the core as an energy absorber.

In the example in Figure 10.15, FOAMCORE, with its shock-absorbing properties and its very smooth and flat surface, provides optimal performance benefits.

10.1.6.4 Backseat Panels

Figure 10.16 shows a backseat panel that is produced using FOAMCORE technology. In this example, the lightweight and shock-absorbing properties of FOAMCORE are highly evaluated.

This product satisfies the requirements for automotive parts (including backseat retention, ECE R17 intrusion performance, etc.).

10.1.6.5 Pallets for Clean Environment

Figure 10.17 shows the pallets that are produced using FOAMCORE technology. In this example, FOAMCORE was selected because it is easy to clean and

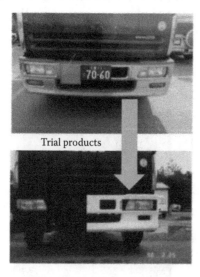

Trial products

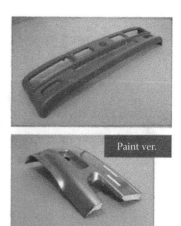

Paint ver.

FIGURE 10.15
Bumper.

FIGURE 10.16
Backseat panels.

chemical resistant and has a high load bearing. This pallet design can withstand a load of up to 1 t.

10.1.6.6 Floats

Figure 10.18 shows the floats that are produced using FOAMCORE technology, which can be used to protect a dam from damage caused by driftwood.

High-density polyethylene is normally used for the shell, and expanded polystyrene is used for the core. This product does not allow for adhesion between the shell and the core.

In this example, the durability, weather resistance, and buoyancy of FOAMCORE provides optimal performance benefits.

10.1.6.7 Emergency Lifeboats

FOAMCORE technology is used for producing the fenders of emergency lifeboats (Figure 10.19). Thermoplastic polyolefin elastomer is used for the shell.

FIGURE 10.17
Pallets for clean environment.

FIGURE 10.18
Float.

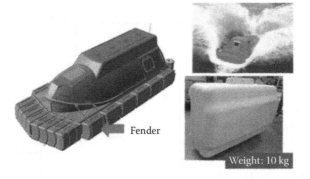

Fender

Weight: 10 kg

FIGURE 10.19
Fenders of emergency lifeboats.

The reason why we use thermoplastic polyolefin elastomer is due to good performance and durability. In this application, the shock-absorbing property provides optimal performance benefits.

10.1.7 Summary

It is expected that the features of FOAMCORE technology such as reduced weight, high rigidity, thermal insulation, shock-absorbing performance, and sound insulation will contribute to energy savings by reducing vehicle weight and allowing for future part recycling for the benefit of society.

10.1.8 Acknowledgments

Many thanks to our colleagues from JSP who have developed many new applications.

10.2 Part II: Extruded Antitermite Polycarbonate Foam

Tatsuyuki Ishikawa

10.2.1 Introduction

JSP has been developing and producing foamed plastics using polystyrene and polyolefin for over 30 years. These products are manufactured by using extrusion foaming or bead foaming processes. The polycarbonate foam reported in this chapter is an extruded foamed product. JSP has three ranges of thicknesses of polycarbonate foam, as shown in the following sections in this part. Here, MIRAPOLYCAFOAM™ is introduced as a heat insulation material with excellent antitermite performance and other unique properties.

(Products)

Product Name	Thickness (mm)	Application
MIRAPOLYCASHEET™	1–3	Frozen food packaging
MIRAPOLYCABOARD™	3–5	Automotive parts
MIRAPOLYCAFOAM™	30–50	Heat insulation material

10.2.2 Manufacturing Process Details

An extrusion process is used for polycarbonate foaming. Two extruders connected in series with a flat die connected in the outlet of second extruder were used.

Polycarbonate resin, which was dried at 120°C for 5 h, was fed into the first extruder. This component was heated to 300°C to be melted and kneaded in the extruders. A physical blowing agent consisting of a hydrocarbon was supplied into the extruder from the injection port, and the mixture was melted and kneaded.

The temperature of expandable and melted resin was adjusted to approximately 210°C, which is suitable for foaming at the outlet of the second extruder.

This melted resin was then extruded from the die lip to the guider, passed through the guider while the resin was foamed. Using this technique, the resin was formed into a board to produce polycarbonate-extruded foam board.

10.2.3 Characteristics of MIRAPOLYCAFOAM™

The range of thickness is 30–50 mm, and that of width is 300–600 mm. MIRAPOLYCAFOAM is normally produced within the range of 30–50 mm thickness and 300–350 mm width.

MIRAPOLYCAFOAM inherits the properties of polycarbonate, which include excellent heat resistance, impact resistance, and toughness. Regarding antitermite protection, MIRAPOLYCAFOAM has been certified as a non-chemical antitermite heat insulation material by the Japan Wood Protection Association for the first time in Japan.

Table 10.1 shows the physical properties of MIRAPOLYCAFOAM and polystyrene foam:

- Antitermite
- Heat resistance
- Impact resistance
- Toughness
- Self-extinguishing flammability properties (limiting oxygen index: 26)
- Thermal insulation
- Recyclable

10.2.4 Antitermite Performance

Antitermite performance was evaluated in an indoor antitermite test and a field antitermite test. Three types of test pieces, MIRAPOLYCAFOAM, polystyrene extruded foam (without antitermite protection), and polystyrene-extruded foam (including antitermite protection), were used for the indoor antitermite test. It was carried out according to JIS K 1571 (2010). Formosan subterranean termite soldier termites (15 heads), worker termites (150 heads), and the test pieces were put into a container. The mass reduction rate of the test piece and the mortality of the termites were measured after 28 days. The mass reduction rate is shown in Equation 10.1. As a result, higher antitermite performance was shown for a test piece with smaller mass reduction rate.

$$L = (m1 - m2)/m1 \times 100 \qquad (10.1)$$

where
L = mass reduction rate (%)
$m1$ = weight before the test (g)
$m2$ = weight after the test (g)

TABLE 10.1

Physical Properties of MIRAPOLYCAFOAM (MPF)

	Units	MPF	Polystyrene Foam
Density	kg/m³	100	28
Thermal conductivity	W/mK	0.04	0.04
Heat distortion temperature	°C	130	80
Flexural strength	N/cm²	200	30
Compressive strength	N/cm²	60	40

Table 10.2 shows test results of antitermite performance.

MIRAPOLYCAFOAM does not kill termites, unlike antitermite protection-treated materials. The mortality rate of termites on MIRAPOLYCAFOAM was almost the same as that on polystyrene foam without antitermite protection. The mass reduction rate of MIRAPOLYCAFOAM was almost the same as antitermite-protected polystyrene foam. The mortality rate of the termites on the polystyrene foam with antitermite protection was 99%.

In other words, MIRAPOLYCAFOAM demonstrates a high antitermite performance even when exposed to an environment with a large number of termites for a long period of time.

Regarding the field antitermite test, three types of test pieces, including MIRAPOLYCAFOAM, polystyrene-extruded foam (without antitermite protection), and polystyrene foam (including antitermite protection), were prepared and buried in the soil of the test field in Kagoshima Prefecture (N31°33′E130°33′) for three years.

Figure 10.20 shows the test pieces after three years. No damage was observed on MIRAPOLYCAFOAM, but some damage by termites was observed on the antitermite-protected polystyrene foam. Thus MIRAPOLYCAFOAM exhibits high antitermite performance without any antitermite additives because of the toughness of polycarbonate.

TABLE 10.2

Antitermite Performance (Indoor Antitermite Test)

Materials	Chemical	Density (kg/m³)	Mass Reduction Rate (%)	Mortality of Termite (%)
MPF	Not used	100	0.8	Under 30
Polystyrene foam	Not used	35	7.6	Under 30
Polystyrene foam	Used	35	1.0	99

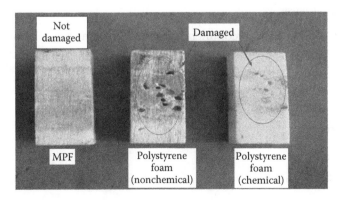

FIGURE 10.20
Antitermite performance (field antitermite test).

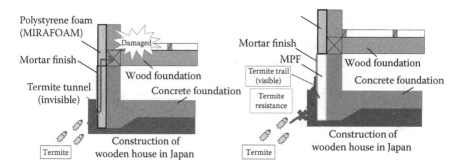

FIGURE 10.21
Antitermite construction method.

10.2.5 Antitermite Construction Method

Figure 10.21 shows the typical insulation structure of a wooden house in Japan. The foundations of the wooden houses in Japan are composed of concrete and wood. Inexpensive polystyrene foam as a heat insulation materials is usually applied on the concrete foundation to insulate the concrete.

However, polystyrene foam is relatively vulnerable to termites, which dig holes in the foam from the ground and reach the wood material of the houses in order to get away from sunlight exposure. Such a penetration into insulation foam and wooden structures is not visible and results in widespread damage to the wooden structure of houses.

On the other hand, when MIRAPOLYCAFOAM is applied on the concrete foundation as a heat insulation material, termites cannot invade into the wooden structure. In rare cases, they make a tunnel through the soil on the surface of the insulating material. Therefore, it is easy to detect the pathway of the termites at an earlier stage and be able to expel them.

10.2.6 Conclusion

MIRAPOLYCAFOAM is recommended as an antitermite insulation material for the foundations of wooden houses. It has a unique advantage because it does not contain antitermite additives but is naturally termite resistant.

References

1. 2004. 40th Anniversary for Japan Styrene Paper (JSP). Pub. by JSP Corp., Tokyo, Japan.
2. 2014. Life cycle assessment and physical properties for expanded foam products. ARPRO EPP, Foamcore. Available at http://www.jsp.com.

3. Sopher, Steven R. 2014. Foamcore structural automotive, rail, and aviation applications for seating and load floors. *SPE Foams 2014*, September, lselin, NJ, Paper #05-2014.
4. JSP Corporation et al., Tokiwa, T. et al. 2013. Method for producing a molded foam and manufacturing equipment. JP Patent 2013-141780.

11

Advanced CAE Technology for Microcellular Injection Molding

Chao-Tsai Huang and Rong-Yeu Chang

CONTENTS

11.1 Introduction

Microcellular technology has been enormously developing and applied in many foaming processes for many industrial polymer manufacturing, since the microcellular batch-processing technology was brought out by Dr. Nam Suh and coworkers at the Massachusetts Institute of Technology (MIT) in the early 1980s [1]. The microcellular application used in reciprocating screw injection molding machines was built by Trexel and Engel in 1998 [2]. Despite the fact that microcellular technology has been developing for many decades and is widely used in current plastic product manufacturing, a reliable computer-aided engineering application is not well developed due to a limited understanding of complex foaming mechanism. Venerus [3] reviewed numerous diffusion-controlled modeling studies of polymer foaming and showed the diffusion-induced bubble growth in viscoelastic liquids numerically having good agreement with experimental data. Taki [4] studied the effects of pressure release rate on bubble density and sizes. Recently, previously developed models of cell foaming have been coupled with three-dimensional (3D) flow motion technology for microcellular injection molding. However, the dynamic features of bubble growth and the integration from a bubble growth mechanism to the final injected parts are not fully developed. In this chapter, we describe the numerical development in microcellular injection molding. Furthermore, the dynamic features of bubble growth, including bubble size and bubble density, are validated with the experimental data done by Turng and coworkers [5]. Moreover, this advanced CAE technology is applied in some cases to realize how and why microcellular process can benefit in detail.

11.2 Numerical Modeling

The details of several model developments can be found in the papers by Taki [4]. We consider that the single-phase fluid is of binary constitution and supercritical fluids (SCFs) dissolved in polymer melt before the onset of bubble nucleation. The polymer melt is assumed to be a generalized Newtonian fluid (GNF). Hence, the nonisothermal 3D flow motion can be mathematically described by the following:

$$\frac{\partial \rho}{\partial t} + \nabla \cdot \rho \mathbf{u} = 0 \tag{11.1}$$

$$\frac{\partial}{\partial t}(\rho \mathbf{u}) + \nabla \cdot (\rho \mathbf{u} \mathbf{u} - \boldsymbol{\sigma}) = \rho \mathbf{g} \tag{11.2}$$

$$\boldsymbol{\sigma} = -p\mathbf{I} + \eta(\nabla \mathbf{u} + \nabla \mathbf{u}^T) \tag{11.3}$$

$$\rho C_P \left(\frac{\partial T}{\partial t} + \mathbf{u} \cdot \nabla T \right) = \nabla \cdot (\mathbf{k} \nabla T) + \eta \dot{\gamma}^2 \tag{11.4}$$

where $\mathbf{u}$ is the velocity vector, T is the temperature, t is the time, p is the pressure, $\boldsymbol{\sigma}$ is the total stress tensor, ρ is the density, η is the viscosity, $\mathbf{k}$ is the thermal conductivity, C_P is the specific heat, and $\dot{\gamma}$ is the shear rate. The finite volume method (FVM), due to its robustness and efficiency, is employed in this study to solve the transient flow field in complex 3D geometry.

The microcellular foaming process happens after the melt is injected into the mold cavity. The 3D numerical simulation is applied for describing the dynamic behavior of bubble growth, which is coupled with macroscopic molten polymer flow. The radius of bubble growth is given as

$$\frac{dR}{dt} = \frac{R}{4\eta} \left(P_D - P_C - \frac{2\gamma}{R} \right) \tag{11.5}$$

where the bubble radius is R, the viscosity is η, the bubble pressure is P_D, the ambient pressure is P_C, and the surface tension is γ. A thin boundary layer condition is assumed, and a dissolved gas concentration profile along the radial direction of the thin shell is described by a diffusion equation, as shown in the following:

$$\frac{\partial c}{\partial t} = D\left[\frac{1}{r^2}\frac{\partial}{\partial r}\left(r^2\frac{\partial c}{\partial r}\right)\right] \tag{11.6}$$

where c is the dissolved gas concentration, and D is the diffusion coefficient.

The dynamic bubble growth behavior is described by the mass transfer at the interface of gas bubble and is proposed by Han and Yoo [6]:

$$\frac{d}{dt}(P_D R^3) = \frac{6D(R_g T)(c_\infty - c_R)R}{-1 + \left\{1 + \frac{2/R^3}{R_g T}\left(\frac{P_D R^3 - P_{D0}R_0^3}{c_\infty - c_R}\right)\right\}^{1/2}} \tag{11.7}$$

The concentration has the following relation:

$$\frac{c_\infty - c}{c_\infty - c_R} = \left(1 - \frac{r - R}{\delta}\right)^2 \tag{11.8}$$

where δ is the concentration boundary thickness.

The dynamic bubble growth behavior is described by the mass transfer at the interface of the gas bubble and is proposed by Han and Yoo [6]. Bubble nucleation, described by Equation 11.9, happens due to the pressure drop of the molten polymer from the sprue to the mold cavity during the filling process. The cell nucleation and bubble growing is a competition mechanism and can be proposed as a classical exponential function that is coupled with a mass conservation of dissolved gas, as given in Equation 11.10.

$$J(t) = f_0\left(\frac{2\gamma}{\pi M_w/N_A}\right)^{1/2}\exp\left(-\frac{16\pi\gamma^3 F}{3k_B T(\bar{c}(t)/k_H - P_C(t)^2)}\right)N_A\bar{c}(t) \tag{11.9}$$

where f_0 and F are the fitting parameters, $\bar{c}$ is the average dissolved gas concentration, N_A is Avogadro's number, k_B is the Boltzmann constant, and M_w is the gas molecular weight.

$$\bar{c}(t)V_{L0} = c_0 V_{L0} - \int_0^t \frac{4\pi}{3}R^3(t - t', t')\frac{P_D(t - t', t')}{R_g T}J(t')V_{L0}\,dt' \tag{11.10}$$

where V_{L0} is the volume of polymer matrix.

The basic injection molding simulations were developed based on the FVM algorithm by Chang and Yang [7]. The 3D modeling of microcellular injection molding simulation technology is further integrated into the basic algorithm to become an advanced CAE technology for the prediction of a cell growth mechanism and the injected product quality. In addition, a comparison of the simulation of microcellular and conventional injection molding is presented for illustrating the benefits of using the microcellular injection molding process.

11.3 Case Studies

11.3.1 Learn Cell Growth Mechanism during Injection Molding

The cell size and the density distribution of microcellular will influence the strength and performance of the final injected product significantly. In this case, the cell growth and density distribution will be discussed. The calculated average values of cell density and cell size versus the dissolved gas amount from the middle cross-section area of a microcellular injection-molded dog-bone sample are compared with experimental data reported by Turng and coworkers [5], as shown in Figures 11.1 and 11.2. For a PP/N$_2$ system, as illustrated in Figure 11.1, the calculated cell size agrees well with experimental data, while the calculated density quantitatively agrees well with the experiment for lower dissolved gas amounts and qualitatively agrees with the experiment for higher gas concentrations. For an LDPE/N$_2$ system, as

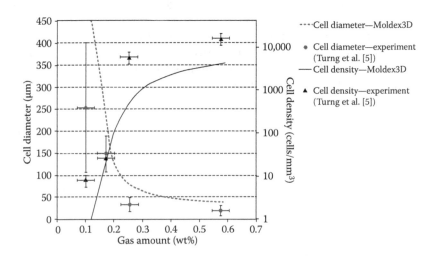

FIGURE 11.1

Comparison of calculated and experimental data of cell sizes and densities for PP/N$_2$ system. (Y-axis error bar is a guess value of 30% of lowest experimental gas concentration 0.1 wt%.)

illustrated in Figure 11.2, the calculated cell size and cell density both have good agreement with the experiments.

The simulation results of two dissolved gas concentrations, 0.24 wt% and 0.57 wt%, for the PP/N_2 system of microcellular injection molded dog-bone samples are analyzed for the cell nucleation and bubble-growing behaviors. In Figure 11.3, the cell density distribution of the molded part surface

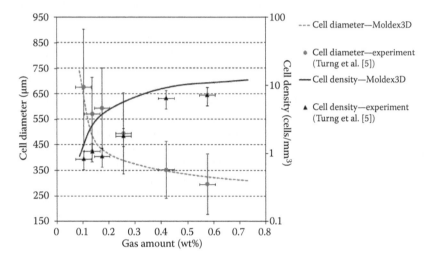

FIGURE 11.2
Comparison of calculated and experimental data of cell sizes and densities for LDPE/N_2 system. (Y-axis error bar is a guess value of 30% of lowest experimental gas concentration 0.1 wt%.)

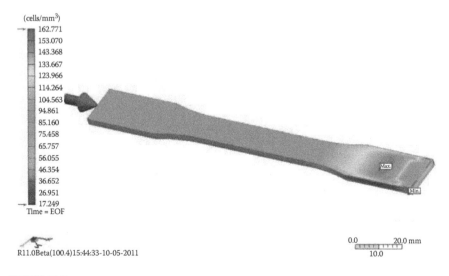

FIGURE 11.3
Cell density distribution of sample surface for 0.24 wt% N_2 SCF dissolved in PP. (Legend scale is 17–163 cells/mm³.)

ranges from 90 to 140 cells/mm³, while that of the whole molded part is from 17 to 163 cells/mm³, and the average value is approximately 111 cells/mm³, which is the mean value calculated from nodes. The maximum value appears at the position where the pressure-driven flow stops flowing, and molten polymer starts to fill the cavity by cell expansion; also, the minimum value happens at the end of cavity where it is last filled by melt. The density distribution of the core section of the molded part, as shown in Figure 11.4, shows a close-density range as that of the surface of molded part. The cell-size distribution results are given in Figures 11.5 and 11.6 for the surface and core section, respectively. The radius of the molded part is from 0.05 to 77.6 µm, and the average radius is 50.09 µm so that one might observe that cells having a size smaller than 10 µm almost appear near the surface (frozen layer). A comparison of Figures 11.5 and 11.6 also tells that the cells near the surface, i.e., frozen layer, are much smaller than the average cell size, which is the phenomena observed in a real microcellular injection-molded part. Calculated results for 0.57 wt% dissolved SCF are illustrated in Figures 11.7 through 11.10. The density distribution of the molded part of 0.57 wt% dissolved SCF is from 2.28×10^3 cells/mm³ to 5.36×10^3 cells/mm³, and the average is 3.71×10^3 cells/mm³; also, the size distribution is from 3.0 to 32.4 µm, and the average is 19.6 µm. In the

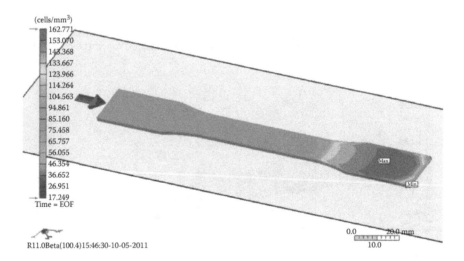

FIGURE 11.4
Cell density distribution along center section of sample for 0.24 wt% N_2 SCF dissolved in PP. (Legend scale is 17–163 cells/mm³.)

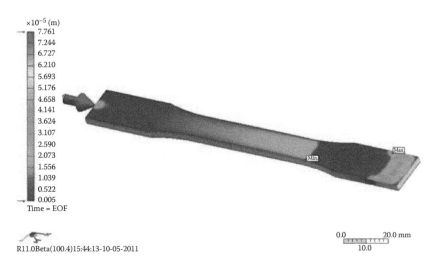

FIGURE 11.5
Cell radius distribution of sample surface for 0.24 wt% N_2 SCF dissolved in PP. (Legend scale is from 0.05 to 77.61 μm.)

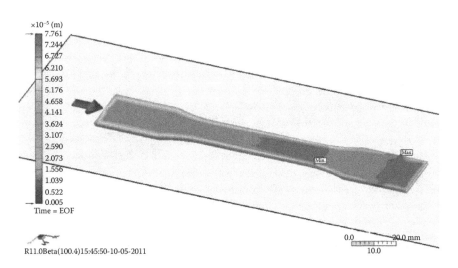

FIGURE 11.6
Cell radius distribution along center section of sample for 0.24 wt% N_2 SCF dissolved in PP. (Legend scale is from 0.05 to 77.61 μm.)

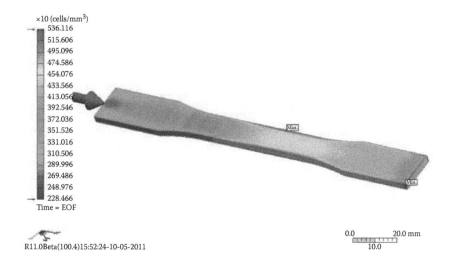

FIGURE 11.7
Cell density distribution of sample surface for 0.57 wt% N_2 SCF dissolved in PP. (Legend scale is $2.28 \times 10^3 \sim 5.36 \times 10^3$ cells/mm^3.)

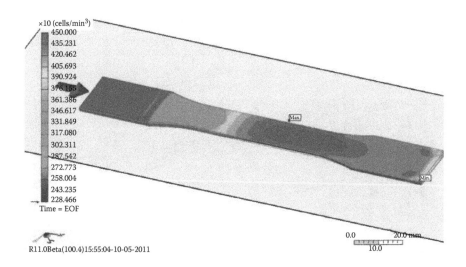

FIGURE 11.8
Cell density distribution along center section of sample for 0.57 wt% N_2 SCF dissolved in PP. (Legend scale is $2.28 \times 10^3 \sim 4.50 \times 10^3$ cells/mm^3.)

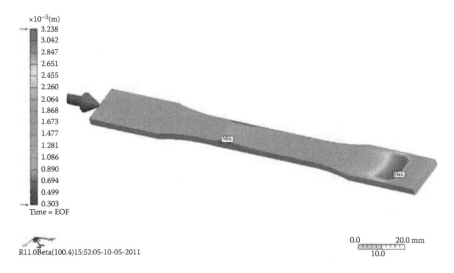

FIGURE 11.9
Cell radius distribution of sample surface for 0.57 wt% N_2 SCF dissolved in PP. (Legend scale is 3.03–32.38 μm.)

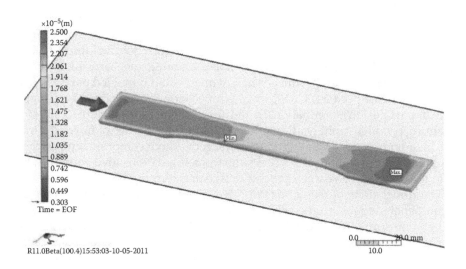

FIGURE 11.10
Cell radius distribution along center section of sample for 0.57 wt% N_2 SCF dissolved in PP. (Legend scale is 3.03–32.38 μm.)

average sense, a higher-dissolved SCF concentration gives a higher cell density and a smaller cell size than that of a lower concentration. The phenomena can be interpreted by the competition behavior between cell nucleation and bubble growing. In addition, lower-dissolved SCF gives a more uniformed cell size distribution.

11.3.2 Advanced CAE for Evaluation of Cell Size and Density in Long Fiber-Reinforced Plastic Injection Molding

Due to the high demand of smart and green technologies, lightweight technologies have become the driving force for automotive and other developments in recent years. Among those technologies, using short and long fiber-reinforced plastics (FRP) to replace some metal components can reduce the weight of an automotive significantly. However, the microstructures of fiber inside the plastic matrix are too complicated to manage and control during the injection molding through the screw, the runner, the gate, and then into the cavity. In addition, despite the fact that microcellular technology has been widely used in current plastic manufacturing, commodities on the market are mainly of sizes like electrical supplies and appliances. Large products, such as automotive instrument panels, were not generally produced from the MuCell process till Ford applied the MuCell process to automotive instrument panel manufacturing in 2011 [8]. The difficulty of the large part-utilizing MuCell process is not only the complicated tool design but also the optimization of molding conditions. A computer-aided engineering application developed for microcellular injection molding process was validated for understanding complex foaming mechanisms and for assisting the process or product design of microcellular molded parts [5]. For a high-tool-cost product, CAE provides an economic virtual mold trial and tool or part design examination. In addition, the microcellular structure of a foaming product strongly affects the mechanical properties of molded parts [5]. In order to reach a weight reduction target without compromising impact strength, microcellular property results from simulations give good insight into the correlation between process design and part quality.

The dimensions of 644.5 * 1415.8 * 562.4 (mm) of an automotive instrument panel model is studied. The mesh was generated by the auto 3D mesh generation software Moldex3D Designer. A MuCell process simulation was performed with 30 wt% glass fiber polypropylene and nitrogen supercritical fluid 0.5 wt%. The molding process conditions were based on typical mold trial conditions of the MuCell process. In this case, comparisons based on typical process conditions for conventional and microcellular injection molding gave a weight reduction of 10.18%, which can be calculated from the simulation results in Table 11.1. The reduced weight is approximately 0.277 kg. A foaming part, in general, has better dimensional stability than a conventional molded part; the improvement of part shrinkage in three directions of a Cartesian coordinate can be found in Table 11.1.

TABLE 11.1

Numerical Results: Comparisons between Conventional Injection
Molding and Microcellular Injection Molding

	Solid Part	MuCell Part
Part weight [g]	2724.2	2446.9
Max. clamping force [ton (m)]	1579	699.5
x-Displacement [mm]	22.13	7.61
y-Displacement [mm]	31.46	14.79
z-Displacement [mm]	21.49	5.43

An instrument panel model with gate locations are shown in Figure 11.11.
The dashed line circle in Figure 11.11 indicates that the molten plastic flow
fulfilled the cavity majorly by foaming instead of by a pressure-driven flow,
which could be clearly observed in Moldex3D flow animation. Corresponding
to this phenomenon, the cell size in this region is relatively bigger than the
average cell size. The flow end region, indicated as a dashed line circle in
Figure 11.11 has a large cell size, which is shown in the dashed line circle in
Figure 11.12. In Figure 11.12, the microcellular cell size on the part surface
shows that regions near gates have a smaller size, which can be interpreted

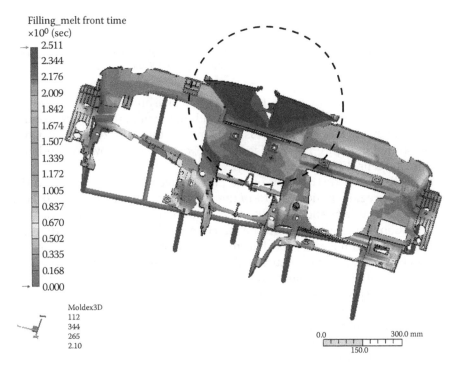

FIGURE 11.11
Gate locations and filling time distribution.

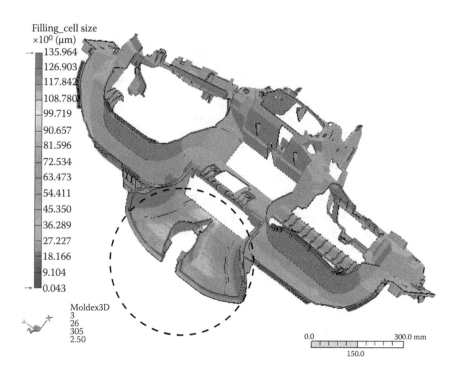

Filling_cell size
×10⁰ (μm)

→	135.964
	126.903
	117.842
	108.780
	99.719
	90.657
	81.596
	72.534
	63.473
	54.411
	45.350
	36.289
	27.227
	18.166
	9.104
→	0.043

Moldex3D
3
26
305
2.50

0.0   300.0 mm
150.0

FIGURE 11.12

Microcellular cell size distribution on part surface; the average cell size of the whole part is 53.8 μm, standard deviation is 12.6 μm, and maximum size is 136 μm.

as cells that are initially nucleated with small size and then taken onto the surface by fountain flow and frozen by cold mold temperature. The average microcellular cell size of the part is 53.8 μm, and the cell size distribution is as illustrated in Figure 11.13. In Figure 11.13, except at the flow end and gate region, the cells distribute uniformly.

The cell density distribution of the part is given in Figure 11.14 and the dashed line region indicated 80% of the range from 0.8×10^6 to 1.2×10^6 cm³. Although the information of cell size distribution and cell number density distribution cannot directly tell the effect on mechanical strength, they are indeed a helpful reference for determining the impact strength. According to previous research, a structure with cell size smaller than 10 μm can have impact strength improvement, and with cells within the same order of 10 μm may improve specific impact strength, i.e., reducing weight without compromising impact strength.

The maximum clamping force reduction is 55.7%, as shown in Table 11.1. The injection sprue pressure is given in Figure 11.15. Due to the fact that the shot size of the MuCell process is smaller than the conventional injection molding, the flow resistance of the MuCell part is smaller than the solid part. Thus, sprue pressure is reduced, further leading to a very small clamping

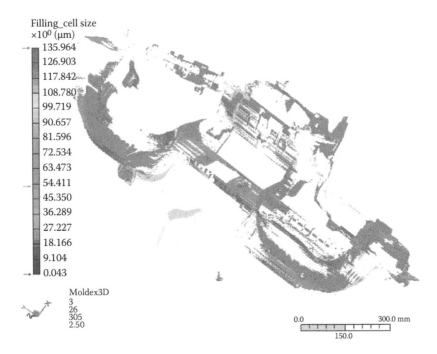

FIGURE 11.13

Distribution of microcellular cell with average size of 53 µm inside the part.

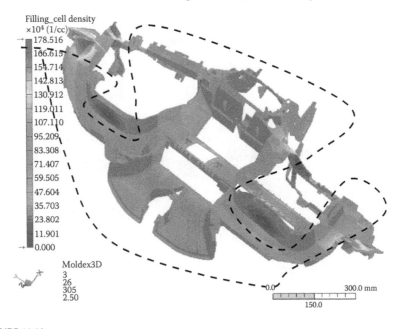

FIGURE 11.14

Cell number density distribution, average is $1.05 \times 10^6 /cm^3$; standard deviation is $2.04 \times 10^5 /cm^3$.

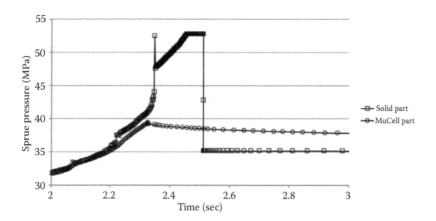

FIGURE 11.15
Sprue pressure comparison between conventional injection molding and microcellular injection molding processes.

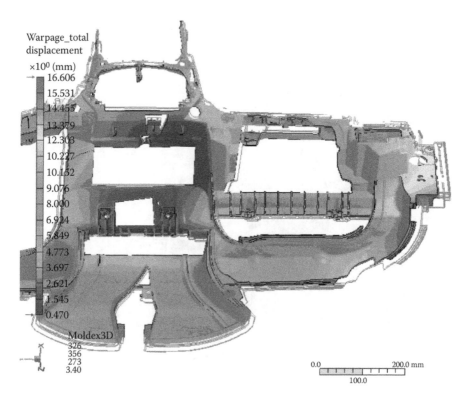

FIGURE 11.16
Total displacement of solid part.

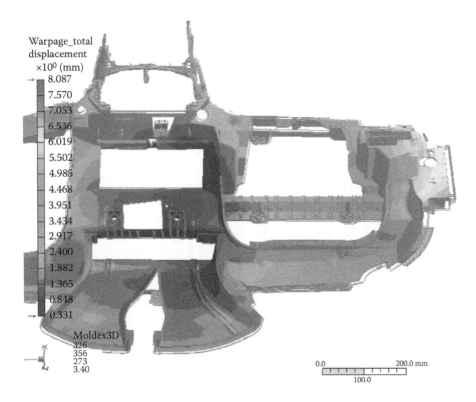

FIGURE 11.17
Total displacement of MuCell part.

force. The total displacements of the solid part and the MuCell part shown in two times the scale are given in Figures 11.16 and 11.17, respectively. The cell growth of MuCell brings the uniform packing under low pressure, which indeed reduces the shrinkage and gives a better directional stability than conventional injection molding.

11.4 Conclusions

This study presents the 3D simulation capability of predicting the dynamic behavior of a microcellular injection molding process. A good quantitative agreement between the simulation and experimental results of cell density distribution and size distribution is found for LDPE, although LDPE has a lower melt strength and shows a bigger error bar of experimental data of cell sizes. The simulation results of microcellular injection molding for nitrogen SCF dissolved in PP show a fairly good agreement with experimental data.

In the industrial application, the validation of simulation results with experimental data meets the criteria of prediction capability for further application of microcellular injection molding simulation for a complex geometrical product.

Acknowledgments

The authors thank Prof. L.-S. Turng for providing us the great research data and endless support. A very special thanks goes to Jeff Webb and Li Qi from the Ford team for encouraging our team. A very special gratitude goes to Dr. Tai-Yi Shiu for his assistance in the preparations of the draft material. We acknowledge with deep appreciation all of our colleagues in CoreTech System (Moldex3D) who helped to make Moldex3D and Mucell modules, especially to Venny, David, Anthony, Dan, Robert, Ethan, Jimmy, and many others.

References

1. N. P. Suh. 1996. Hanser/Garden Publications, Cincinnati, Chapter 3.
2. J. Xu. 2010. John Wiley & Sons, New Jersey.
3. D. C. Venerus. 2003. Modeling diffusion-induced bubble growth in polymer liquids. *Cellular Polymers*, 22, 89–101.
4. K. Taki. 2008. Experimental and numerical studies on the effects of pressure release rate on number density of bubbles and bubble growth in a polymeric foaming process. *Chemical Engineering Science*, 63, 3643–3653.
5. J. Lee, L.-S. Turng, E. Dougherty, and P. Gorton. 2011. A novel method for improving the surface quality of microcellular injection molded parts. *Polymer*, 52, 1436–1446.
6. C. D. Han and H. J. Yoo. 1981. Studies on structural foam processing 4. Bubble-growth during mold filling. *Polymer Engineering and Science*, 21, 518–533.
7. R.-Y. Chang and W.-H. Yang. 2001. Numerical simulation of mold filling in injection molding using a three-dimensional finite volume approach. *International Journal for Numerical Methods in Fluids*, 37, 125–148.
8. D. Smock. 2011. *Plastics Today*. Available at http://www.plasticstoday.com/.
9. T.-Y. Shiu, Y.-J. Chang, C.-T. Huang, D. Hsu, and R.-Y. Chang. 2012. Dynamic behavior and experimental validation of cell nucleation and growing mechanism in microcellular injection molding process. *SPE ANTEC Tech. Paper*.

Index